United States Nuclear Regulatory Commission

Protecting People and the Environment

NUREG/CR-7143
SAND-2007-2270

I0482633

Characterization of Thermal-Hydraulic and Ignition Phenomena in Prototypic, Full-Length Boiling Water Reactor Spent Fuel Pool Assemblies After a Postulated Complete Loss-of-Coolant Accident

Office of Nuclear Regulatory Research

AVAILABILITY OF REFERENCE MATERIALS
IN NRC PUBLICATIONS

NRC Reference Material

As of November 1999, you may electronically access NUREG-series publications and other NRC records at NRC's Public Electronic Reading Room at http://www.nrc.gov/reading-rm.html. Publicly released records include, to name a few, NUREG-series publications; *Federal Register* notices; applicant, licensee, and vendor documents and correspondence; NRC correspondence and internal memoranda; bulletins and information notices; inspection and investigative reports; licensee event reports; and Commission papers and their attachments.

NRC publications in the NUREG series, NRC regulations, and Title 10, "Energy," in the *Code of Federal Regulations* may also be purchased from one of these two sources.
1. The Superintendent of Documents
   U.S. Government Printing Office Mail Stop SSOP
   Washington, DC 20402–0001
   Internet: bookstore.gpo.gov
   Telephone: 202-512-1800
   Fax: 202-512-2250
2. The National Technical Information Service
   Springfield, VA 22161–0002
   www.ntis.gov
   1–800–553–6847 or, locally, 703–605–6000

A single copy of each NRC draft report for comment is available free, to the extent of supply, upon written request as follows:
Address: U.S. Nuclear Regulatory Commission
          Office of Administration
          Publications Branch
          Washington, DC 20555–0001
E-mail: DISTRIBUTION.RESOURCE@NRC.GOV
Facsimile: 301–415–2289

Some publications in the NUREG series that are posted at NRC's Web site address http://www.nrc.gov/reading-rm/doc-collections/nuregs are updated periodically and may differ from the last printed version. Although references to material found on a Web site bear the date the material was accessed, the material available on the date cited may subsequently be removed from the site.

Non-NRC Reference Material

Documents available from public and special technical libraries include all open literature items, such as books, journal articles, transactions, *Federal Register* notices, Federal and State legislation, and congressional reports. Such documents as theses, dissertations, foreign reports and translations, and non-NRC conference proceedings may be purchased from their sponsoring organization.

Copies of industry codes and standards used in a substantive manner in the NRC regulatory process are maintained at—
          The NRC Technical Library
          Two White Flint North
          11545 Rockville Pike
          Rockville, MD 20852–2738

These standards are available in the library for reference use by the public. Codes and standards are usually copyrighted and may be purchased from the originating organization or, if they are American National Standards, from—
          American National Standards Institute
          11 West 42nd Street
          New York, NY 10036–8002
          www.ansi.org
          212–642–4900

Legally binding regulatory requirements are stated only in laws; NRC regulations; licenses, including technical specifications; or orders, not in NUREG-series publications. The views expressed in contractor-prepared publications in this series are not necessarily those of the NRC.

The NUREG series comprises (1) technical and administrative reports and books prepared by the staff (NUREG–XXXX) or agency contractors (NUREG/CR–XXXX), (2) proceedings of conferences (NUREG/CP–XXXX), (3) reports resulting from international agreements (NUREG/IA–XXXX), (4) brochures (NUREG/BR–XXXX), and (5) compilations of legal decisions and orders of the Commission and Atomic and Safety Licensing Boards and of Directors' decisions under Section 2.206 of NRC's regulations (NUREG–0750).

NUREG/CR-7143
SAND-2007-2270

United States Nuclear Regulatory Commission

Protecting People and the Environment

Characterization of Thermal-Hydraulic and Ignition Phenomena in Prototypic, Full-Length Boiling Water Reactor Spent Fuel Pool Assemblies After a Postulated Complete Loss-of-Coolant Accident

Manuscript Completed: October 2012
Date Published: March 2013

Prepared by
E. R. Lindgren and S. G. Durbin

Sandia National Laboratory
Albuquerque, NM 87185

G. A. Zigh, Technical Advisor
A. Velazquez-Lozada, Project Manager

NRC Job Code Y6758

Prepared for
Division of Systems Analysis
Office of Nuclear Regulatory Research
U.S. Nuclear Regulatory Commission
Washington, DC 20555-0001

ABSTRACT

The NRC regulates the operation of the civilian nuclear power plant fleet by establishing and enforcing regulatory requirements for their design, construction, and operation. To protect the health and safety of the public and environment, the NRC requires all nuclear power plants to have a spent fuel pool where the used reactor fuel assemblies are allowed to cool for a number of years before being moved to interim or permanent storage. Spent fuel pools (SFP) are robust structures with an extremely low likelihood of a complete loss of coolant under traditional accident scenarios. However, in the wake of the terrorist attacks of September 11, 2001, the SFP accident progression was reconsidered and reevaluated using best-estimate accident codes.

In 2001, the NRC staff performed an evaluation of the potential accident risk in a SFP at decommissioning plants in the United States. This evaluation is documented on NUREG-1738, "Technical Study of Spent Fuel Pool Accident Risk at Decommissioning Nuclear Power Plants," (ML010430066). The study described a modeling approach of a typical decommissioning plant with design assumptions and industry commitments, thermal-hydraulic analyses performed to evaluate spent fuel stored in the SFP at decommissioning plants, risk assessment of SFP accidents, consequence calculations and the implications for decommissioning regulatory requirements. Some assumptions in the accident progression were known to be necessarily conservative, especially the estimation of the fuel damage. Consequently, the NRC continued SFP accident research by applying best-estimate computer codes to predict the severe accident progression following various postulated accident initiators. These code studies identified various modeling and phenomenological uncertainties that prompted a need for experimental confirmation. The present experimental program was undertaken to address thermal-hydraulic issues associated with complete loss-of-coolant accidents in boiling water reactor SFPs. All of the experiments and numerical simulations described in this report were performed at Sandia National Laboratories (SNL) in Albuquerque, New Mexico.

The objective of this project was to provide basic thermal-hydraulic data associated with a postulated SFP complete loss-of-coolant accident. The accident conditions of interest for the SFP were simulated in a full-scale prototypic fashion (electrically-heated, prototypic assemblies in a prototypic SFP rack) so that the experimental results closely represent actual fuel assembly responses. A major impetus for this work was to facilitate code validation (primarily MELCOR) and reduce modeling uncertainties within the code.

The research summarized in this report achieved the stated objective and resolved several unexpected technical challenges related with thermocouple attachments and the choice of appropriate input power in the heated design test that would cause the ignition. The close coupling of the experimental and numerical programs allowed for rapid validation and improvement of the MELCOR whole pool calculations. Because of the success of this approach, this project will be used as a model for subsequent studies.

CONTENTS

ABSTRACT iii

LIST OF FIGURES ix

LIST OF TABLES.. xvii

EXECUTIVE SUMMARY xix

ABBREVIATIONS/DEFINITIONS xxi

1   INTRODUCTION..1
    1.1   Background..1
    1.2   Objectives and Scope of Testing 1
    1.3   Test Plan Synopsis..1
    1.4   Report Organization...3

2   HEATER DESIGN TEST 1 5
    2.1   Objective..5
    2.2   Experimental..5
        2.2.1   Electric Heater Design 5
        2.2.2   Initial Experimental Assembly 7
        2.2.3   Final Experimental Assembly 14
        2.2.4   Additional Sampling 15
        2.2.5   Initial Test Operation 16
        2.2.6   Initial Test Analysis 19
        2.2.7   Final Test Operation 19
    2.3   Data Analysis...23
        2.3.1   Ignition Details...23
        2.3.2   Smoke Sampling Results..31
    2.4   Discussion..32
        2.4.1   Time of Ignition and Burn Rate 32
        2.4.2   Post-Test MELCOR Analysis 32
    2.5   Technical Issues...33
        2.5.1   TC Attachment 33
        2.5.2   Molten Zirconium-Magnesium Oxide (Zr-MgO) Reaction 34
    2.6   Conclusion...35

3   HEATER DESIGN TEST 2 37
    3.1   Background...37
    3.2   Objective..37
    3.3   Electric Heaters..38
    3.4   Experimental..38
        3.4.1   Additional Sampling 47
        3.4.2   Test Operation..48
        3.4.3   Ignition Details...50

	3.4.4	Post-Test MELCOR Analysis	55
	3.4.5	Post-Test Inspection	56
	3.4.6	Smoke Sampling Results	61
3.5	Issues		61
3.6	Conclusion		61
4	**HYDRAULIC ANALYSIS OF THE SPENT FUEL POOL EXPERIMENT**		**63**
4.1	Experimental Apparatus and Procedures		63
4.2	Hydraulic Analysis		66
	4.2.1	Blocked Water Rod Results	67
	4.2.2	Unblocked Water Rod Results	72
	4.2.3	Water Rod Flow Rate	74
	4.2.4	Bypass Hole Flow Rate	78
4.3	Summary		79
	4.3.1	Application of Experimental Results to a MELCOR Model	79
5	**THERMAL–HYDRAULIC ANALYSIS OF THE SPENT FUEL POOL EXPERIMENT**		**83**
5.1	Experimental Apparatus and Procedures		83
	5.1.1	Hardware Components	83
	5.1.2	Thermocouple Layout	83
	5.1.3	Flow Measurements and Calibrations	85
5.2	Modeling Efforts		92
	5.2.1	COBRA-SFS (Coolant Boiling in Rod Arrays-Spent Fuel Storage)	92
	5.2.2	MELCOR – Severe Accident Analysis	94
5.3	Results		97
	5.3.1	Transient Thermal Response	97
	5.3.2	"Steady State" Peak Cladding Temperature	102
	5.3.3	Steady State Natural Induction Flow Rates	102
5.4	Summary		103
6	**SUMMARY OF THE 1×4 INCOLOY SHORT STACK THERMAL RADIATION EXPERIMENTS OF THE SFP PROJECT**		**105**
6.1	Assembly Design		105
	6.1.1	TC Layout	106
6.2	MELCOR Analysis Methodology		108
	6.2.1	MELCOR Code	108
	6.2.2	Short 1×4 Assembly Model	109
6.3	COBRA 1×4 Model		112
6.4	Results		114
6.5	Model Sensitivity to Emissivity		120
	6.5.1	Additional COBRA Model Modifications	128
7	**FULL-SCALE ZIRCALOY PRE-IGNITION AND IGNITION RESULTS OF THE SFP EXPERIMENT**		**129**
7.1	Apparatus		129
7.2	Pre-Ignition Testing and Analysis		135
	7.2.1	Comparison of Thermal Results	135
	7.2.2	Comparison of Air Flow Results	136

	7.3	Ignition Test and Analysis	138
		7.3.1 Thermal Data	138
		7.3.2 Oxygen Data	139
		7.3.3 Comparison of Air Flow Results	142
		7.3.4 Comparison of Thermal Results	145
	7.4	MELCOR Post-test Comparison Summary	147
	7.5	Discussion	153
8		1×4 ZIRCALOY PRE-IGNITION AND IGNITION RESULTS OF THE SFP EXPERIMENT	155
	8.1	Assembly Design	155
		8.1.1 General Construction	155
		8.1.2 TC Layout	158
		8.1.3 Oxygen Monitors	161
	8.2	Ignition Test Results	163
		8.2.1 Input Parameters	163
		8.2.2 Oxygen Profiles	165
		8.2.3 Axial Temperature Profiles	167
		8.2.4 Radial Temperature Gradients	171
		8.2.5 Post-Mortem of the Test Apparatus	176
	8.3	MELCOR Baseline Case Comparison	177
		8.3.1 Model Input Parameters	177
		8.3.2 Comparison of Results	180
	8.4	MELCOR Sensitivity Study	184
		8.4.1 Modification to Kinetics Model	185
		8.4.2 Comparison of Results	187
	8.5	Summary	191
9		REPORT SUMMARY	193
	9.1	Heater Design Tests	193
	9.2	Separate Effects Tests	194
	9.3	Integral Effects Tests	194
	9.4	Key Findings	195
	9.5	Uncertainties and Potential Mitigating Factors	196
10		REFERENCES	197
		APPENDIX A. SFP HEATER DESIGN TEST 1 DATA SUMMARY	A-1
		APPENDIX B. ERROR ANALYSIS	B-1
		APPENDIX C. TSI HOT WIRE ANEMOMETER MEASUREMENTS	C-1
		APPENDIX D. ADDITIONAL DATA AND FINAL POST-TEST MELCOR RESULTS FOR THE FULL-SCALE ZIRCALOY IGNITION TEST	D-1

LIST OF FIGURES

Figure 1.1   Illustration of the closely-coupled relationship between the various testing elements and the supporting MELCOR modeling..3

Figure 2.1   Cumulative power distribution along a typical spent fuel rod and the electric heater rod..7

Figure 2.2   Specific heat of electric heaters compared to spent nuclear fuel..........................7

Figure 2.3   Schematic of the heater design test..8

Figure 2.4   Bundle of heater rods..9

Figure 2.5   TC attachment detail..11

Figure 2.6   Quartz tube installed over heater bundle..11

Figure 2.7   Stainless steel foil in place over quartz tube..12

Figure 2.8   Insulation and side-viewing slit..13

Figure 2.9   Initial test setup..14

Figure 2.10   More insulated final test setup..15

Figure 2.11   Temperature of air entering the bottom of the assembly and temperature of the air leaving the top of the assembly during the initial test operation..................16

Figure 2.12   History of the electrical current supplied to each of the twelve heater rods during the initial test operation..17

Figure 2.13   Axial temperature profile developed during the initial test operation..................17

Figure 2.14   Radial temperature profiles developed during the initial test operation..............18

Figure 2.15   Axial differential temperature change measured during the initial test operation..18

Figure 2.16   First derivative of axial temperature profile with major experimental events noted..21

Figure 2.17   Close-up of chimney during ignition initiation..22

Figure 2.18   Smoke filled CYBL vessel during burn..23

Figure 2.19   Oxygen concentration of off gas during time of Zircaloy ignition..................24

Figure 2.20   Heater currents during time of Zircaloy ignition..24

Figure 2.21   Axial temperature profile during time of ignition..25

Figure 2.22   First derivative of axial temperature profile during time of ignition..................25

Figure 2.23   Radial temperature profile at the 36-in. elevation during ignition..................26

Figure 2.24   First derivative of radial temperature profile at the 36-in. elevation during ignition..26

Figure 2.25   Radial temperature profile at the 24-in. elevation during ignition..................27

Figure 2.26   First derivative of radial temperature profile at the 24-in. elevation during ignition..27

Figure 2.27   Radial temperature profile at the 12-in. elevation during ignition..................28

Figure 2.28   First derivative of radial temperature profile at the 12-in. elevation during ignition..28

Figure 2.29   Temperature on the outside of insulation at the 42-in. level..................31

Figure 2.30   Heater Design Test 1 PCT with and without post-breakaway oxidation kinetics..33

Figure 2.31   TC attachment options, (a) small hood and (b) full wrap..................................34

Figure 2.32   Photo image of burnt heater rod cross sections, (a) lower rod segment end detail, and (b) upper rod segment end detail..34

Figure 2.33   Electron microprobe images of burnt heater rod section, (a) scanning electron microscopy (SEM) image, (b) Zirconium analysis, and (c) magnesium analysis...35

Figure 3.1   Schematic of the second heater design test.......................................37

Figure 3.2   Bundle of heater rods. 39

Figure 3.3   TC attachment detail. 42

Figure 3.4   Quartz tube installed over heater bundle. 43

Figure 3.5   Stainless steel foil in place over quartz tube and ceramic positioning block is in place...44

Figure 3.6   Installation of quartz wall TCs and light pipes into ceramic positioning block. ...45

Figure 3.7   Detail of quartz tube wall TCs and light pipe installation. 46

Figure 3.8   Initial test setup for Heater Design Test 2. 47

Figure 3.9   Heater currents during time of Zircaloy ignition with major experimental events noted. 50

Figure 3.10   Voltage applied to heater rods. 50

Figure 3.11   Oxygen concentration of off gas with major experimental events noted. 51

Figure 3.12   Axial temperature profile with major experimental events noted. 52

Figure 3.13   First derivative of axial temperature profile with major experimental events noted...52

Figure 3.14   Radial temperature profile at the 12-in., 24-in., and 36-in. elevations. 53

Figure 3.15   Axial temperature profile along quartz tube. 54

Figure 3.16   Close-up of light pipes during burn. 54

Figure 3.17   Composite axial rod temperature profile using corrected wall temperature measurements...55

Figure 3.18   Peak cladding temperature data and MELCOR calculation with and without post-breakaway oxidation kinetics...56

Figure 3.19   Outer insulation jacket after burn. 57

Figure 3.20   Inlet gas temperature, exit gas temperature in chimney, and other miscellaneous temperatures. 58

Figure 3.21   Post-test quartz tube and inner thermal radiation shield after removal of insulation...59

Figure 3.22   Detail of heater rod rubble. 60

Figure 3.23   Detail of heater rod rubble at the 36-in. level after bundle removal from test stand. 61

Figure 4.1   Experimental apparatus showing as-built port locations. 64

Figure 4.2   Pressure traces recorded during the blocked water rod testing for measurements across 2–8 and 8–17...65

Figure 4.3   Acrylic sleeves are inserted over the top of the water rods to block the flow. 66

Figure 4.4   Pressure drop as a function of bundle velocity for the SFP assembly with blocked water rods. 68

Figure 4.5   Hydraulic loss coefficients as a function of the upper, measured experimental flow range. 69

Figure 4.6   Cumulative pressure drop as a function of axial location for flow rates of 100, 200, 300, 400, 500, and 600 slpm...72

Figure 4.7   Pressure drop as a function of bundle velocity for the SFP assembly with unblocked water rods. 73

Figure 4.8 Total air flow rate through the assembly as a function of pressure drop across 2–17 for blocked (closed) and unblocked (open) water rods..................................75

Figure 4.9 Fraction of flow rate through the water rods as a function of total flow rate as determined from the 2–17 pressure drop data...76

Figure 4.10 Pressure drop as a function of bundle velocity for the SFP assembly with unblocked water rods. 77

Figure 4.11 Fraction of flow rate through the bypass holes as a function of total flow rate as determined from 1–B pressure drop data. 79

Figure 5.1. Experimental mock fuel assembly showing internal, as-built (a) axial and (b) lateral thermocouple (TC) locations...84

Figure 5.2 Arrangement of insulation, radiation barriers, and external TCs......................85

Figure 5.3 Schematic showing the layout of the assembly inlet hot wires (hot wires 2 and 4). 86

Figure 5.4 Hot wire 4 velocity traverse across the assembly inlet pipe for input volumetric flow rates of 50 (red) and 125 slpm (blue). 87

Figure 5.5 Inlet hot wire calibrations for hot wire 2 (HW2) (diamonds) and hot wire 4 (HW4) (squares) on 1/20/06 (closed) and 1/31/06 (open).............................88

Figure 5.6 Schematic showing layout of the drain hole hot wire (HW1 and HW3). 88

Figure 5.7 Drain hole hot wire calibrations for hot wire 1 (HW1) (diamonds) and hot wire 3 (HW3) (squares) on 1/20/06 (closed) and 1/31/06 (open).....................89

Figure 5.8. Schematic of the measurement of the pressure drop across the bypass holes for (a) calibration and (b) heated testing. 90

Figure 5.9. Annular flow rate as a function of the bypass pressure drop for the hydraulic assembly (diamonds), analytic solution (line), and the external bottom tie plate calibration (squares)..91

Figure 5.10. Schematic of the COBRA model layout...93

Figure 5.11. Detail view of the COBRA bundle layout. 93

Figure 5.12. Repeating patterns used in the MELCOR separate effects models. 95

Figure 5.13. MELCOR nodalization of the full-length assembly experiment. 97

Figure 5.14 Transient thermal response at different axial locations for a power input of 1370 W and closed bypass/open drains flow configuration. 98

Figure 5.15 Transient thermal response at different axial locations for a power input of 1370 W and open bypass/closed drains flow configuration. 99

Figure 5.16. Comparison of the thermal transient response with the MELCOR and COBRA codes for a power input of 1370 W and open bypass/closed drains flow configuration. 100

Figure 5.17 Comparison of experimental (symbols) to MELCOR (solid lines) and COBRA (dashed lines) temperatures as a function of axial location at times of 3 (♦), 6 (■), 9 (▲), and 12 hours (●). 101

Figure 5.18 Similar to the figure above for the open bypass/closed drains configuration......101

Figure 5.19 PCT as a function of assembly input power for experimental (symbols), MELCOR (blue lines), and COBRA (red lines)..102

Figure 5.20 Volumetric flow rates as a function of assembly input power for experimental (symbols) and MELCOR (lines)..103

Figure 6.1 Schematic and pictures of the assembly design...105

Figure 6.2 Photographs showing the final insulated assembly. 106

Figure 6.3   Overall TC layout. 107
Figure 6.4   Center assembly TC locations and naming conventions. 107
Figure 6.5   Peripheral assemblies TC locations and naming conventions. 108
Figure 6.6   Schematic of the 1×4 configuration. 109
Figure 6.7   MELCOR nodalization of the full-length assembly experiment. 110
Figure 6.8   Slab layout for the 1×4 assembly. 113
Figure 6.9   Individual rod array illustration. 114
Figure 6.10   Symmetric peripheral bundle temperatures as a function of time for the
              powered test conducted on 3/21/06. 115
Figure 6.11   Power test 2/27/06 – 1.5 kW center assembly, 26-in. level. 116
Figure 6.12   Power test 3/15/06 – 3 kW center assembly, 11-in. level. 116
Figure 6.13   Power test 3/15/06 – 3 kW center assembly, 26-in. level. 117
Figure 6.14   Power test 3/15/06 – 3 kW center assembly, 41-in. level. 117
Figure 6.15   Power test 3/21/06 – 3 kW center assembly, 11-in. level. 118
Figure 6.16   Power test 3/21/06 – 3 kW center assembly, 26-in. level. 118
Figure 6.17   Power test 3/21/06 – 3 kW center assembly, 41-in. level. 119
Figure 6.18   Power test 3/23/06 – 5 kW center assembly, 11-in. level. 119
Figure 6.19   Power test 3/23/06 – 5 kW center assembly, 26-in. level. 120
Figure 6.20   Power test 3/23/06 – 5 kW center assembly, 41-in. level. 120
Figure 6.21   MELCOR baseline emissivity for power test 3/21/06 – 3 kW center assembly,
              11-in. level. 121
Figure 6.22   MELCOR baseline emissivity for power test 3/21/06 – 3 kW center assembly,
              26-in. level. 122
Figure 6.23   MELCOR baseline emissivity for power test 3/21/06 – 3 kW center assembly,
              41-in. level. 122
Figure 6.24   COBRA baseline emissivity for power test 3/21/06 – 3 kW center assembly,
              11-in. level. 123
Figure 6.25   COBRA baseline emissivity for power test 3/21/06 – 3 kW center assembly,
              26-in. level. 123
Figure 6.26   COBRA baseline emissivity for power test 3/21/06 – 3 kW center assembly,
              41-in. level. 124
Figure 6.27   Effective two-surface emissivity as a function of the Zircaloy emissivity for
              stainless steel emissivities of 0.33, 0.38, and 0.43. 124
Figure 6.28   Maintain effective emissivity for power test 3/21/06 – 3 kW center assembly,
              11-in. level. 125
Figure 6.29   Maintain effective emissivity for power test 3/21/06 – 3 kW center assembly,
              26-in. level. 125
Figure 6.30   Maintain effective emissivity for power test 3/21/06 – 3 kW center assembly,
              41-in. level. 126
Figure 6.31   Increase Zircaloy emissivity to 0.80 and set stainless steel emissivity to 0.43
              for power test 3/21/06 – 3 kW center assembly, 11-in. level. 126
Figure 6.32   Increase Zircaloy emissivity to 0.80 and set stainless steel emissivity to 0.43
              for power test 3/21/06 – 3 kW center assembly, 26-in. level. 127
Figure 6.33   Increase Zircaloy emissivity to 0.80 and set stainless steel emissivity to 0.43
              for power test 3/21/06 – 3 kW center assembly, 41-in. level. 127

Figure 6.34   Peripheral temperature profiles at 4.1 hours elapsed test time for the powered test conducted on 3/21/06. 128

Figure 7.1   Schematic of externally mounted sensors. 130

Figure 7.2   *In situ* oxygen sensor seal and flow schematic. 131

Figure 7.3   Plan view schematic of oxygen sensor installation. 132

Figure 7.4   Experimental mock fuel assembly showing internal, as-built (a) axial and (b) lateral TC locations. 133

Figure 7.5   Schematic showing the layout of the assembly inlet hot wires 2 and 4. 134

Figure 7.6   Inlet hot wire calibrations for hot wire 2 (diamonds) and hot wire 4 (squares) ...134

Figure 7.7   Temperature as a function of time for the pre-ignition test with a power input of 2250 W. 135

Figure 7.8   MELCOR predictions for temperature as a function of time for the pre-ignition test with a power input of 2250 W. 136

Figure 7.9   Flow rate and pressure drop across the bypass as a function of time for hot wire 2 (HW2) (blue), hot wire 4 (HW4) (red), MELCOR (green), and pressure gage (black). 137

Figure 7.10   Comparison of pre-ignition MELCOR predictions and experimentally measured assembly inlet flow rates as a function of bundle power 138

Figure 7.11   Temperature as a function of time at different axial levels during the ignition test. 139

Figure 7.12   Signal and temperature response for the oxygen sensor located at $z = 144$ in. 140

Figure 7.13   Comparison of the measured oxygen concentrations measured by the *in situ* oxygen sensor and the flow-through oxygen monitor. 141

Figure 7.14   Measured oxygen concentration from the monitor devices. 142

Figure 7.15   Comparison of MELCOR pre-test prediction with and without sensor leakage versus the measured assembly inlet flow rate. 143

Figure 7.16   Comparison of MELCOR post-test prediction with additional sensor leakage versus the measured assembly inlet flow rate. 144

Figure 7.17   Calculated leakage flow from the post-test enhanced leakage model. 145

Figure 7.18   Calculated temperature response from the posttest enhanced leakage model. 146

Figure 7.19   Measured median temperature response. 147

Figure 7.20   Flow rate as a function of time for the experiment hot wires and the final post-test MELCOR model. 148

Figure 7.21   PCT for both the ignition experiment and the final post-test MELCOR model..149

Figure 7.22   Breakaway lifetime function as a variable of time in the MELCOR ignition model. 149

Figure 7.23   Cladding temperatures at $z = 24$ in. for both experiment and MELCOR. 150

Figure 7.24   Cladding temperatures at $z = 48$ in. for both experiment and MELCOR. 150

Figure 7.25   Cladding temperatures at $z = 72$ in. for both experiment and MELCOR. 151

Figure 7.26   Cladding temperatures at $z = 96$ in. for both experiment and MELCOR. 151

Figure 7.27   Cladding temperatures at $z = 119$ in. for both experiment and MELCOR. 152

Figure 7.28   Cladding temperatures at $z = 144$ in. for both experiment and MELCOR. 152

Figure 7.29   Postmortem of the assembly after ignition. 153

Figure 8.1   Left: Expanded schematic of the apparatus construction. Upper right: Photo of the center assembly under construction. Lower right: Photo of the undercarriage of the apparatus. 156

Figure 8.2   Flow diagram of mass flow controllers, air heaters, and assemblies.....................157
Figure 8.3   Photographs of the final assembly with insulation applied around (a) the
               perimeter and (b) the lower portion. 158
Figure 8.4   Overall type K (73K to 1523K temperature range) TC layout. 159
Figure 8.5   Center assembly TC locations and naming conventions. 159
Figure 8.6   TC locations and naming conventions in the peripheral assemblies. 160
Figure 8.7   TC locations and naming conventions in the channel canisters and pool cells. ...161
Figure 8.8   AMI Inc. Model 65 oxygen monitors. 162
Figure 8.9   Oygen sampling locations for each assembly as shown in (a) the as-built
               apparatus and (b) a schematic with sampling tube penetrations indicated in
               red. 162
Figure 8.10   Center assembly input power for the ignition test. 164
Figure 8.11   Air flow rates for the different channels inside the apparatus during the
               ignition test...164
Figure 8.12   Air inlet temperatures for the different channels inside the apparatus during
               the ignition test...165
Figure 8.13   Response of all oxygen monitors as a function of time............................166
Figure 8.14   Overhead images of the test apparatus at elapsed test times of (a) 5 hours and
               (b) 7 hours. 167
Figure 8.15   Temperature of the GX rod in the center assembly as a function of time for
               different axial locations..168
Figure 8.16   Temperature of the EZ and EQ rods in the west and east assemblies,
               respectively, as a function of time for different axial locations..........................169
Figure 8.17   Detailed temperature of the peripheral assembly as a function of time at $z =$
               10.6 in. 169
Figure 8.18   Temperature across the central portion of the apparatus as a function of time
               at $z = 10.6$ in..172
Figure 8.19   Temperature across the peripheral portion of the apparatus as a function of
               time at $z = 10.6$ in...172
Figure 8.20   Temperature across the central portion of the apparatus as a function of time
               at $z = 25.6$ in..173
Figure 8.21   Temperature across the peripheral portion of the apparatus as a function of
               time at $z = 25.6$ in...173
Figure 8.22   Temperature across the central portion of the apparatus as a function of time
               at $z = 40.6$ in..174
Figure 8.23   Temperature across the peripheral portion of the apparatus as a function of
               time at $z = 40.6$ in...174
Figure 8.24   Post-ignition test photographs of the apparatus with the insulation removed. 176
Figure 8.25   Post-ignition overhead photographs of (a) the apparatus and (b) the inlet air
               flow boxes..177
Figure 8.26   Comparison of air flow rates for the different channels inside the test
               apparatus and the MELCOR model..178
Figure 8.27   Comparison of air inlet temperatures for the different channels inside the test
               apparatus and the MELCOR model..179
Figure 8.28   Default emissivity model in MELCOR with high temperature correction..........180

Figure 8.29   Comparison of oxygen concentrations for the different assemblies inside the test apparatus and the MELCOR model. 181

Figure 8.30   Comparison of temperatures inside the center assembly and MELCOR model for $z = 4$ to 14 in. 182

Figure 8.31   Comparison of temperatures inside the center assembly and MELCOR model for $z = 28$ to 38 in. 183

Figure 8.32   Comparison of temperatures inside the center assembly and MELCOR model for $z = 37$ to 47 in. 183

Figure 8.33   Comparison of temperatures inside the peripheral assembly and MELCOR model. 184

Figure 8.34   Comparison of CTE prediction with Kearns' model for Zircaloy in the circumferential direction. 186

Figure 8.35   Modification of the default MELCOR oxidation kinetics model. 186

Figure 8.36   Average and PCT as a function of time in the center assembly at $z = 37.6$ in. ...187

Figure 8.37   Comparison of oxygen concentrations for the different assemblies inside the test apparatus and the MELCOR sensitivity model. 188

Figure 8.38   Comparison of the maximum and average temperatures in the center assembly to the MELCOR baseline and sensitivity models. 189

Figure 8.39   Comparison of the maximum and average temperatures in the peripheral canister to the MELCOR baseline and sensitivity models. 189

Figure 8.40   Comparison of the maximum and average temperatures in the peripheral assemblies to the MELCOR baseline and sensitivity models. 190

Figure 8.41   Effect on temperature response in the full-scale Zircaloy MELCOR model to the modified oxidation kinetics model. 190

LIST OF TABLES

Table 1.1   Summary of the testing elements in the experimental program. 2
Table 2.1   Composition of heater rods. 6
Table 2.2   Correlation of assembly and rod power versus the required voltage. 6
Table 2.3   Heater Design Test 1 instrument layout. 10
Table 2.4   Chronology of events during experiment. 20
Table 3.1   Composition of heater rods. 38
Table 3.2   Heater Design Test 2 instrument layout. 40
Table 3.3   Chronology of events during experiment. 49
Table 4.1   Dimensions of assembly components in the 9×9 BWR. 63
Table 4.2   Summary of the flow areas and hydraulic diameters of the SFP experiment. 66
Table 4.3 S_{LAM} and k coefficient analysis data for pressure drops between 2–8, 8–17, and 2–17 for blocked water rods. 68
Table 4.4 S_{LAM} and k values for different ranges of experimental air flow rate. 69
Table 4.5 S_{LAM} and k coefficients for the blocked water rod assembly. 70
Table 4.6   Spacer S_{LAM} and k coefficients calculated with the appropriate spacer hydraulic diameter and flow area. 70
Table 4.7   Calculation of effective S_{LAM} and k values for different span lengths. 71
Table 4.8 S_{LAM} and k coefficient analysis data for pressure drops between 2–8, 8–17, and 2–17 for unblocked water rods. 73
Table 4.9 S_{LAM} and k coefficients for the unblocked water rod assembly assuming all flow passes through the bundle. 74
Table 4.10 S_{LAM} and k coefficients for the unblocked water rod assembly assuming equal air velocities in the bundle and water rods. 74
Table 4.11 S_{LAM} and k coefficient analysis data for pressure drops between 2–8, 8–17, and 2–17 for unblocked water rods. 77
Table 4.12 S_{LAM} and k coefficients for the unblocked water rod assembly using the experimentally determined bundle flow rate ($Q_{tot} - Q_{wr}$). 78
Table 5.1   Dimensions of assembly components in the 9×9 BWR. 83
Table 5.2 S_{LAM} and k analysis across 1–B and 17–B for Incoloy and stainless steel assemblies. 85
Table 5.3   Integrated flow rates from the hot wire velocity profiles. 87
Table 5.4   Summary of all thermal-hydraulic tests for the SFP experiment. 104
Table 6.1   Summary of all 1×4 thermal radiation testing on the Incoloy *short stack* assembly. 114
Table 7.1   Elapsed experimental time to reach 15% oxygen concentration for the oxygen sensors and monitors. 142
Table 8.2   Elapsed experimental time to reach reduced oxygen concentrations for the oxygen monitors. 166
Table 8.3   Measured burn rate at different axial heights based on an ignition criterion of 1220 K. 170
Table 8.4   Measured temperatures across the apparatus at 4.82 hours. 175
Table 9.1   Summary of the testing elements in the experimental program. 193

EXECUTIVE SUMMARY

Spent fuel pools (SFP) are robust structures with an extremely low likelihood of a complete loss of coolant under traditional accident scenarios. However, in the wake of the terrorist attacks of September 11, 2001, the SFP accident progression was reevaluated using best-estimate accident codes. In 2001, the United State Nuclear Regulatory Commission (NRC) staff performed an evaluation of the potential accident risk in a spent fuel pool (SFP) at decommissioning plants in the United States [1]. This evaluation is documented on NUREG-1738, "Technical Study of Spent Fuel Pool Accident Risk at Decommissioning Nuclear Power Plants," (ML010430066). The best-estimate computer code studies identified various modeling and phenomenological uncertainties that prompted a need for experimental confirmation. This experimental program was undertaken to address thermal-hydraulic issues associated with complete loss-of-coolant accidents in boiling water reactor (BWR) SFPs.

The objective of this project was to provide basic thermal-hydraulic data associated with a SFP complete loss-of-coolant accident. The accident conditions of interest for the SFP were simulated in a full-scale prototypic fashion (electrically-heated, prototypic assemblies in a prototypic SFP rack) so that the experimental results closely represent actual fuel assembly responses. A major impetus for this work was to facilitate code validation (primarily MELCOR) and reduce modeling uncertainties within the code.

The Spent Fuel Pool Heatup and Propagation Phenomena project was conducted from April 2004 until November 2006 over which time seven unique experimental apparatuses were tested at Sandia National Laboratories in Albuquerque, New Mexico. In addition to these experiments, extensive simulation efforts were undertaken with the MELCOR severe accident modeling code to understand and predict the behavior observed in the tests. The key findings from this integrated experimental and simulation program are:

- Electrically heated spent fuel rod simulators can be fabricated with Zircaloy cladding to accurately represent the decay heat, thermal mass and Zircaloy reactivity of a prototypic spent fuel rod.

- The measured form and friction loss coefficients of a prototypic BWR assembly were significantly different from generally accepted values. Use of the measured coefficients was vital for accuracy when calculating (with MELCOR) the naturally induced flow in a heated, prototypic BWR assembly.

- Incorporation of "breakaway" Zircaloy oxidation kinetics into MELCOR was vital for accurately capturing the Zircaloy heat-up to ignition and oxygen consumption.

- For the full length ignition test, the MELCOR model predicted the peak cladding temperature (PCT) of the assembly to within 40 K at all times and the time of ignition to within 5 minutes.

- For the 1×4 ignition experiment, the standard MELCOR model predicted ignition in the center and peripheral assemblies to within 30 and 15 minutes, respectively. The error in ignition timing between the simulations and experiment is approximately 10%. The difference in timing is likely due to the inability of the lumped parameter approach used in MELCOR to account for steep radial temperature gradients.

- Post-mortem examination of the integral test assemblies revealed gross distortion of the pool rack and channel box, rubblization of the tubing bundle and accumulation of debris on the bottom tie plate that resulted in flow blockage. Flow blockage was also evident from molten aluminum (originating from Boral plates built into the pool rack) that collected on and below the bottom tie plates.

ABBREVIATIONS/DEFINITIONS

%LEL	percent lower explosive limit
%O$_2$	percent oxygen
ANL	Argonne National Laboratory
BE	best estimate
BWR	boiling water reactor
COBRA	Coolant Boiling in Rod Arrays: thermal-hydraulic and radiative analysis code
COBRA-SFS	Coolant Boiling in Rod Arrays-Spent Fuel Storage
COR	CORE – fuel response model in MELCOR
Cr	chromium
CTE	coefficient of thermal expansion
CVH	control volume hydrodynamics model in MELCOR
CYBL	cylindrical boiling
DAQ	data acquisition
EPA	error propagation analysis
Fe	iron
HS	heat structure
HW	hot wire
ID	inside diameter
IR	infrared
LEL	lower explosive limit
MELCOR	severe accident analysis code
MgO	magnesium oxide
Ni	nickel
NPT	national pipe thread
NRC	Nuclear Regulatory Commission
O$_2$	oxygen
OD	outside diameter
PCT	peak cladding temperature
PWR	pressurized water reactor
RADGEN	view factor input generation code for COBRA
RMS	root mean square
scfm	standard cubic feet per minute (standard defined at 0°C and 1 atm)
SEM	scanning electron microscopy
SFP	spent fuel pool
slpm	standard liters per minute (standard defined at 0°C and 1 atm)
Sn	tin
SNL	Sandia National Laboratories
TC	thermocouple
vol%	volume percent
wt%	weight percent
Zr	Zirconium

1 INTRODUCTION

1.1 Background

In 2001, the United States Nuclear Regulatory Commission (NRC) staff performed an evaluation of the potential accident risk in a spent fuel pool (SFP) at decommissioning plants in the United States [1]. NUREG-1738 was prepared to provide a technical basis for decommissioning rulemaking for permanently shutdown nuclear power plants. The study described a modeling approach of a typical decommissioning plant with design assumptions and industry commitments, thermal-hydraulic analyses performed to evaluate spent fuel stored in the SFP at decommissioning plants, risk assessment of SFP accidents, consequence calculations and the implications for decommissioning regulatory requirements. It was known that some of the assumptions in the accident progression in NUREG-1738 were necessarily conservative, especially the estimation of the fuel damage. Furthermore, the NRC desired to expand the study to include accidents in the SFPs of operating power plants. Consequently, the NRC continued SFP accident research by applying best-estimate computer codes to predict the severe accident progression following various postulated accident initiators. The best-estimate computer code studies identified various modeling and phenomenological uncertainties that prompted a need for experimental confirmation [2]. The present experimental program was undertaken to address thermal-hydraulic issues associated with complete loss-of-coolant accidents in boiling water reactor (BWR) SFPs.

1.2 Objectives and Scope of Testing

The objective of this project was to provide basic thermal-hydraulic data associated with a SFP complete loss-of-coolant accident. The accident conditions of interest for the SFP were simulated in a full-scale prototypic fashion (electrically-heated, prototypic assemblies in a prototypic SFP rack) so that the experimental results closely represent actual fuel assembly responses. A major impetus for this work was to facilitate code validation (primarily MELCOR) and reduce questions about buoyancy driven flow, heat transfer modes, oxidation reaction kinetics and with the interpretation of the experimental results. It was necessary to simulate a cluster of assemblies to represent a higher decay (younger) assembly surrounded by older, lower power assemblies. Specifically, this program provided data and analysis confirming: (1) MELCOR modeling of inter-assembly radiant heat transfer, (2) flow resistance modeling and the natural convective flow induced in a fuel assembly as it heats up in air, (3) the potential for and nature of thermal transient (i.e., Zircaloy fire) propagation, and (4) mitigation strategies concerning fuel assembly management.

1.3 Test Plan Synopsis

Three configurations of fuel assemblies were utilized in the experimental testing. Initial tests were conducted in a small configuration in order to assess the performance and suitability of Zirconium clad electrically-heated spent fuel rod simulators. Two such tests were conducted in order to demonstrate that the heater design was capable of initiating a Zirconium fire. The second configuration examined a single, full-length highly prototypic fuel assembly (stainless steel, Zircaloy, and Incoloy versions) inside a prototypical pool rack cell. The stainless steel rod assembly was unheated and only used for the hydraulic characterization. The Incoloy fuel assemblies were used to conduct high temperature tests while minimizing complications of oxidizing Zircaloy surfaces. The Zircaloy version of the full-length fuel assembly was used for

1

tests that were taken to conditions of high temperature oxidation and ignition. The final configuration was five Zircaloy short (1/3 length) assemblies in a 3×3 pool rack. The short array of assemblies was designed to simulate a slice from the middle to upper portion of an array of full-length assemblies. This was accomplished in the final ignition test by accurately controlling both the flow rate and temperature of air introduced into the bottom of each partial assembly.

Two types of testing were conducted: (1) separate effects tests and (2) integral effects tests. In the separate effects tests, the experiments were designed to investigate a specific heat transfer or flow phenomenon such as thermal radiative coupling or induced natural convective flow. These tests were non-destructive and involved some non-prototypic materials (e.g., stainless steel and Incoloy). This phase of testing involved conducting a greater number of less complicated tests. In the integral effects tests, all hydraulic flow and heat transfer phenomena were investigated simultaneously. These tests were specified with boundary conditions which led to the destruction of the experimental apparatus and thus involved a limited number of more complex experiments. Table 1.1 summarizes in chronological order the phased tests conducted as part of this experimental program.

Table 1.1   Summary of the testing elements in the experimental program

Description	Purpose	Assembly	Rod material
Heater Design	Test electrical heater performance, preliminary data on zircaloy fire	12 rod bundle	Zircaloy
Separate Effects	Hydraulics – determine form loss and laminar friction coefficients	Prototypic	Stainless Steel
Separate Effects	Thermal hydraulics – Determine input conditions for partial length experiments	Prototypic	Incoloy
Separate Effects	Thermal radiation – radiation coupling in a 1×4 arrangement	Prototypic – Partial length	Incoloy
Integral Effects	Axial Ignition – temperature profiles, induced flow, axial O_2 profile, nature of fire	Prototypic	Zircaloy
Integral Effects	Radial Propagation – Determine nature of radial fire propagation	Prototypic – Partial length	Zircaloy

A unique aspect of this project was the deliberate close coupling of the experiments with numerical analysis. The principal code used was the severe accident code MELCOR. At each step in the experimental program, MELCOR was used (1) as a tool for the experimental design, (2) for the pre-test results prediction, and (3) for post-test analysis of the calculated and measured responses. The post-test analysis helped identify important response parameters, which often led to improvements in the conduct of the next phase of testing and improvements in the modeling approach. The experimental and modeling findings and improvements from post-test analysis of the previous experimental program tests were subsequently used to design and predict the next phase of experiments, and so on. For example, the pressure drops measured in the hydraulic testing did not initially match the MELCOR predictions based on best-estimate hydraulic parameters. When MELCOR was updated with the experimentally derived hydraulic parameters, the MELCOR predictions of the next thermal-hydraulic testing phase were in excellent agreement with the data.

Figure 1.1 illustrates the closely coupled relationship between the various experiments and the MELCOR modeling. This figure also stresses the overall objective of this program to validate MELCOR with full-scale prototypic experimental measurements to reduce uncertainties in whole pool accident analyses. The phased, multi-scale test program and integrated relationships between the test program and the analytical modeling efforts were critical to the comprehensive understanding of the complex accident phenomena.

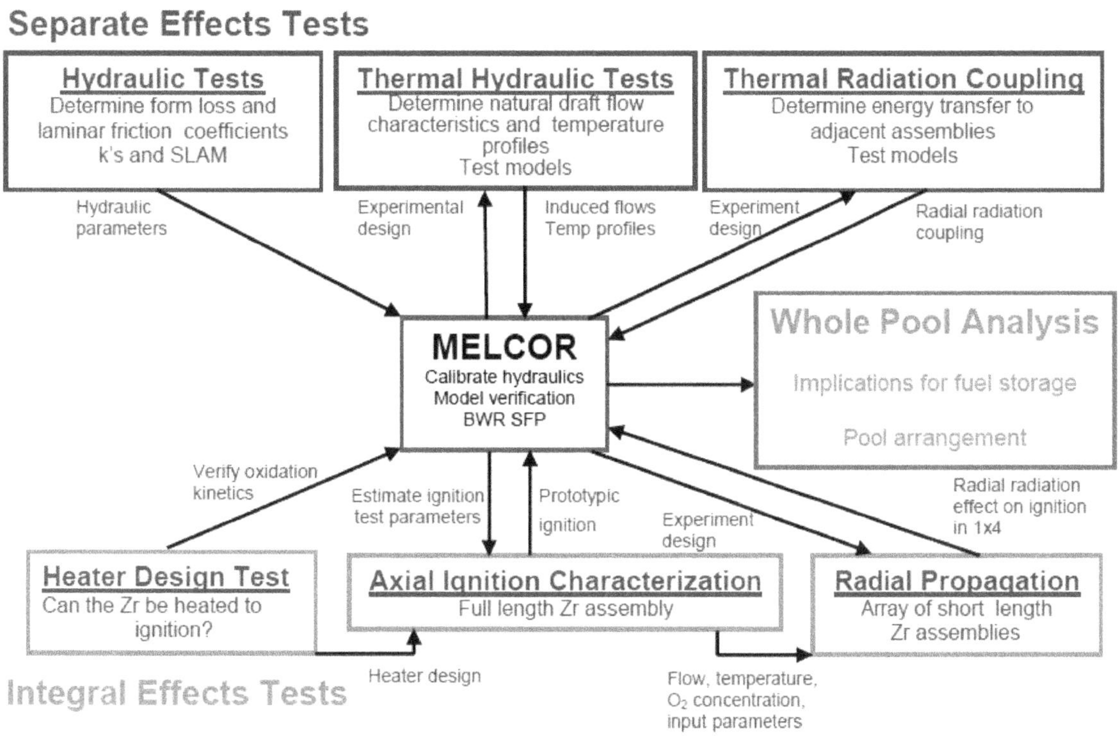

Figure 1.1   Illustration of the closely-coupled relationship between the various testing elements and the supporting MELCOR modeling

1.4   Report Organization

This report is organized in a series of sections with each describing a particular experiment or phase of experiments. Section 2 describes the first heater design test where the design and physical properties of the heaters are described. Both the full oxygen test checkout test and Section 3 describe the second heater design test, which was conducted under reduced oxygen conditions. Section 4 describes the hydraulic characterization of the highly prototypic 9×9 BWR assembly. Next, Section 5 describes the thermal-hydraulic testing where the naturally induced flows inside a heated assembly are characterized. Section 6 describes the thermal radiative coupling experiments; Section 7 describes the ignition testing of a full-length Zircaloy assembly. Section 8 describes the ignition testing of the 1×4 array of short length Zircaloy assemblies. Finally, Section 9 summarizes the key findings.

2 HEATER DESIGN TEST 1

2.1 Objective

The main objective of the first heater design test was to verify that the heater design would produce high enough surface temperatures for long enough to ignite the Zircaloy cladding before the heater element fails. A secondary objective was to gain experience with monitoring a Zirconium fire and to identify important issues.

The surrogate fuel assemblies used in this program consisted of a Zircaloy tubing jacket with a central Nichrome heater element surrounded by compacted magnesium oxide. The heaters were assembled by Watlow Electric Manufacturing Company (Watlow), a large manufacturer of industrial electric heaters. Watlow has considerable experience in manufacturing reliable heater rods. The Nichrome element melts when it reaches a temperature of about 1400°C (1673 K). Typically, heaters of this type are rated for intermediate (~4 hours) duty at surface temperatures of 1000°C (1273 K). This allows for a 400 K temperature drop between the Nichrome element and the surface of the heater jacket. The actual temperature drop between the jacket surface and the Nichrome element is highly dependent on the power output of the heater. The lower the power output, the lower the temperature drop, hence, the higher the temperature the heater jacket surface can reach before element failure. The power output of the heaters used in this study was at most 25 W/ft, which is considered very low (an order of magnitude lower than typical). Thus, surface temperatures of at least 1225°C (1500 K) and perhaps as high as 1350°C (1623 K) were expected before heater element failure. Depending upon the heat losses, the oxidation layer thickness, the ventilation characteristics (i.e., flow rate, inlet gas temperature, oxygen content), as well as other factors, 1225°C could be above or below the temperature where rapid oxidation (i.e., ignition) occurs.

2.2 Experimental

2.2.1 Electric Heater Design

The design of the Zircaloy jacketed heater was typical of those made by Watlow. Each heater was comprised of a central Nichrome heater element, compressed magnesium oxide powder, and an outer Zircaloy 2 jacket. The composition of the heater rods is listed in Table 2.1. The final magnesium oxide powder density was 2.720 g/cm^3 corresponding to a solids fraction of 0.759.

Table 2.1   Composition of heater rods

Component		Nominal wt%	Mass/rod (lb)	Mass/rod (g)	# Rods	Total mass (g)	Mwt (g/mol)	Total moles (mol)
Zircaloy 2			0.368					
	Zr	98.40%		164.2513	12	1971.015	91.224	21.60632
	Sn	1.30%		2.169986	12	26.03983	118.71	0.219357
	Fe	0.18%		0.30046	12	3.605515	55.845	0.064563
	Cr	0.10%		0.166922	12	2.003064	51.9961	0.038523
	Ni	0.07%		0.116845	12	1.402145	58.6934	0.023889
		100.05%						
MgO			0.377	171.0043	12	2052.052	40.3044	50.91385
Nichrome			0.046	20.86525	12	250.383		

The diameter of the heater rods used in the heater design test bundle was smaller than prototypic due to the availability of Zircaloy tubing. The tubing used was 0.440-in. diameter, 0.028-in. wall Zircaloy 2 tubing for a boiling water reactor (BWR) assembly. This tubing was made into 48-in. long, 0.375-in. diameter heaters by Watlow in a process whereby the 0.440-in. tubing was drawn through a die that reduced the diameter to 0.375 in. and compressed the magnesium oxide powder considerably. Compression of the magnesium oxide powder is required to establish good thermal contact between the Nichrome heater element and the outer metallic sheath.

Consideration was given to how well the electric heater would represent a spent fuel rod from a thermal perspective. The power produced by a spent fuel rod depends primarily on the age of the spent fuel rod, which is defined as the time since it was removed from the reactor. Table 2.2 shows a typical BWR spent fuel assembly power for various ages along with the corresponding power per rod and linear foot . The electric heater rods were designed to produce 25 W/ft at 120 volts. With this design, spent fuel assemblies with ages from as short as three days old to over two years old could easily be simulated.

Table 2.2   Correlation of assembly and rod power versus the required voltage

Time (days)	Assembly peak power (kW)	Total rod power (W)	Linear rod power (W/ft)	Voltage (V)
3	23.93	323.40	25.87	122.07
10	15.01	202.84	16.23	96.68
100	5.17	69.80	5.58	56.72
365	2.30	31.03	2.48	37.81
730	1.33	17.91	1.43	28.73

In order to simplify the heater rod fabrication, the power was distributed linearly along the heated length of the rod. A typical spent fuel rod produces a chopped cosine power distribution with 130% peak power at the center and 60% power at the ends. The difference between the cosine power distribution of a typical spent fuel rod and the linear power distribution of the electric heater rod is shown in Figure 2.1. The discrepancy is not great and can be accounted for in the model validation effort.

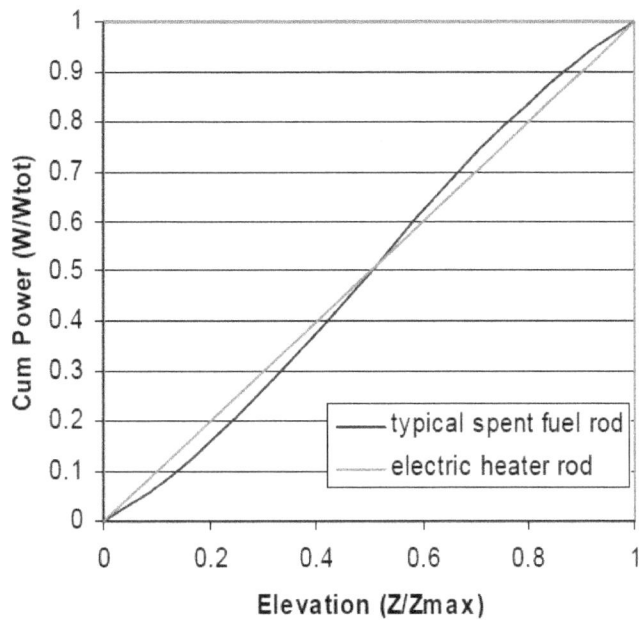

Figure 2.1   **Cumulative power distribution along a typical spent fuel rod and the electric heater rod**

Another important thermal property of the electric heater rod is the thermal mass. This is best characterized by the specific heat (on a volume basis) and is shown in Figure 2.2 for spent fuel, compacted magnesium oxide (based on the as-built compacted magnesium oxide density), and as aluminum oxide. The agreement between magnesium oxide and spent nuclear fuel is very close over the entire temperature range making the magnesium oxide an ideal surrogate.

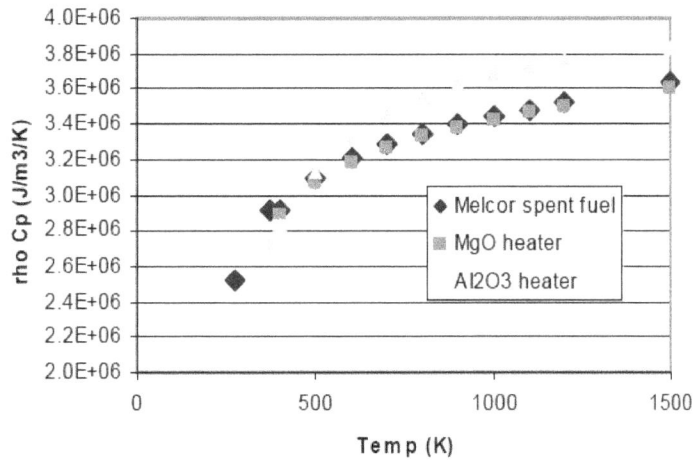

Figure 2.2   **Specific heat of electric heaters compared to spent nuclear fuel**

2.2.2 Initial Experimental Assembly

The experiment was constructed and operated in a cylindrical boiling (CYBL) vessel located in Sandia National Laboratories (SNL) in Albuquerque, New Mexico. Twelve 4-ft. long, 0.375-in. diameter heater rods with a pitch of 0.539 in., as shown in Figure 2.3, were assembled into a bundle (Figure 2.4). The heater rods were instrumented with 0.032-in. ungrounded Incoloy sheathed type K thermocouples (TCs) in a configuration as shown in Table 2.3. Table 2.3 also lists other instrumentation used in the test and was collected by a computerized data acquisition system at nominally 1-second intervals. The TCs were attached to the Zircaloy tubing by spot welding a small piece of stainless steel foil across the tip, as shown in Figure 2.5. A 57-mm ID (2.244-in.) quartz tube was placed around the heater bundle, and the tube was surrounded with reflective foil (Figure 2.6 and Figure 2.7). Initially, the bundle was insulated with 3 in. of Kaowool™. A small axial slit through the insulation and foil was made to permit viewing (see Figure 2.3 and Figure 2.8). However, in initial testing of this setup, the temperature rise rate was low indicating that the heat loss was too high. The initial test setup is shown in Figure 2.9.

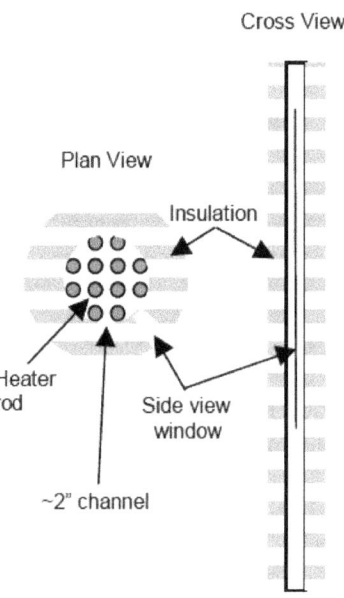

Figure 2.3   **Schematic of the heater design test**

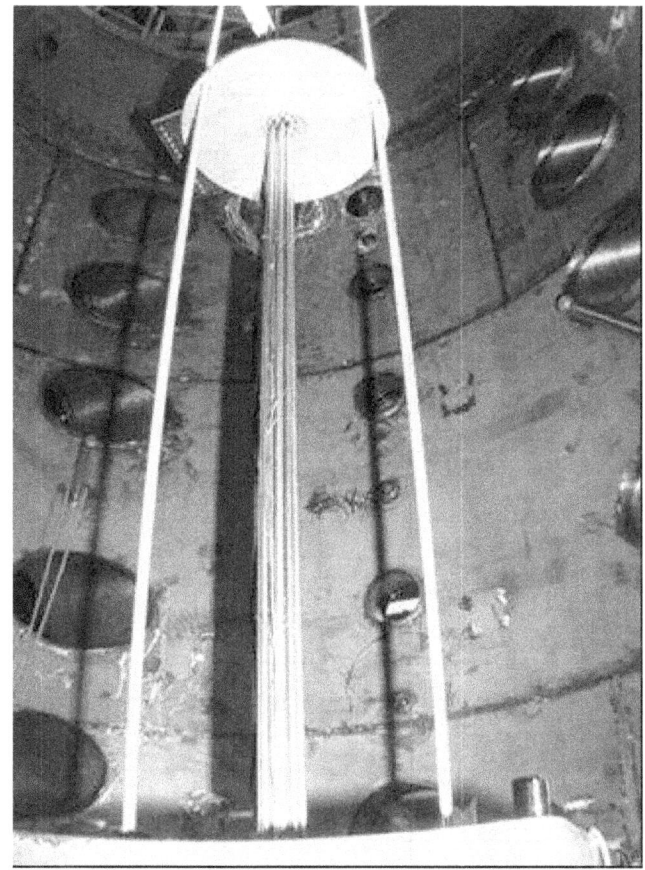

Figure 2.4   Bundle of heater rods

Table 2.3   Heater Design Test 1 instrument layout

Heater Design Test Instrumentation

Chan #	Name	Description	Instrument type	Max Volt	Units
0	D0	Heater Axial Elevation 0"	K TC		
1	D6	Heater Axial Elevation 6"	K TC		
2	D12	Heater Axial Elevation 12"	K TC		
3	D18	Heater Axial Elevation 18"	K TC		
4	D24	Heater Axial Elevation 24"	K TC		
5	D30	Heater Axial Elevation 30"	K TC		
6	D36	Heater Axial Elevation 36"	K TC		
7	D42	Heater Axial Elevation 42"	K TC		C
8	D48	Heater Axial Elevation 48"	K TC		C
9	W12	W Heater Elevation 12"	K TC		C
10	X12	X Heater Elevation 12"	K TC		C
11	Y12	Y Heater Elevation 12"	K TC		C
12	Z12	Z Heater Elevation 12"	K TC		C
13	W24	W Heater Elevation 24"	K TC		C
14	X24	X Heater Elevation 24"	K TC		C
15	Y24	Y Heater Elevation 24"	K TC		C
16	Z24	Z Heater Elevation 24"	K TC		C
17	W36	W Heater Elevation 36"	K TC		C
18	X36	X Heater Elevation 36"	K TC		C
19	Y36	Y Heater Elevation 36"	K TC		C
20	Z36	Z Heater Elevation 36"	K TC		C
21	F24	F Heater Elevation 42"	K TC		C
22	INLET	Air inlet temperature	K TC		C
23	OVEN	Oven temperature	K TC		C
24	Chimney	Chimney Air temperature	K TC		C
25	A current	Current of Heater A	Current xducer	5	ampere
26	B current	Current of Heater B	Current xducer	5	ampere
27	C current	Current of Heater C	Current xducer	5	ampere
28	D current	Current of Heater D	Current xducer	5	ampere
29	E current	Current of Heater E	Current xducer	5	ampere
30	F current	Current of Heater F	Current xducer	5	ampere
31	G current	Current of Heater G	Current xducer	5	ampere
32	H current	Current of Heater H	Current xducer	5	ampere
33	W current	Current of Heater W	Current xducer	5	ampere
34	X current	Current of Heater X	Current xducer	5	ampere
35	Y current	Current of Heater Y	Current xducer	5	ampere
36	Z current	Current of Heater Z	Current xducer	5	ampere
37	Oxygen	Oxygen in chimney	Oxygen		percent
38*	Air Flow	Mass air flow into oven	MKS Flow	10	slpm
39**	Out 42	Outer skin temp 42"	K TC		C

*Not used
**Final assembly only

Key
5° SLIT
(A B / C D E F / W X Y Z / G H)

Figure 2.5　　TC attachment detail

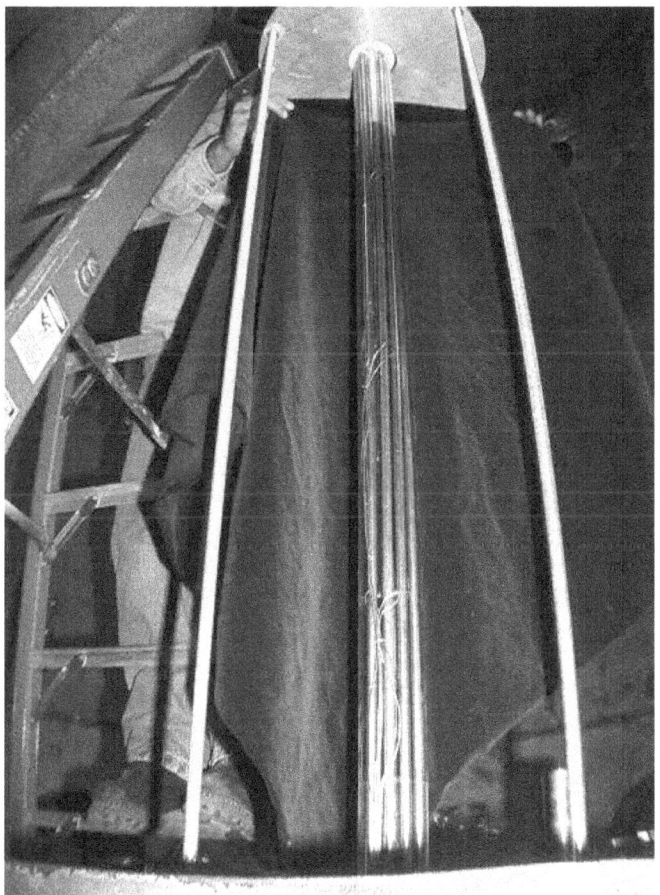

Figure 2.6　　Quartz tube installed over heater bundle

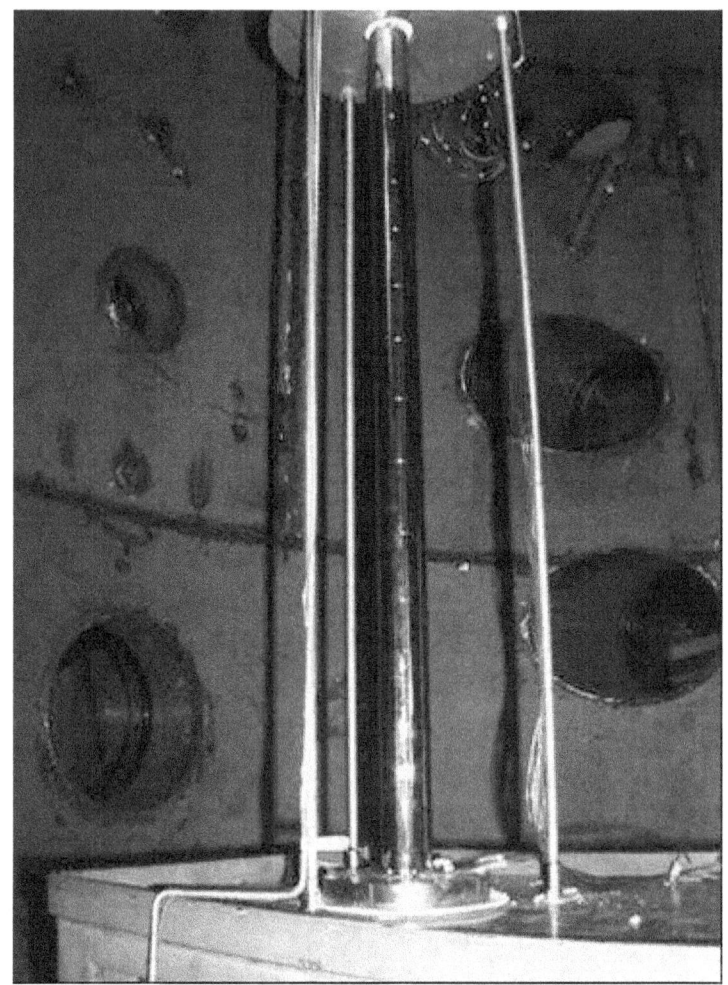

Figure 2.7   Stainless steel foil in place over quartz tube

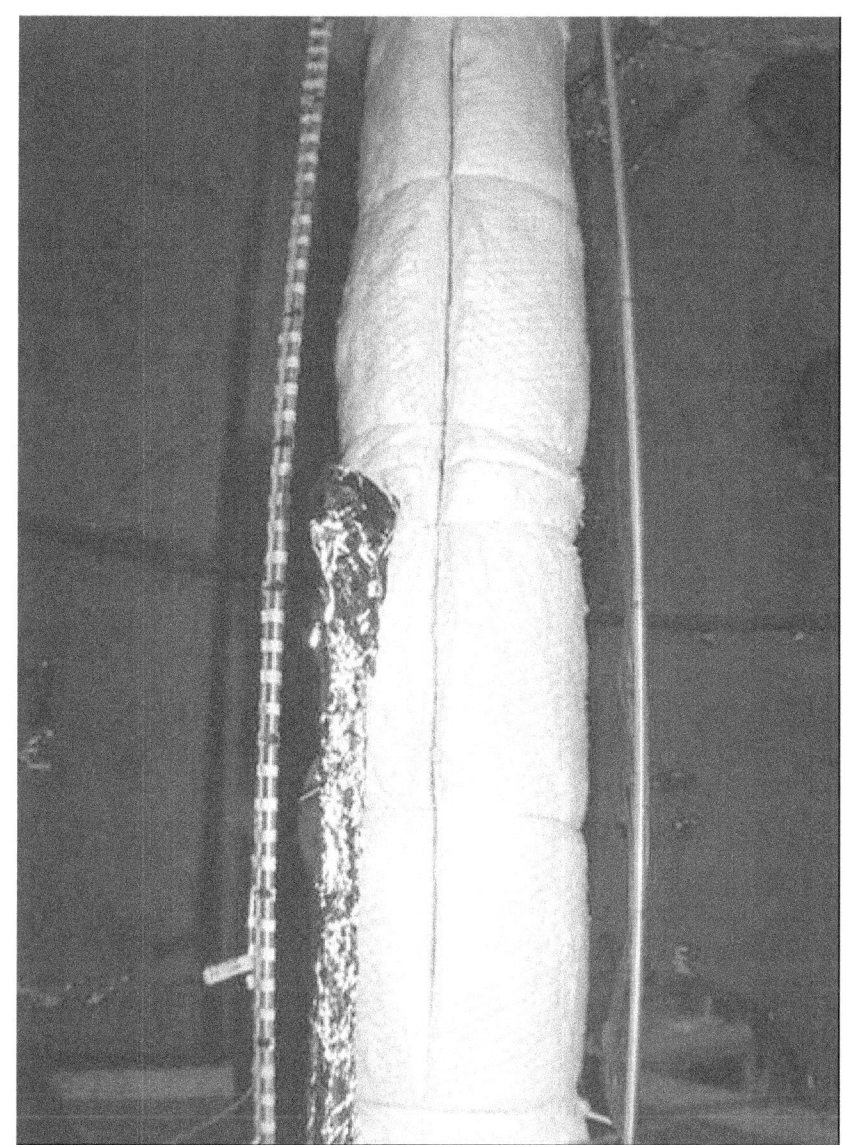

Figure 2.8   Insulation and side-viewing slit

Figure 2.9   Initial test setup

2.2.3   Final Experimental Assembly

After the initial heat test indicated a heat-up rate too low, the test was terminated, and the assembly was allowed to cool. It was decided to increase the insulation and abandon the side-viewing slit. The viewing slit was filled with insulation and a stainless steel foil wrap was placed around the initial 3-in. layer of insulation. An additional 3-in. wrap of insulation was added and a final stainless steel sheet metal wrap was added. In addition, the top plate and chimney were also insulated. The final test setup is shown in Figure 2.10.

Figure 2.10   More insulated final test setup

2.2.4   Additional Sampling

In addition to the data collected by the instrumentation shown in Table 2.3, three other sampling efforts collected data for this experiment. The first effort was to monitor the oxygen concentration of the exit gas, the second effort was the sampling of smoke particulates for metals analysis, and the third was video recording of the experiment from a number of perspectives.

Oxygen: The oxygen concentration of the gas leaving the test bundle was monitored using an ITX (S/N: 0307010-286) portable atmosphere-monitoring instrument. The instrument measured both oxygen concentration and the percent lower explosive limit (%LEL). The sample was collected from the mid point of the chimney through stainless steel tubing that was routed through an ice bath and outside the CYBL vessel where the oxygen monitor was located. The sample was drawn using the instrument's internal sample pump. The instrument's sampling system included a small particulate filter that plugged shortly after smoke was emitted from the test bundle. The percent oxygen ($\%O_2$) and %LEL data were collected internally along with the time at 5-second intervals.

Smoke: Health physics personnel installed two particulate sampling devices inside the CYBL vessel to analyze the total metal content of the smoke generated by the experiment. One device was located about 3 ft. above the chimney and the other was located about midway between the top of the chimney and the upper lip of the CYBL vessel. The sample filters were analyzed for the metals added to Zirconium to make the Zircaloy 2 alloy, which include tin, iron, nickel, and chromium. The samples were also analyzed for magnesium, a principal constituent of the

15

magnesium oxide insulation used inside the heaters. Table 2.1 presents the initial mass of each of these constituents that was initially present in the test bundle.

Video: Video perspectives included a close-up looking down the chimney, a side view of the experimental setup, and a distant top view from the top opening of the CYBL vessel.

2.2.5 Initial Test Operation

The initial experimental setup was tested on December 9, 2004. The air preheater (and data collection system) was started at 11:06 a.m., and the air flow of 1.0 scfm was started at 11:13 a.m. The preheat air temperature set point was 600 K (327°C). The power to the heater rods was applied at 1 p.m. (114 minutes into experiment). The initial power was a total of 800 W, which corresponds to about 16.6 W/ft. At 2:24 p.m. (198 minutes) the total power was increased to 1200 W or 25 W/ft. At 2:45 p.m., the power was terminated concluding the initial test.

The data collected during this initial testing follows. Figure 2.11 shows the inlet and exit air temperatures. Figure 2.12 shows the current history of the test bundle. Figure 2.13 and Figure 2.14 show the axial and radial temperature profiles. Figure 2.15 shows the rate of change of the axial temperature profile.

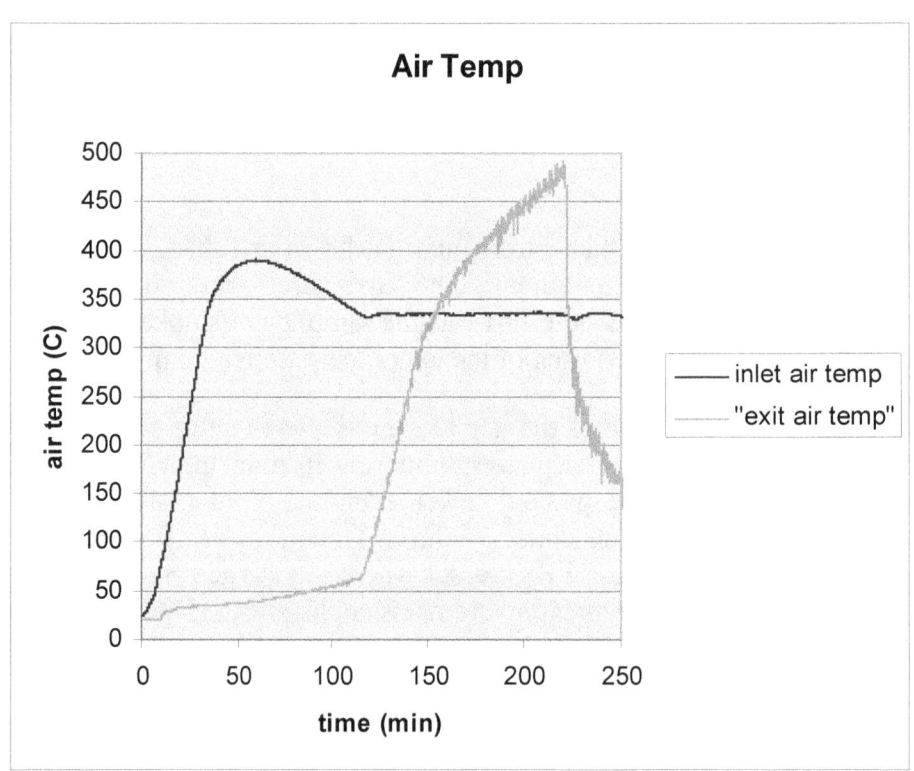

Figure 2.11   **Temperature of air entering the bottom of the assembly and temperature of the air leaving the top of the assembly during the initial test operation**

16

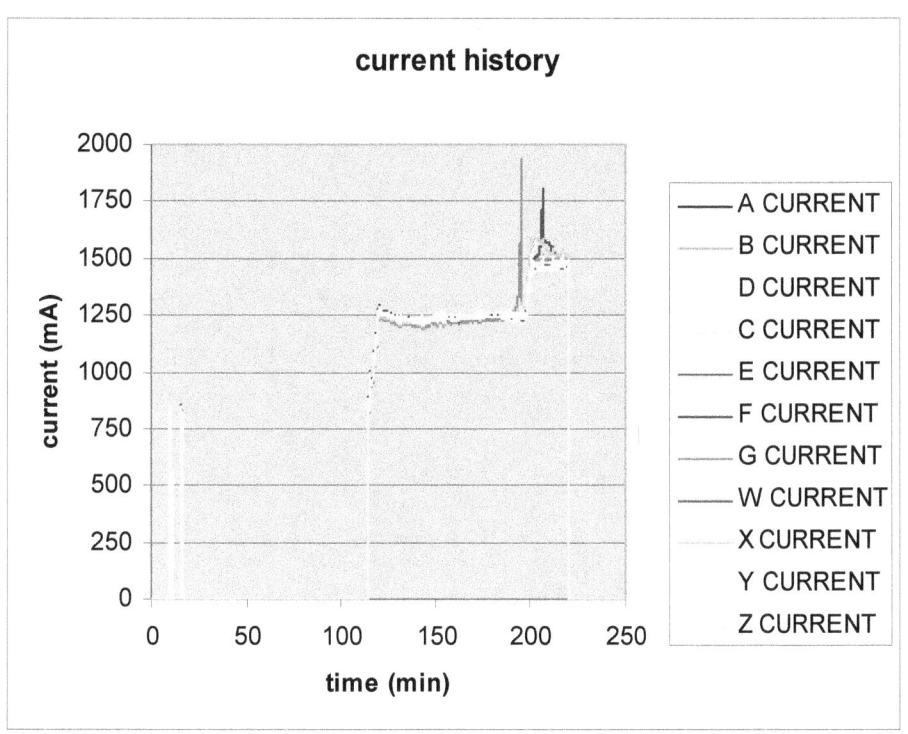

Figure 2.12   History of the electrical current supplied to each of the twelve heater rods during the initial test operation

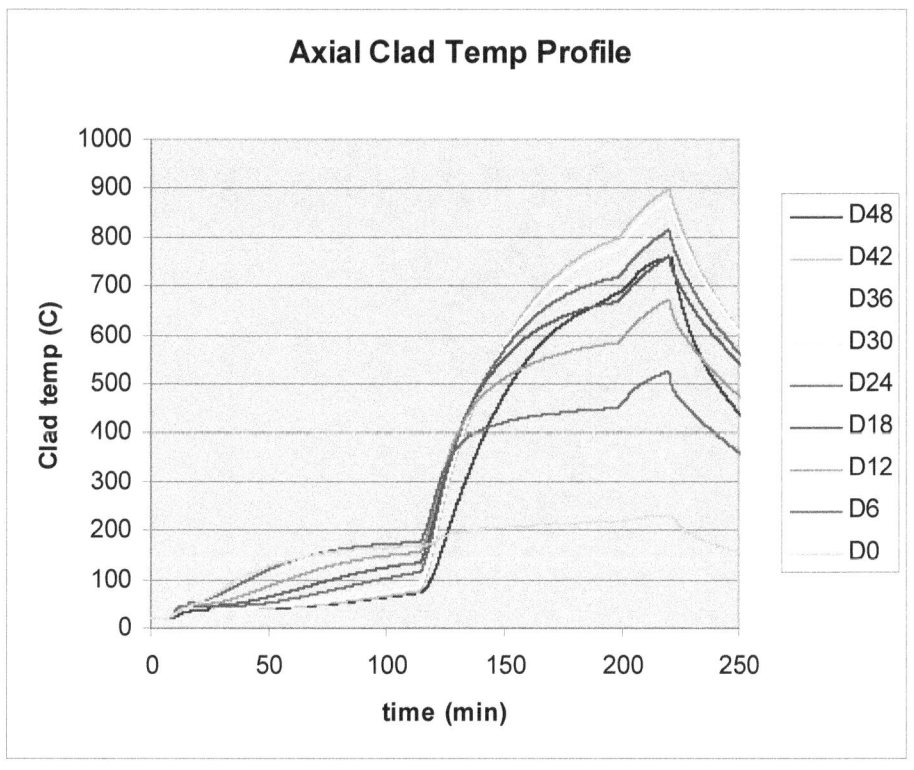

Figure 2.13   Axial temperature profile developed during the initial test operation

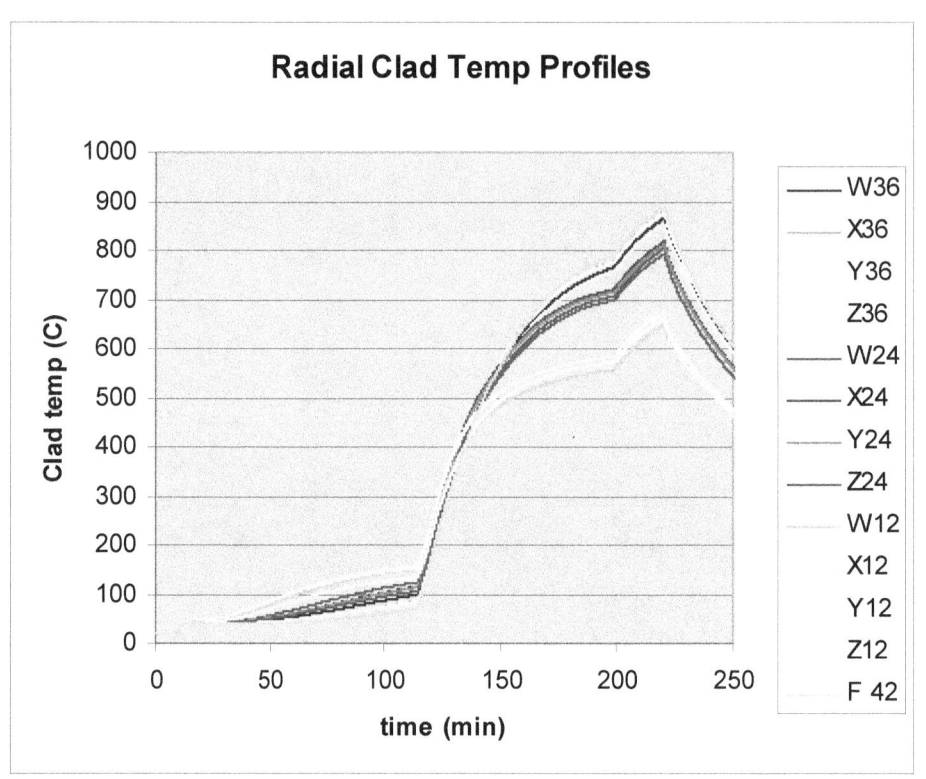

Figure 2.14   Radial temperature profiles developed during the initial test operation

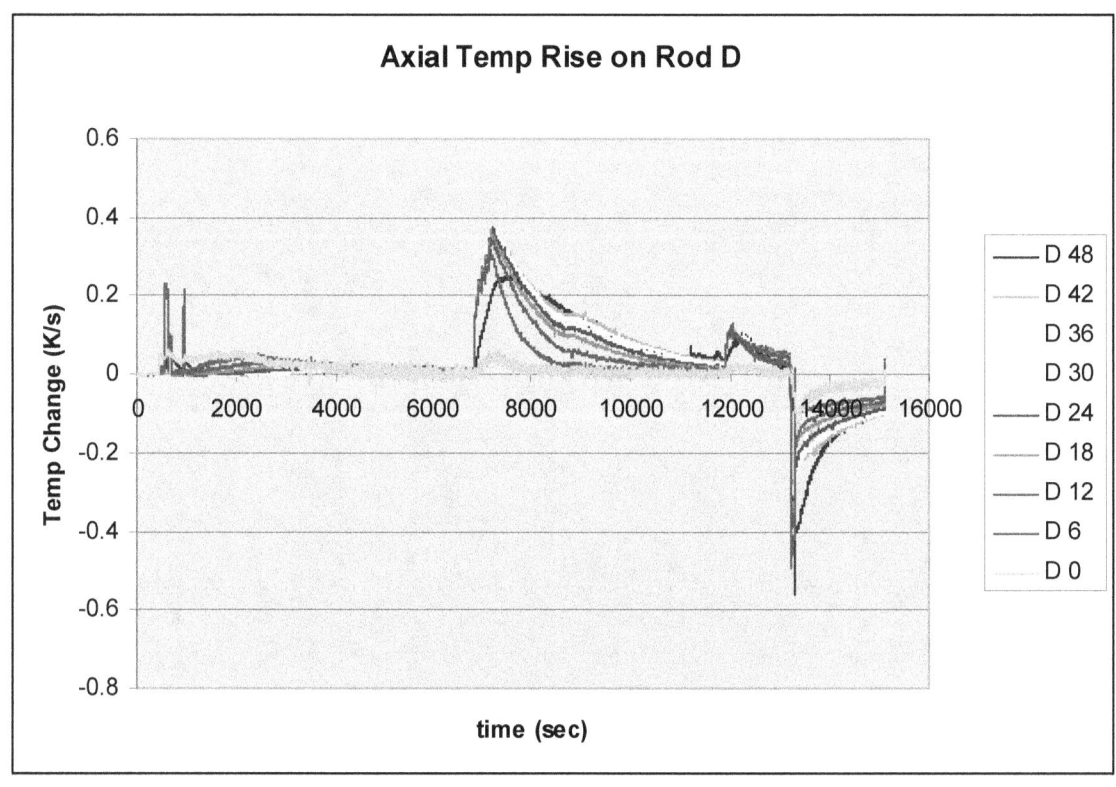

Figure 2.15   Axial differential temperature change measured during the initial test operation

2.2.6   Initial Test Analysis

In order to ensure ignition of the Zirconium cladding, the best available MELCOR model indicated the clad temperature needed to reach 1000°C at a rate of at least 0.1 K/s. If the heat-up rate was not high enough, a Zirconium oxide layer could grow such that ignition would not be possible at the maximum power rating of the heaters. Figure 2.13 shows that the clad peak temperature was approaching 800°C at 198 minutes into the test. Figure 2.15 shows that the heat-up rate at this time (11,000 seconds) was only about 0.04 K/s, far below the 0.1 K/s needed. At this point, it was decided to increase the power to 1200 W. This increased the heat-up rate to 0.12 K/s; but as the clad temperature approached 900°C, the heat-up rate fell back to 0.04 K/s. Thus, ignition was not ensured so the test was terminated. The results of this initial testing clearly indicated that the heat losses were too great to ensure ignition of the heater rods.

2.2.7   Final Test Operation

The initial testing, though terminated, brought the Zirconium heater rods to sufficient temperature long enough to produce a significant oxide layer. This oxide layer would make igniting the rods more difficult than before. A number of measures were taken to ensure ignition during the final test even with this pre-established oxide layer. As described above, the insulation around the test bundle was more than doubled, and the side-viewing slit was abandoned. It was decided that the inlet air temperature would be increased to 750 K (477°C) and that the heater power could be increased to 120% of rating without compromising the heaters. Finally, convective losses would be reduced and further oxide buildup minimized by reducing the air flow until sufficient temperatures were reached.

Table 2.4 lists the chronology of events during the final test. Many of the main events are noted in Figure 2.16, which shows the temperature change of the axial profile. The data shown on this plot is the calculated first derivative of the temperature data over a moving 10-sample (nominally 10-second) window. Presenting the temperature data as the first derivative most clearly denotes the temperature response to the events listed in Table 2.4. A complete record of all the data collected during the final test can be found in Appendix A, SFP2 Data Summary.

Table 2.4   Chronology of events during experiment

Time (s)	Event Value	Event Description
0		Inlet temperature set to 600 K, flow 1.0 scfm
4813.5		Inlet temperature increased to 750 K
6433.50		O_2 data on (12:50 on O_2 data file, 3 min difference)
6485.70	800	W on heater rods (16.7 W/ft)
7445.70	0.23	K/s
8405.70	0.18	K/s
9125.70	0.126	K/s
9339.40	0.1	scfm air inlet (13:39:42 O_2 time)
9639.40	0.11	K/s
9809.20	1200	W on heater rods (25 W/ft)
10049.20	20.6	% O_2
10169.20	0.17	K/s (20.5% O_2)
10889.20	20.4	%O_2
11309.20	20.3	%O_2
11549.20	20.2	%O_2
11909.20	20	%O_2
11969.20	19.9	%O_2
12022.50	1440	W (94 min experiment time)
12082.50	19.8	%O_2
12142.50	19.7	%O_2
12308.50	1	scfm air inlet IGNITION (14:28:02 O_2 time)
12628.50	8	%O_2
12688.50	7.5	%O_2 SMOKE (14:34 O_2 time)
12808.50	4.8	%O_2 (filter may be plugged)
12868.50	4.1	%O_2 (restart timer)
12988.50	2.8	%O_2 (filter may be plugged)
13948.50	20.5	%O_2 (this sample w/o filter)
15557.80	0	scfm air inlet (15:22:56 video time)

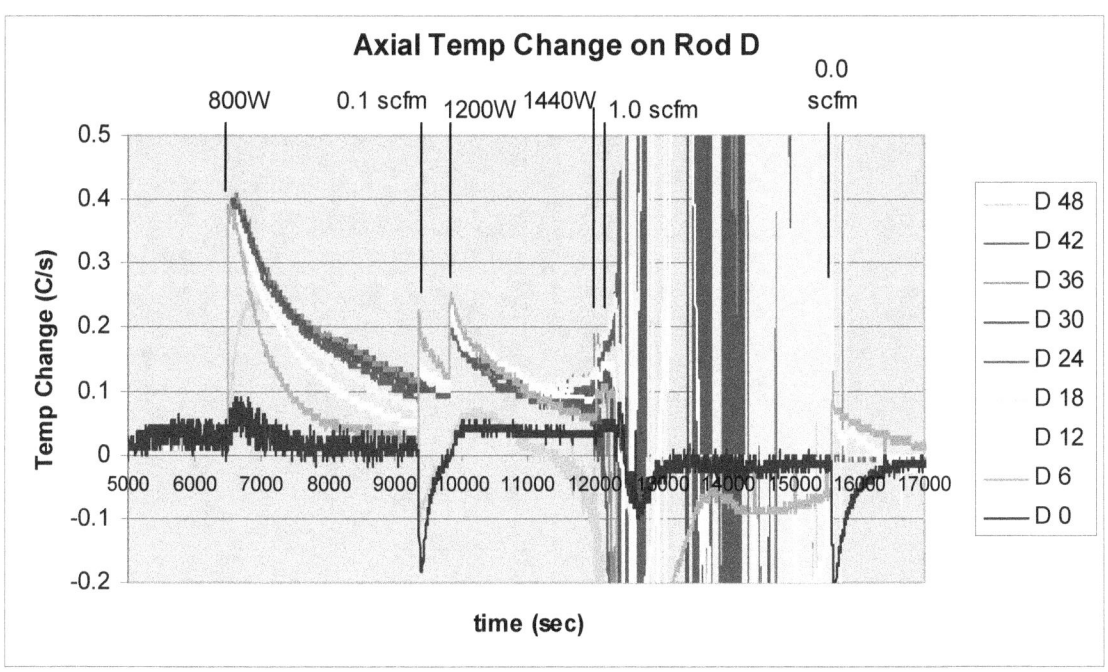

Figure 2.16   First derivative of axial temperature profile with major experimental events noted

The test was started on December 10, 2004, at 11 a.m. The inlet air flow was 1.0 scfm and 750 K (477°C). Power was applied to the heater rods at a level of 800 W at about 12:50 p.m. At 1:30 p.m., peak temperatures were approaching 700°C, but heat rates were dropping to 0.1 K/s. At this time, the air flow rate was reduced to 0.1 scfm. This had only a minimal effect on the heat rate of the hotter rods; therefore, at about 1:45 p.m., the power was increased to 1200 W. At about 2:20 p.m., the peak temperatures were approaching 1000°C, but the heat rate had dropped to less than 0.1 K/s. At this point, the power was increased to 1440 W, 120% of power rating. Five minutes later, with clad temperatures at 1000°C, the air was increased back to 1 scfm. Ignition was immediately evident as a bright glow in the chimney, and at about 2:30 p.m., smoke was first seen (Figure 2.17). The burn continued until about 3:20 p.m., when the air flow was terminated and the test was allowed to cool off (Figure 2.18).

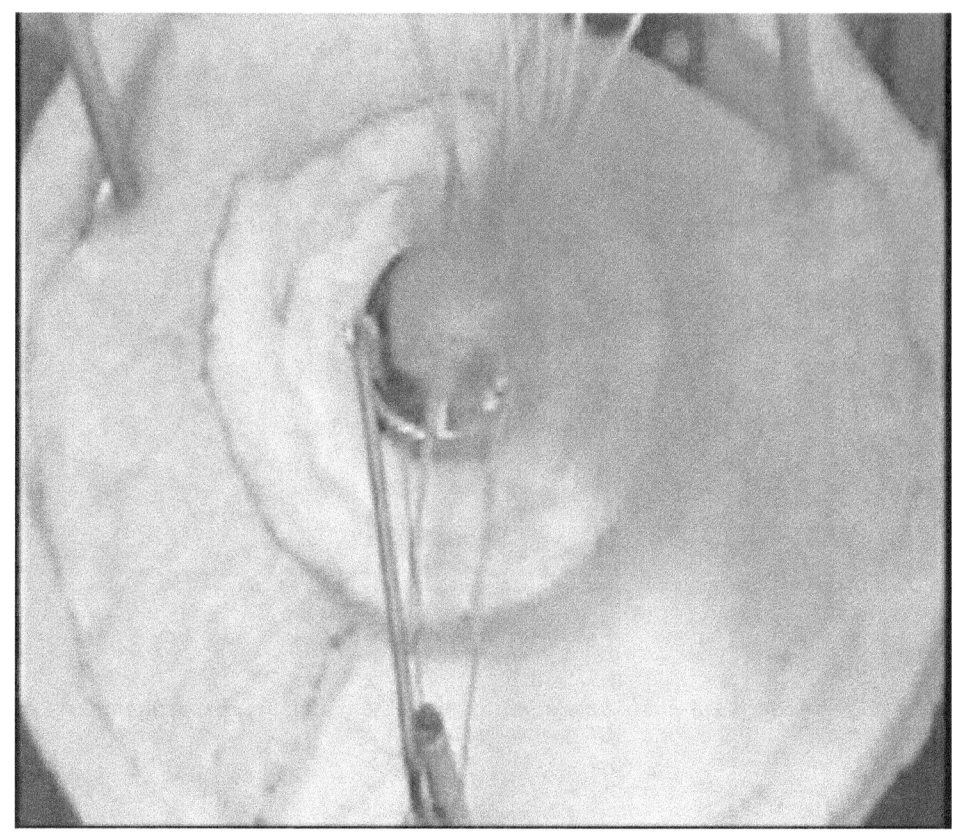

Figure 2.17   Close-up of chimney during ignition initiation

Figure 2.18   **Smoke filled CYBL vessel during burn**

2.3   Data Analysis

2.3.1   Ignition Details

As shown in Figure 2.16, the temperature data after ignition and during the burn period is very complicated. Figure 2.19 through Figure 2.28 show 1000-second intervals of the data at the time of ignition for various locations in the bundle. From this detailed examination of the data, it can be determined when and where the ignition occurred and when the burn reached various thermocouples, as well as some insights gained into some complicating events that occurred during thc burn.

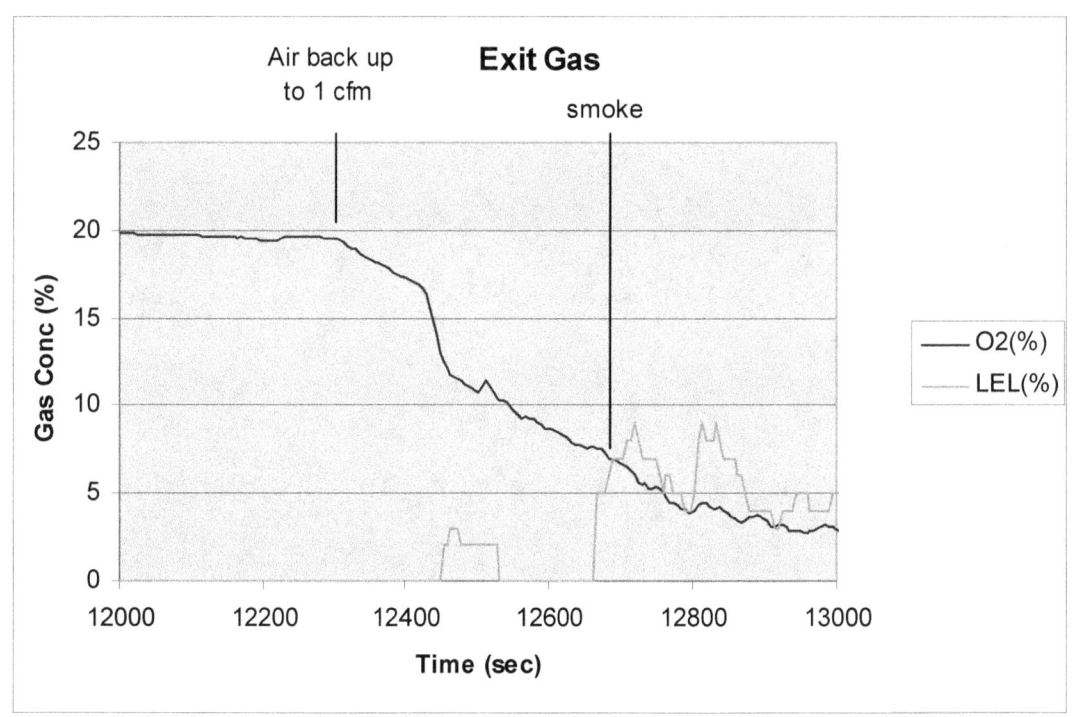

Figure 2.19 Oxygen concentration of off gas during time of Zircaloy ignition

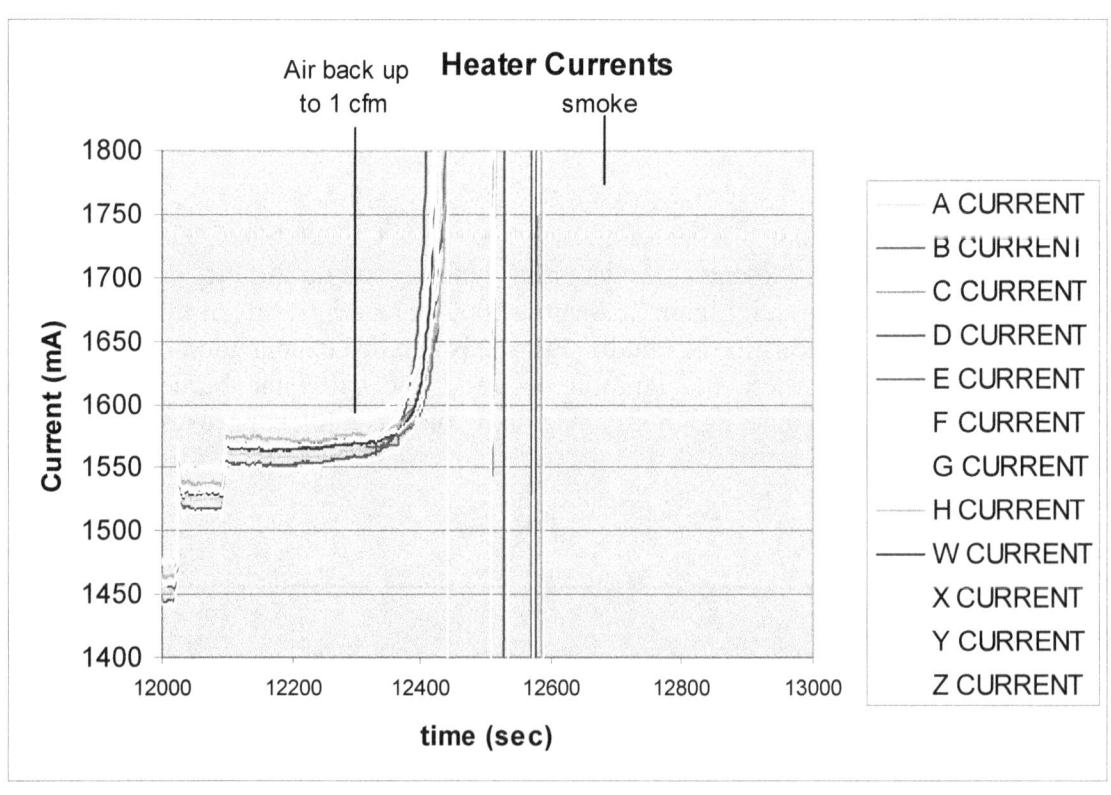

Figure 2.20 Heater currents during time of Zircaloy ignition

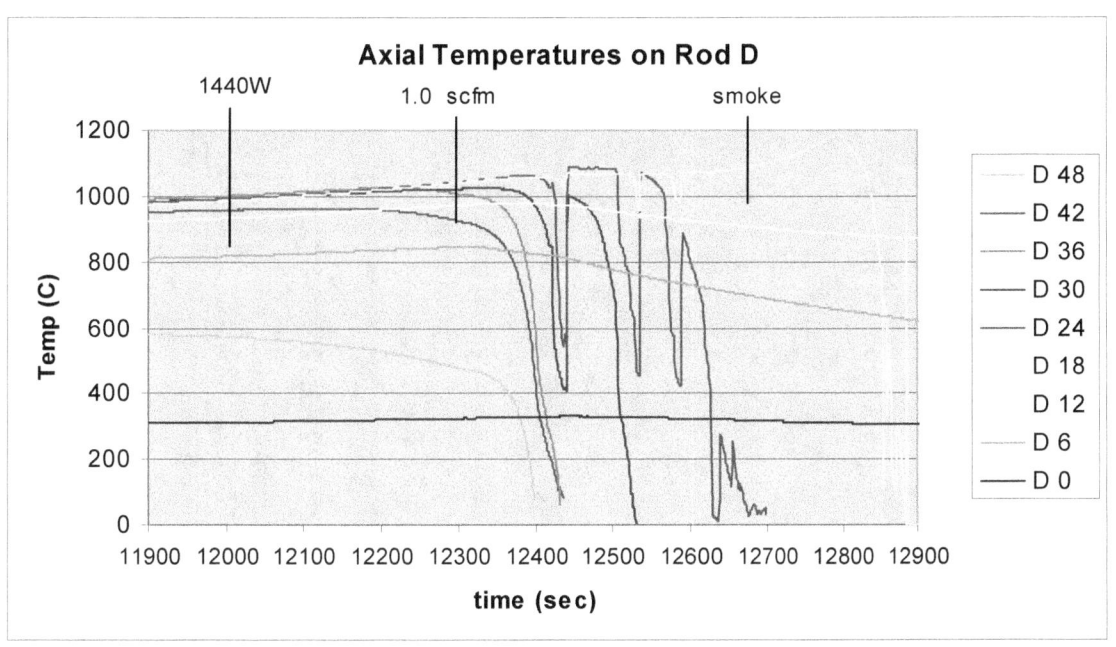

Figure 2.21   Axial temperature profile during time of ignition

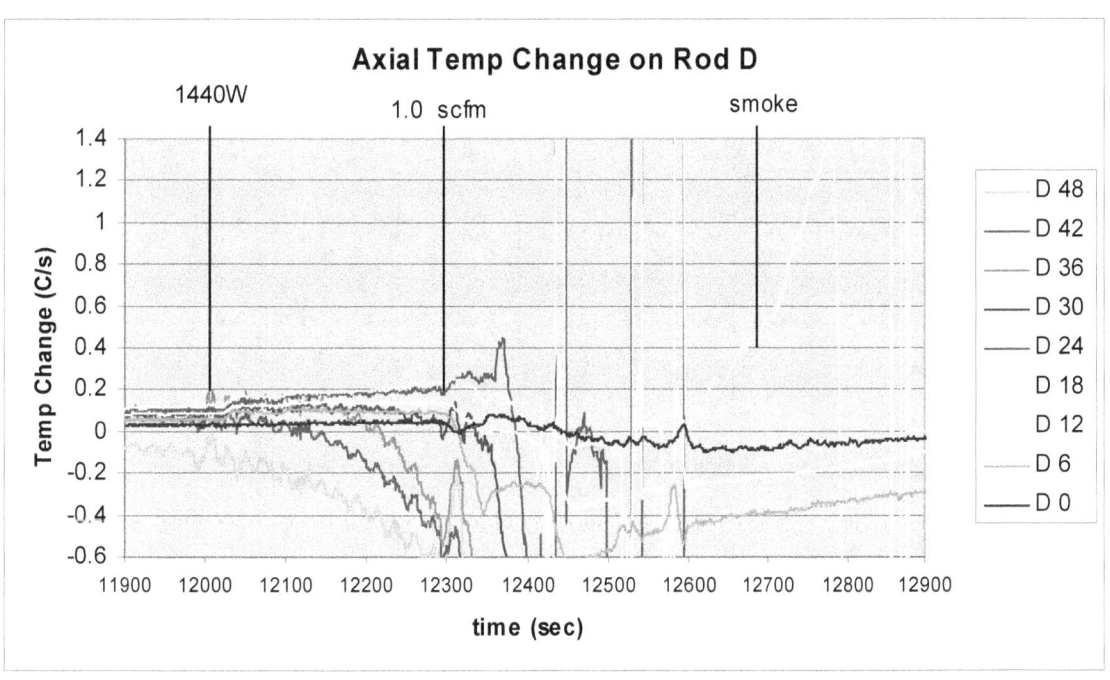

Figure 2.22   First derivative of axial temperature profile during time of ignition

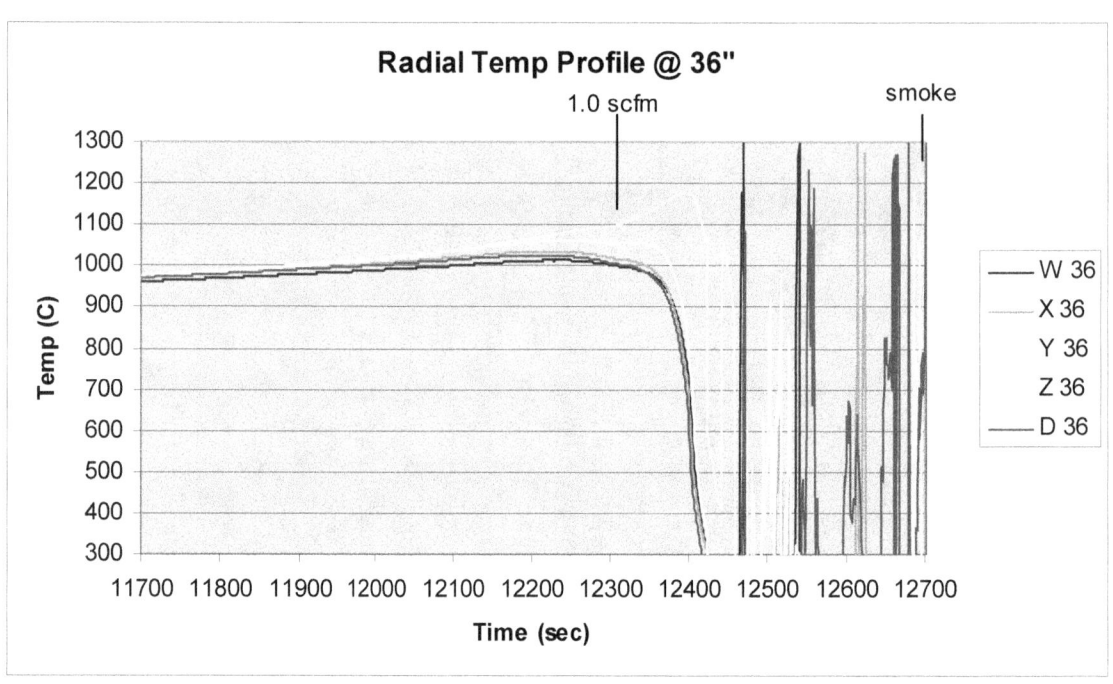

Figure 2.23   Radial temperature profile at the 36-in. elevation during ignition

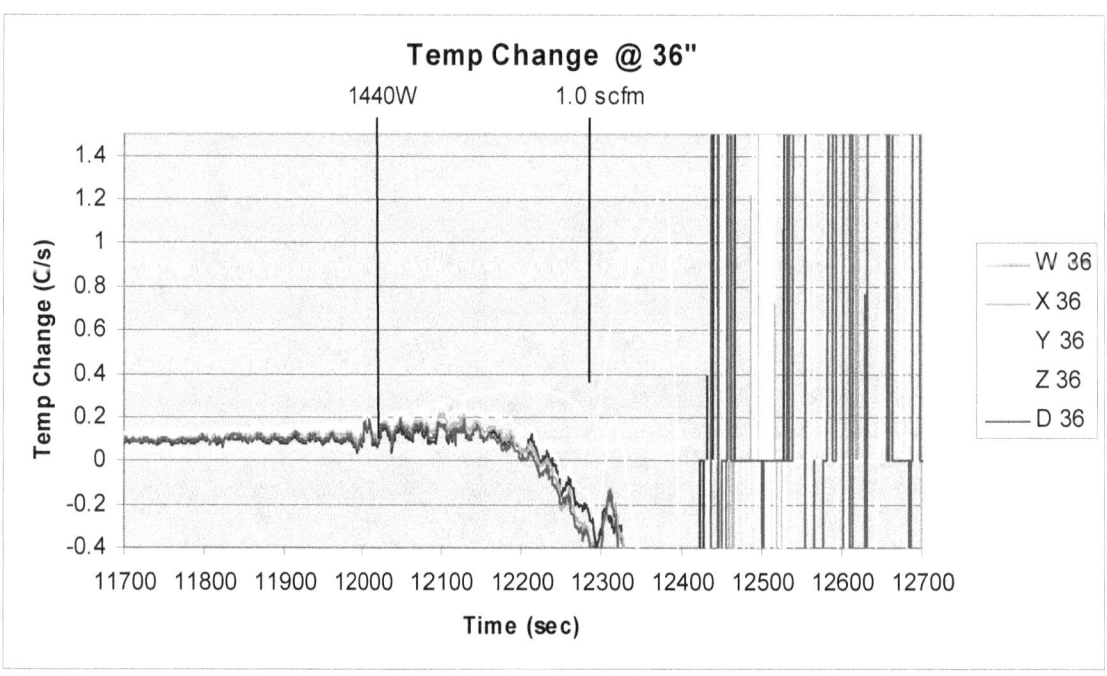

Figure 2.24   First derivative of radial temperature profile at the 36-in. elevation during ignition

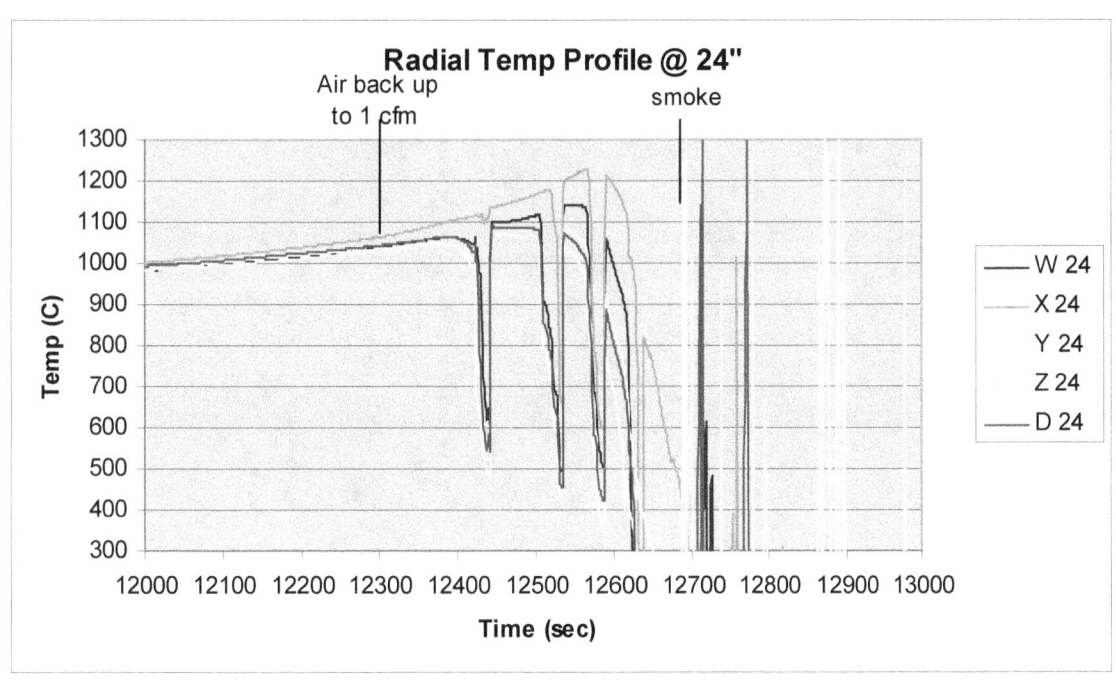

Figure 2.25   Radial temperature profile at the 24-in. elevation during ignition

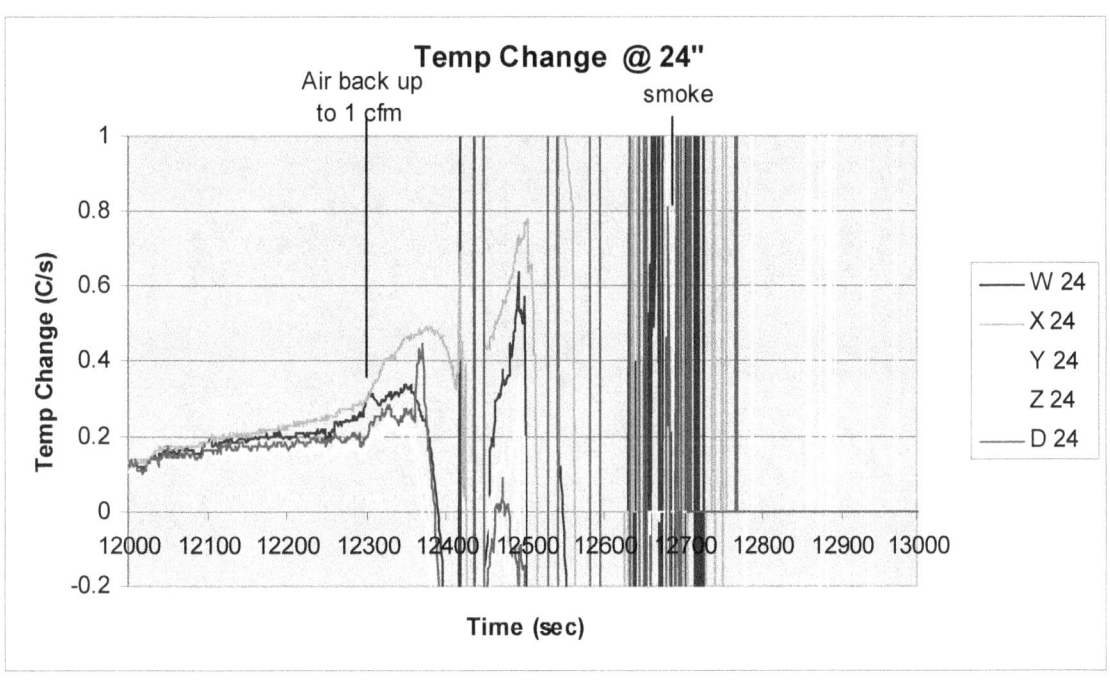

Figure 2.26   First derivative of radial temperature profile at the 24-in. elevation during ignition

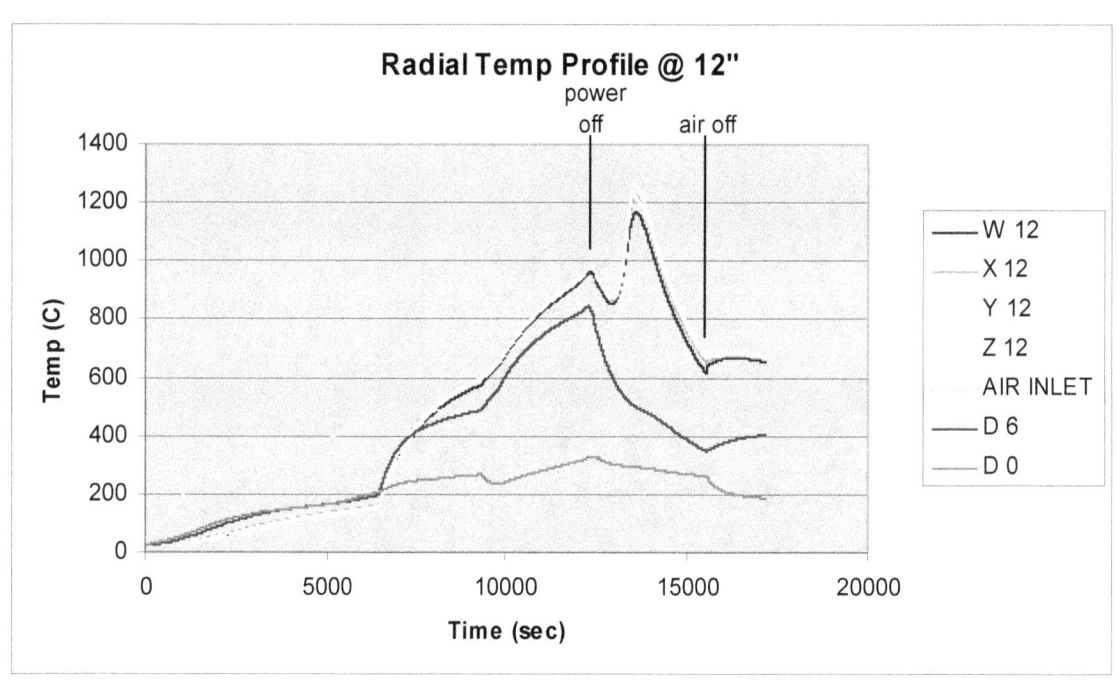

Figure 2.27   Radial temperature profile at the 12-in. elevation during ignition

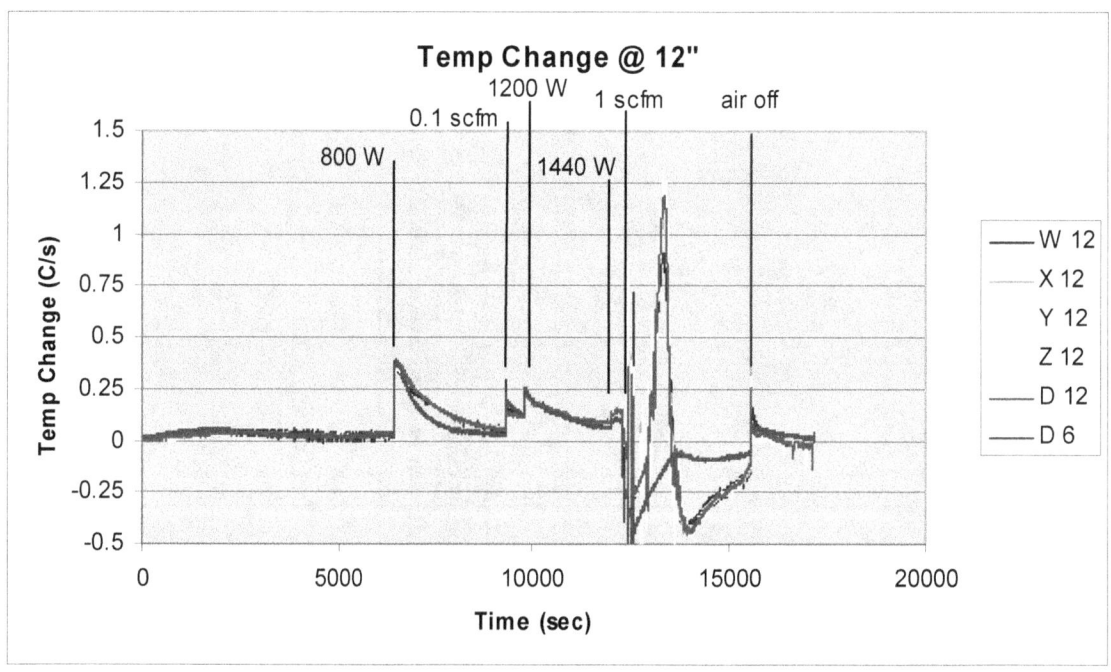

Figure 2.28   First derivative of radial temperature profile at the 12-in. elevation during ignition

Time of Ignition: The time of ignition can be seen in the oxygen content measurement. Figure 2.19 shows the oxygen content during this ignition period. As soon as the air flow was increased, the oxygen content decreased sharply, clearly indicating the onset of Zirconium ignition. Some time shortly after smoke was emitted from the test bundle, the filter on the oxygen sampling system plugged, and the data was compromised. It is not known what constituent in the off gas stream could have resulted in the lower explosive limit (%LEL) response.

Heater Short: Figure 2.20 shows the heater currents during the ignition and initial burn. The first complicating event occurred within 1 minute of returning the full air flow to the bundle when the heater shorted out and tripped the power breaker. This was unexpected, as heater failure was expected to occur as an open circuit when the heater element melted. The breaker was reset twice, but each time it tripped again within 20 seconds.

It is postulated that the heaters shorted out when the Zircaloy cladding melted and penetrated the magnesium oxide insulation. This evidently greatly compromised the electrical insulating properties of the magnesium oxide and allowed a short from the heater element to the Zirconium cladding. A short to ground was provided by the TC sheaths, which were tied off to the electrically grounded steel test stand. The involvement of the TCs in the shorting process is also evident in the temperature data and will be discussed further. Electrically isolating the TC sheaths from ground may prevent the circuit breaker tripping problem and allow the heater to operate longer into the burn.

Temperature Response: Figure 2.21 and Figure 2.22 show the axial temperature profile and the temperature change during the ignition period. Note that in general, this data shows that the temperatures were cooler at the bottom (e.g., D0 and D6) and hotter at the top (e.g., D42). The noted exception is D48, which requires some explanation. The heater rods were built with a 1.5-in. no heat zone on each end. Thus, TC D48 at the top and D0 at the bottom are attached to the no heat zones and therefore measure lower temperatures than otherwise would be expected. The temperature of D0 is further lowered by its proximity of the relatively large thermal mass of the lower heater connection plate.

Ignition and Burn Region: Ignition should be evident by a rapid increase in temperature leading to TC failure at which point the data becomes erratic. In most of the temperature data collected during this experiment, this expected behavior is not clearly evident. Instead, the temperature data drops sharply followed by TC failure. It is believed that this is due to a couple of reasons. The first is the shorting through the TC sheath discussed earlier. In Figure 2.21 , this effect is most clearly seen in the traces of TCs D18 and D24. The blips in the data correspond to the heater rod shorts. Note that the further down the bundle, the effect of the heater short is less evident. TCs D0, D6, and D12 show no effect of the short. TC D18 shows only a minor effect and recovers along its previous trend indicating the measurement was not compromised. TC D24 is more seriously affected but recovers after the first two shorts. However, after the second short, the temperature indicated by D24 drops sharply, is then affected by the third short, and recovers again only to continue to drop precipitously and fail. TC D30 exhibits a similar behavior after the first heater short. TCs D36, D42, and D48 never recovered from the first short. Note that at the time of the first short, D42 was indicating a lower temperature than D36. Similar TC responses are found on the radial profiles at 36 in. and 24 in. as shown in Figure 2.23 and Figure 2.25, respectively. While the heater current spikes complicated the TC data, careful inspection does reveal important information.

The cleanest TC failure due to Zirconium combustion for TC D18 is seen in Figure 2.21. As mentioned before, this TC was little affected by the heater power spikes. As the burn front reached the 18-in. level, the temperature indicated by this TC rose toward 1200°C and then sharply dropped off and became erratic. The point of failure is more easily seen as occurring at

12,779 seconds on Figure 2.22, which shows the first derivative of temperature. The time between failure and obvious erratic behavior is about 1 minute.

The next cleanest failure is seen on TC Z36 (as shown in Figure 2.23), one of the TCs making up the radial profile at the 36-in. level. Like D18, the temperature rose toward 1200°C at 12,380 seconds and then sharply fell off with about 1 minute between failure and erratic behavior. This occurred just after the air flow rate was turned up and just before the first heater power spike. Again, the temperature change plot shown in Figure 2.24 clearly demarks the time of Z36 failure.

Figure 2.24 also shows some subtle but unusual behavior in the other TCs located at the 36-in. level. All the other TCs were cooling off prior to the air flow increase. There are no good physical explanations as to why or how the heater rod clad surface could begin to cool down even a small amount. This behavior is more likely due to loss of thermal contact with the cladding surface at the sensor tip or high temperature damage at a point below the sensor tip. Note that all of these TCs responded to the air flow increase as seen by the positive spike in the temperature change and then failed before the current spike. The temperatures indicated by these TCs were between 1000 and 1060°C, quite a bit lower that that reached by Z36 before failure. This cooling behavior for TCs D48, D42, and D36 is also seen in Figure 2.22, but it is not seen at the 24-in. level in Figure 2.26 .

Below Burn Region: The burn did not make it all the way down to the 12-in. level. Figure 2.27 shows the temperature history for the 12-in. level along with the inlet air temperature and the TCs D0 and D6 for comparison. When the heater power was lost, the peak temperatures at the 12-in. level were approaching 1000°C, but Zirconium ignition did not occur. Rather, the rods began to cool down due to convective cooling of the inlet air. The same trends are evident on TC D0 but at lower temperatures. About 8 minutes after the heater power went off, the temperatures at the 12-in. level began to rise again due to the approach of the burn from above. The temperatures peaked about 20 minutes after the heater power was lost, resulting in the highest temperature detected, 1264°C on Z12, but TC failure did not occur. Rather, the temperatures began to cool off at about the same rate as after the heater power was lost. The temperature change is shown in Figure 2.28. When the air was finally turned off, the temperatures at the 12-in. level increased slightly due to the loss of convective cooling, supporting the conclusion that the TCs survived. The same trend is evident on TC D6.

Bottom Plate Heat Loss: The temperature rise of TC D6 when the air flow was stopped is counter-intuitive when compared to the inlet air temperature. Because the air temperature was greater than D6 when the flow was stopped, it seems the temperature at D6 should have dropped faster, not risen. Looking at the temperature response of TC D0 provides the needed insight. The temperature at D0 was always over 100° lower than the inlet air temperature, and when the air flow stopped, the temperature here did drop at a faster rate. It seems the heat losses at the bottom plate of the bundle assembly were quite large. The temperature of the air entering the bundle was probably closer to the temperature indicated by TC D0 than by the inlet air temperature.

Balance of Burn: After the approach of the burn to the 12-in. level at 13,350 seconds, it did not go out but returned to burn on the upper levels above 36 in., which were previously oxygen starved. This return is evident in the video records as a return of intense glow out the top of the chimney. No TC survived in the upper portion of the bundle to provide any temperature data, but the intensity of the glow seems to indicate an even more intense burn. This is substantiated by TC *Out 42* located on the outside of the insulation jacket at the 42-in. level (Figure 2.29). After the peak temperature was recorded at the 12-in. level, the temperature here rose dramatically. The intensity of the burn was subsiding when the air flow was terminated at 15,558 seconds. Data collection was terminated at 17,180 seconds.

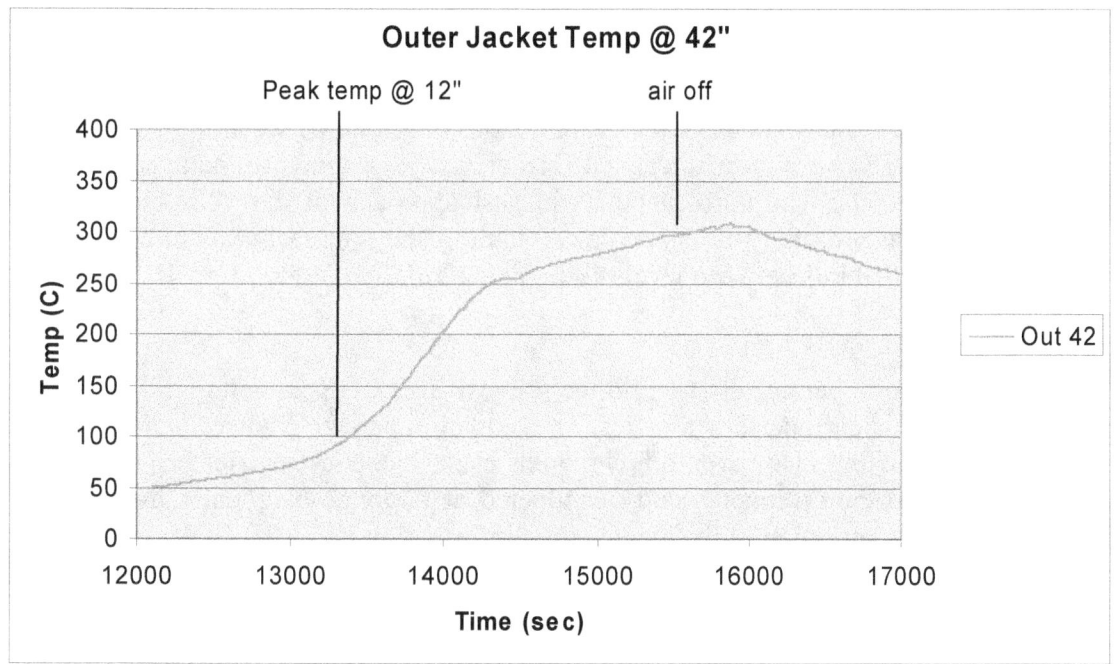

Figure 2.29 **Temperature on the outside of insulation at the 42-in. level**

2.3.2 Smoke Sampling Results

The most surprising product of this experiment was the analytical results of the smoke sampling. Magnesium was found instead of the tin that was expected. The sample taken close to the chimney collected more particulates than the sample taken further up in the CYBL vessel as was expected. The sample close to the chimney collected 580 μg of magnesium (5.0 μg detection limit) but no tin (5.0 μg detection limit), nickel (1.0 μg detection), or iron (5.0 μg detection). Chromium was detected at a level of 2.1 μg, just above the detection limit of 2.0 μg, which leaves this result questionable. The other sample yielded 320 μg of magnesium and nothing else. The metals analysis provides no information on the speciation of the metals detected, but any magnesium sampled would most certainly be in the form of magnesium oxide.

2.4 Discussion

2.4.1 Time of Ignition and Burn Rate

The time of ignition can be set by the oxygen data as occurring when the air rate is increased to 1.0 scfm at 12,300 seconds. The TCs in the upper portion of the bundle (D48, D42, D36, Y36, X36, and W36) exhibited a slight but unusual cooling trend 2 to 3 minutes prior to this. The first clear indication of ignition by TCs occurred at Z36 at 12,380 seconds. TC D30 failed about the same time indicating that ignition occurred between the 30-in. and 36-in. levels and compromised all TCs above this except Z36. The burn reached the TCs at the 24-in. level between 12,500 and 12,560 seconds. The apparent axial burn rate between the 30-in. and 24-in. level was 0.040 in/s. The burn clearly reached D18 at 12,778 seconds so the average axial burn rate slowed to 0.022 in/s. The burn made its closest approach to the 12-in. level at 13,350 seconds, but its actual proximity is uncertain. Assuming the approach was close, the average axial burn rate was 0.010 in/s between levels 18 in. and 12 in.

2.4.2 Post-Test MELCOR Analysis

The Zirconium oxidation parameters used in the pre-test MELCOR modeling did not accurately explain the observed oxidation characteristics. The oxidation parameters used were thought to be the best estimate at the time. However, a few months prior to conducting Heater Design Test 1, a new Zircaloy 4 air oxidation kinetic study conducted at Argonne National Laboratory (ANL) was published [2]. Incorporation of this new kinetic information into MELCOR resulted in much better agreement with the experimental data. Figure 2.30 shows the peak cladding temperature (PCT) measured during the experiment along with the MELCOR prediction using the new kinetics. The new ANL kinetic study identified a breakaway phenomena in the oxidation of Zircaloy after which the kinetic rate of oxidation greatly increased. Thus, the study provides two oxidation kinetic parameters, one for pre-breakaway and one for post-breakaway, and empirical time-at-temperature criteria for the occurrence of breakaway. No mechanistic explanation of breakaway was identified. Incorporation of the pre- and post-breakaway kinetics into the MELCOR model resulted in excellent agreement with the data. The aborted initial attempt for conducting this experiment resulted in a thermal history too complex to test ANL's breakaway criteria; therefore in Figure 2.30, the post-breakaway kinetics was triggered at 1350 K. The oxidation kinetic rate used in the pre-test modeling was midway between the pre- and post-breakaway kinetic rates. This resulted in the model over predicting the oxidation energy before breakaway and under predicting the oxidation energy after breakaway.

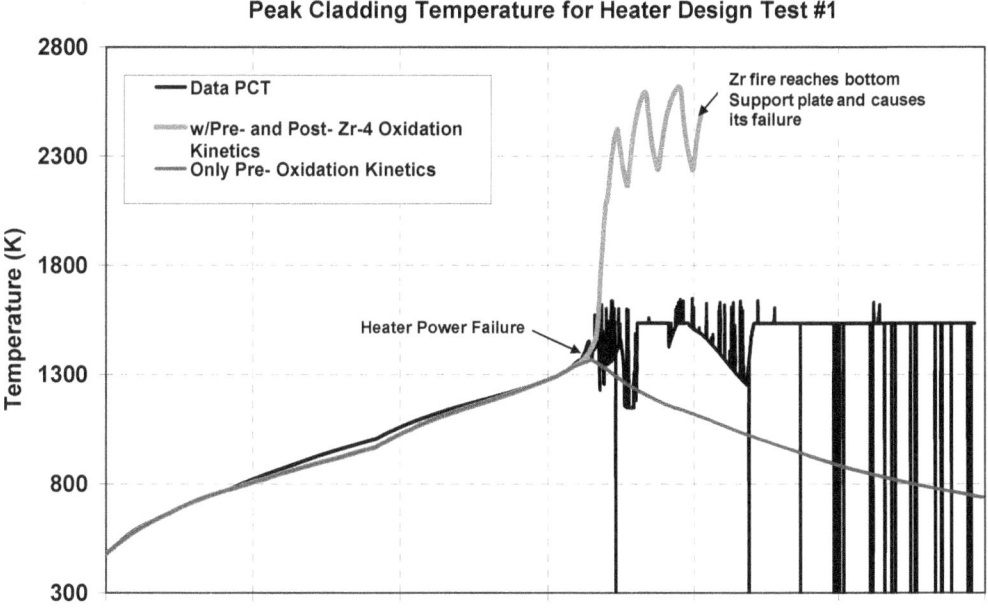

Figure 2.30   Heater Design Test 1 PCT with and without post-breakaway oxidation kinetics

2.5   Technical Issues

A number of important issues regarding the experimental approach have been identified. These issues will be identified and the actions taken will be briefly discussed here.

2.5.1   TC Attachment

A contributing factor to the cooling trend shown by the TCs in the upper portion of the bundle may be due to loss of thermal contact between the TC tip and the heater rod surface. This could occur in at least two ways:

1) The TC tips were held in contact with the Zircaloy surface by spot welding a stainless steel foil to the Zircaloy tube across the tip. Zirconium and iron form a Zirconium rich (88 wt% Zr) eutectic that melts at temperatures as low as 928°C. If a portion of the spot weld melted at this temperature, thermal contact could be lost.

2) The other possibility is that the formation of the Zirconium oxide layer eventually weakened the spot weld attachment.

It was suspected that the TC tip attachment method was responsible for this problem. Figure 2.31(a) shows how the TC was attached using a small hood of stainless steel foil. It is believed that this attachment method would be susceptible to either failure method described earlier. Both failure mechanisms could be prevented by wrapping the foil all the way around the rod and spot welding it to itself and the rod as shown in Figure 2.31(b). Thus, if the weld to the rod is lost, the weld to itself may hold.

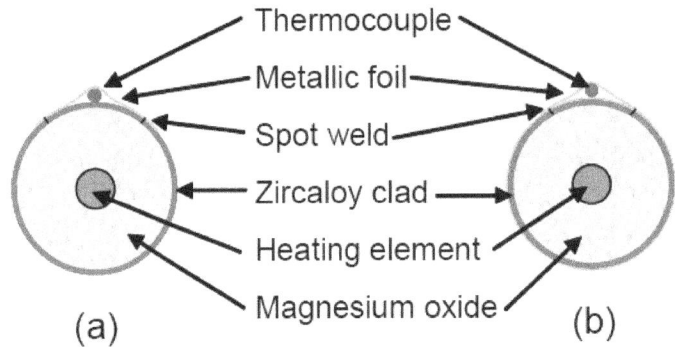

Thermocouple
Metallic foil
Spot weld
Zircaloy clad
Heating element
Magnesium oxide

(a)   (b)

Figure 2.31   TC attachment options, (a) small hood and (b) full wrap

2.5.2   Molten Zirconium-Magnesium Oxide (Zr-MgO) Reaction

The predominance of magnesium in the smoke sampling filters strongly suggests there was a reaction between molten Zirconium and magnesium oxide. The likely reaction is

$$Zr(liq) + 2MgO => ZrO2 + 2Mg(gas) \hspace{2cm} \textbf{Reaction (1)}$$

where molten Zirconium reacts with the magnesium oxide packing and forms Zirconium oxide and magnesium gas. The reaction is endothermic, requiring about 350 kJ/mole, and the free energy of this reaction becomes negative (favored) above 1350°C. The magnesium gas that is produced will enter the gas stream in the bundle and react exothermically (1450 kJ/mole) with oxygen wherever it is encountered, forming magnesium oxide.

Cross sections of the burnt heater rods were examined to determine the extent of zirconium penetration. Figure 2.32 (a) and (b) show an end view of a heater rod segment taken from the lower and upper section of the test bundle, respectively. Many different crystalline phases are evident throughout the cross section. Note that there are three roughly concentric circular layers containing different solid materials. The outermost layer is whitish in color, the middle layer is dark grey and is thicker in the upper sample, and the inner central layer is lighter grey.

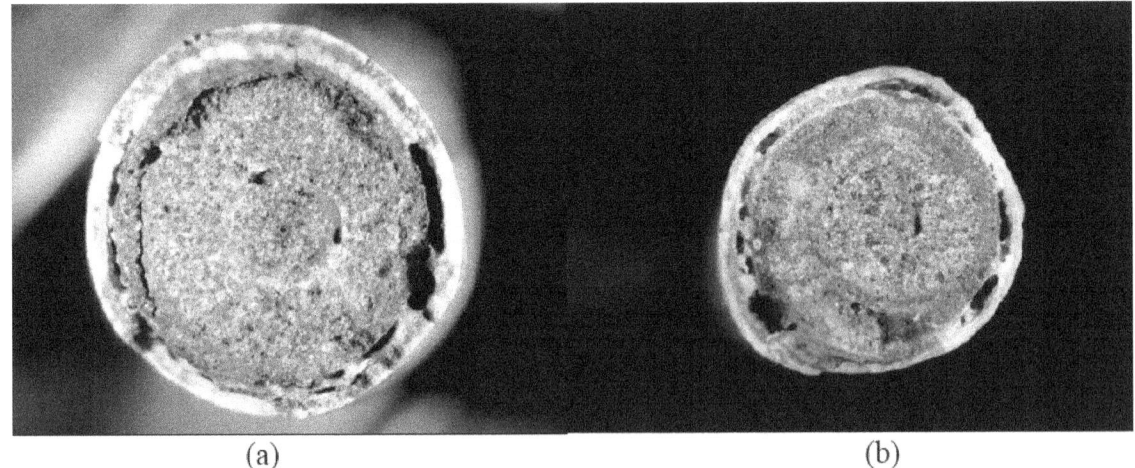

(a)   (b)

Figure 2.32   Photo image of burnt heater rod cross sections, (a) lower rod segment end detail, and (b) upper rod segment end detail

Figure 2.33 shows electron microprobe analyses of these three layers that indicate that the outer layer is pure Zirconium oxide as expected from the oxidation of the Zirconium metal in air. The middle grey is Zirconium oxide with a small fraction of magnesium. This layer is likely formed by the reaction of molten zirconium and magnesium oxide as shown in Reaction 1. The inner central region is still primarily magnesium oxide with a complex mix of minerals formed by the interaction of magnesium oxide with components of the molten Nichrome wire. There are small amounts of Zirconium associated with some of these minerals. This analysis helps to confirm that Reaction 1 was taking place. This analysis also shows that the Zirconium oxide formed by Reaction 1 is a small fraction of the Zirconium oxide formed by air oxidation. Given that the complicating reaction may not occur for an hour after ignition occurs, most if not all the objectives of the overall Spent Fuel Pool Heat-up and Propagation Experiment project can be achieved. It is recommended that no modifications to the basic heater design are needed to mitigate this issue.

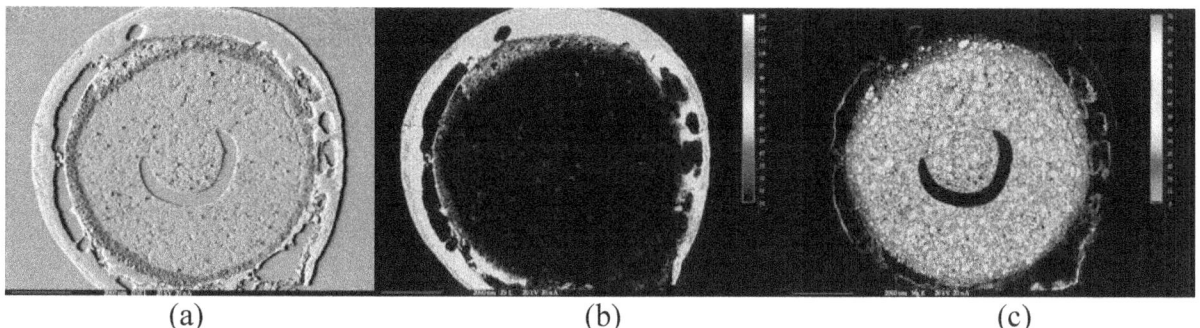

(a)   (b)   (c)

Figure 2.33   **Electron microprobe images of burnt heater rod section, (a) scanning electron microscopy (SEM) image, (b) Zirconium analysis, and (c) magnesium analysis**

2.6   Conclusion

The main objective of the test was achieved. The Zircaloy clad was successfully ignited using the Watlow heater design. A number of issues were also identified. Issues of TC attachment and electrical grounding were identified and can likely be addressed through design improvements. An issue of high temperature Zirconium reaction with magnesium oxide used in the heater design was identified. The impact of this interaction on the suitability of the heater design was determined to be minimal and no design changes are required.

3   HEATER DESIGN TEST 2

3.1   Background

The first heater design test was successfully performed on December 17, 2004, and demonstrated that the Watlow heater design will result in the auto ignition of the Zircaloy cladding when heated under simulated spent fuel conditions in air. As documented in Section 2 a couple of issues with this experiment were identified including concerns about TC attachment and reaction of molten zirconium with the magnesium oxide used to fill the electric heater rods.

3.2   Objective

The objective of the second heater design test was to ignite the simulated fuel rod bundle under reduced oxygen conditions that might better represent the situation near the top of a full-length assembly. In a full-length assembly, oxidation of the Zirconium cladding in the lower, cooler region depletes the oxygen concentration of the air reaching the upper, hotter portion where ignition first occurs. The burn in this test was expected to be slower and cooler (making it easier to observe) and perhaps better represent the initial fire in a prototypic assembly.

This experiment also served to test some instrumentation improvements suggested by the first test. In the first test, the type K TCs attached to the heater rods failed in a manner that made it difficult to determine with certainty where the burn front was located. One possible problem was detachment of the TC tip at high temperatures. To mitigate this possibility, an improved method for attaching the TCs to the heater rods was used (using a full wrap of shim). To provide more data on the location of the burn front, an array of 14 platinum TCs was added to the instrumentation. This array monitored the temperature axially along the outside surface of the quartz tube. Between each of these TCs was a 3-mm quartz light pipe, which was monitored visually and recorded on videotape (Figure 3.1).

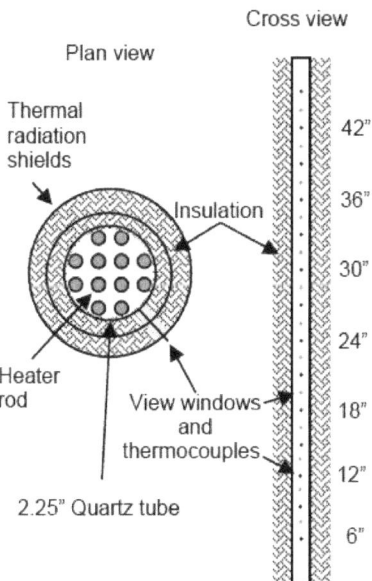

Figure 3.1   Schematic of the second heater design test

3.3 Electric Heaters

The heater design was the same as that used in the first heater design test. The Zircaloy jacketed electric heaters were manufactured by Watlow Electric Manufacturing Company. The design of the heater was typical of those made by Watlow. Each heater was comprised of a central Nichrome heater element, compressed magnesium oxide powder, and an outer Zircaloy 2 jacket. The composition of the heater rods is shown in Table 3.1. The final magnesium oxide powder density was 2.720 g/cm^3, corresponding to a solids fraction of 0.759.

Table 3.1   Composition of heater rods

Component	Nominal wt%	Mass/rod (lb)	Mass/rod (g)	# Rods	Total mass (g)	Mwt (g/mol)	Total moles (mol)
Zircaloy 2		0.368					
Zr	98.40%		164.2513	12	1971.015	91.224	21.60632
Sn	1.30%		2.169986	12	26.03983	118.71	0.219357
Fe	0.18%		0.30046	12	3.605515	55.845	0.064563
Cr	0.10%		0.166922	12	2.003064	51.9961	0.038523
Ni	0.07%		0.116845	12	1.402145	58.6934	0.023889
	100.05%						
MgO		0.377	171.0043	12	2052.052	40.3044	50.91385
Nichrome		0.046	20.86525	12	250.383		

The diameter of the heater rods used in the heater design test bundle was smaller than prototypic due to the availability of Zircaloy tubing. The tubing used was 0.440-in. diameter, 0.028-in. wall Zircaloy 2 tubing for BWR assembly. This tubing was made into 48-in. long, 0.375-in diameter heaters by Watlow in a process whereby the 0.440-in. tubing was drawn through a die that reduced the diameter to 0.375 in. and compressed the magnesium oxide powder considerably. Compression of the magnesium oxide powder is required to establish good thermal contact between the Nichrome heater element and the outer metallic sheath.

3.4 Experimental

The experiment was constructed and operated in the CYBL vessel located in Building 6585C in Technical Area III of SNL in Albuquerque, New Mexico. Twelve 4-ft long, 0.375-in. diameter heater rods with a pitch of 0.539 in. (as shown in Figure 3.1) were assembled into a bundle (Figure 3.2). The heater rods were instrumented with 0.032-in. ungrounded Inconel type K TCs in a configuration as shown in Table 3.2. Table 3.2 also lists other instrumentation used in the test and was collected by a computerized data acquisition system at nominally 2-second intervals.

Figure 3.2   **Bundle of heater rods**

Table 3.2   Heater Design Test 2 instrument layout

Chan #	Name	Description	Instrument type	Max Volt	Units
0	D0	Heater Axial Elevation 0"	K TC		C
1	D6	Heater Axial Elevation 6"	K TC		C
2	D12	Heater Axial Elevation 12"	K TC		C
3	D18	Heater Axial Elevation 18"	K TC		C
4	D24	Heater Axial Elevation 24"	K TC		C
5	D30	Heater Axial Elevation 30"	K TC		C
6	D36	Heater Axial Elevation 36"	K TC		C
7	D42	Heater Axial Elevation 42"	K TC		C
8	D48	Heater Axial Elevation 48"	K TC		C
9	W12	W Heater Elevation 12"	K TC		C
10	X12	X Heater Elevation 12"	K TC		C
11	Y12	Y Heater Elevation 12"	K TC		C
12	Z12	Z Heater Elevation 12"	K TC		C
13	W24	W Heater Elevation 24"	K TC		C
14	X24	X Heater Elevation 24"	K TC		C
15	Y24	Y Heater Elevation 24"	K TC		C
16	Z24	Z Heater Elevation 24"	K TC		C
17	W36	W Heater Elevation 36"	K TC		C
18	X36	X Heater Elevation 36"	K TC		C
19	Y36	Y Heater Elevation 36"	K TC		C
20	Z36	Z Heater Elevation 36"	K TC		C
21	F24	F Heater Elevation 24"	K TC		C
22	Oven Out	Oven Outlet Temperature	K TC		C
23	Oven Wall	Oven Wall Temperature	K TC		C
24	Chimney	Chimney Air Temperature	K TC		C
25	Base Plate	Base Plate Temperature	K TC		C
26	Bottom Atm	Gas Inlet Temperature	K TC		C
27	Conn 21	In 21 Connector Temperature	K TC		C
28	Out 12	Outer Insulation Temp 12"	K TC		C
29	Out 36	Outer Insulation Temp 36"	K TC		C
30	In 6	Inner Skin 6"	S TC		C
31	In 9	Inner Skin 9"	B TC		C
32	In 12	Inner Skin 12"	S TC		C
33	In 15	Inner Skin 15"	B TC		C
34	In 18	Inner Skin 18"	S TC		C
35	In 21	Inner Skin 21"	B TC		C
36	In 24	Inner Skin 24"	S TC		C
37	In 27	Inner Skin 27"	B TC		C
38	In 30	Inner Skin 30"	S TC		C
39	In 33	Inner Skin 33"	B TC		C
40	In 36	Inner Skin 36"	S TC		C
41	In 39	Inner Skin 39"	B TC		C
42	In 42	Inner Skin 42"	S TC		C
43	In 45	Inner Skin 45"	B TC		C

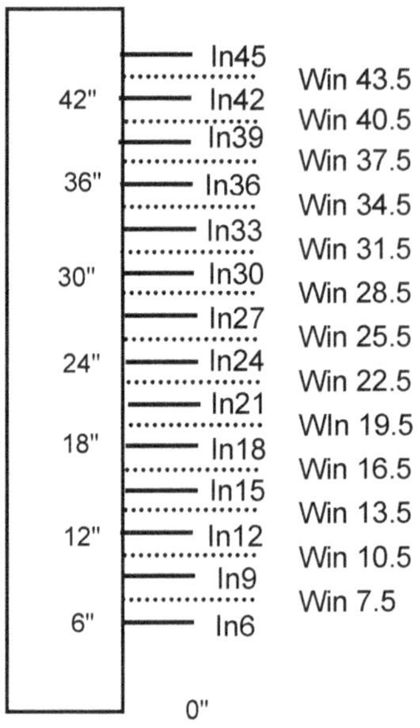

Diagram (right side):

```
        In45        Win 43.5
42"     In42        Win 40.5
        In39
                    Win 37.5
36"     In36
                    Win 34.5
        In33
30"     In30        Win 31.5
        In27        Win 28.5
24"     In24        Win 25.5
        In21        Win 22.5
18"     In18        WIn 19.5
        In15        Win 16.5
12"     In12        Win 13.5
        In9         Win 10.5
6"      In6         Win 7.5

             0"
```

Heater Nomenclature

```
        A   B
    C   D   E   F     3-mm quartz
    W   X   Y   Z     windows and type
        G   H         S/B TCs
```

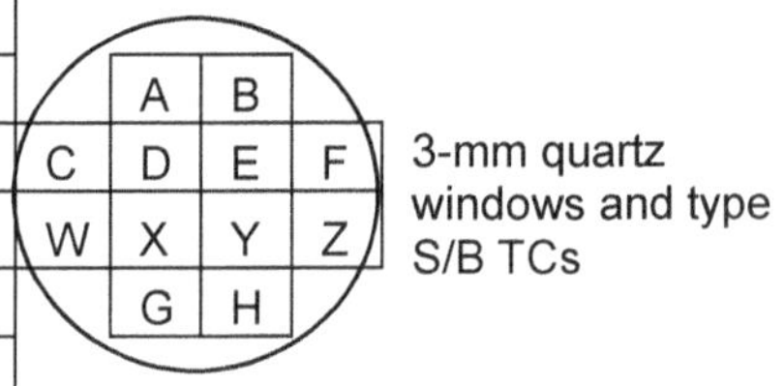

Table 3.3   Heater Design Test 2 instrument layout (continued)

Chan #	Name	Description	Instrument type	Max Volt	Units
44	A current	Current of Heater A	Current xducer	5	ampere
45	B current	Current of Heater B	Current xducer	5	ampere
46	C current	Current of Heater C	Current xducer	5	ampere
47	D current	Current of Heater D	Current xducer	5	ampere
48	E current	Current of Heater E	Current xducer	5	ampere
49	F current	Current of Heater F	Current xducer	5	ampere
50	G current	Current of Heater G	Current xducer	5	ampere
51	H current	Current of Heater H	Current xducer	5	ampere
52	W current	Current of Heater W	Current xducer	5	ampere
53	X current	Current of Heater X	Current xducer	5	ampere
54	Y current	Current of Heater Y	Current xducer	5	ampere
55	Z current	Current of Heater Z	Current xducer	5	ampere
56	Mass Flow	Mass Air Flow into Oven	MKS Flow	10	slpm
57	Heater Volts	Voltage Applied to Heaters	Voltage xducer	10	volts

Inner skin types B and S TCs are exposed junction with bead in contact with the quartz tube.
Outer insulation TC is attached to the outer metal wrap opposite (approximate) the type S and B TCs.
The connector TC is located on the mid point of the ceramic connectors at the end of the type B TC.

The TCs attachment method used to attach the TC to the Zircaloy tubing was slightly different than that used in the first heater test. In the first test, the TCs were attached by spot welding a small piece of stainless steel shim or foil across the tip. In the second heater test, the TC was attached using a full wrap of the stainless foil (Figure 3.3). The full wrap provides a more secure attachment that is less prone to failure due to the oxidation of the Zircaloy or the melting of the nickel-Zirconium eutectic.

Figure 3.3   TC attachment detail
Note: The tips of the TCs are attached with a full wrap of shim stock.

Another experimental improvement was the addition of a heater on the bottom brass base plate. In the first heater design test, it appeared that the thermal mass of this plate cooled the incoming gas flow such that the bottom heater rod temperature never approached the inlet air temperature. With this additional heater, the temperature of the base plate could be controlled to match the inlet air temperature.

A 57-mm ID (2.244-in.) quartz tube was placed around the heater bundle (see Figure 3.4), and the tube was surrounded with reflective foil. The bundle was insulated with 5 in. of Kaowool™. Two additional steel radiation shields were used. One was located on the outside of the insulation, and the other was located midway through the insulation. An axial array of 14 platinum TCs (types S and B) was installed at 3-in. intervals in contact with the outside surface of the quartz tube. In between these TCs was a similarly installed quartz light pipe to serve as a visual indicator of the burn front location. The TCs and light pipes were held in place by a ceramic positioning block as pictured in Figure 3.5, Figure 3.6 and Figure 3.7. The final test setup is pictured in Figure 3.8.

Figure 3.4   Quartz tube installed over heater bundle

Note: The *mouse hole* in the bottom of the quartz tube is used to route the TCs away from the bundle.

Figure 3.5   Stainless steel foil in place over quartz tube and ceramic positioning block is in place

Figure 3.6 **Installation of quartz wall TCs and light pipes into ceramic positioning block**

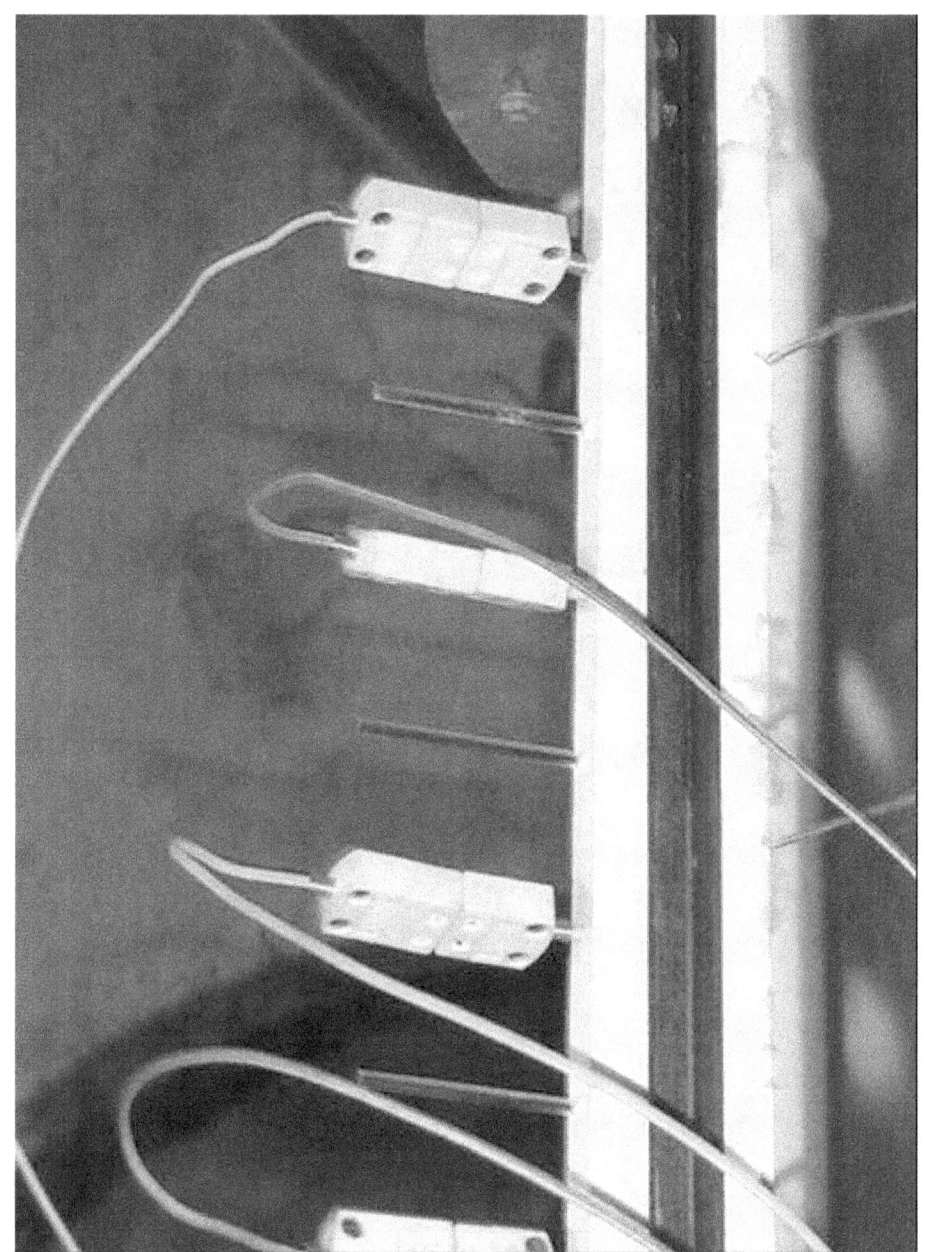

Figure 3.7   Detail of quartz tube wall TCs and light pipe installation

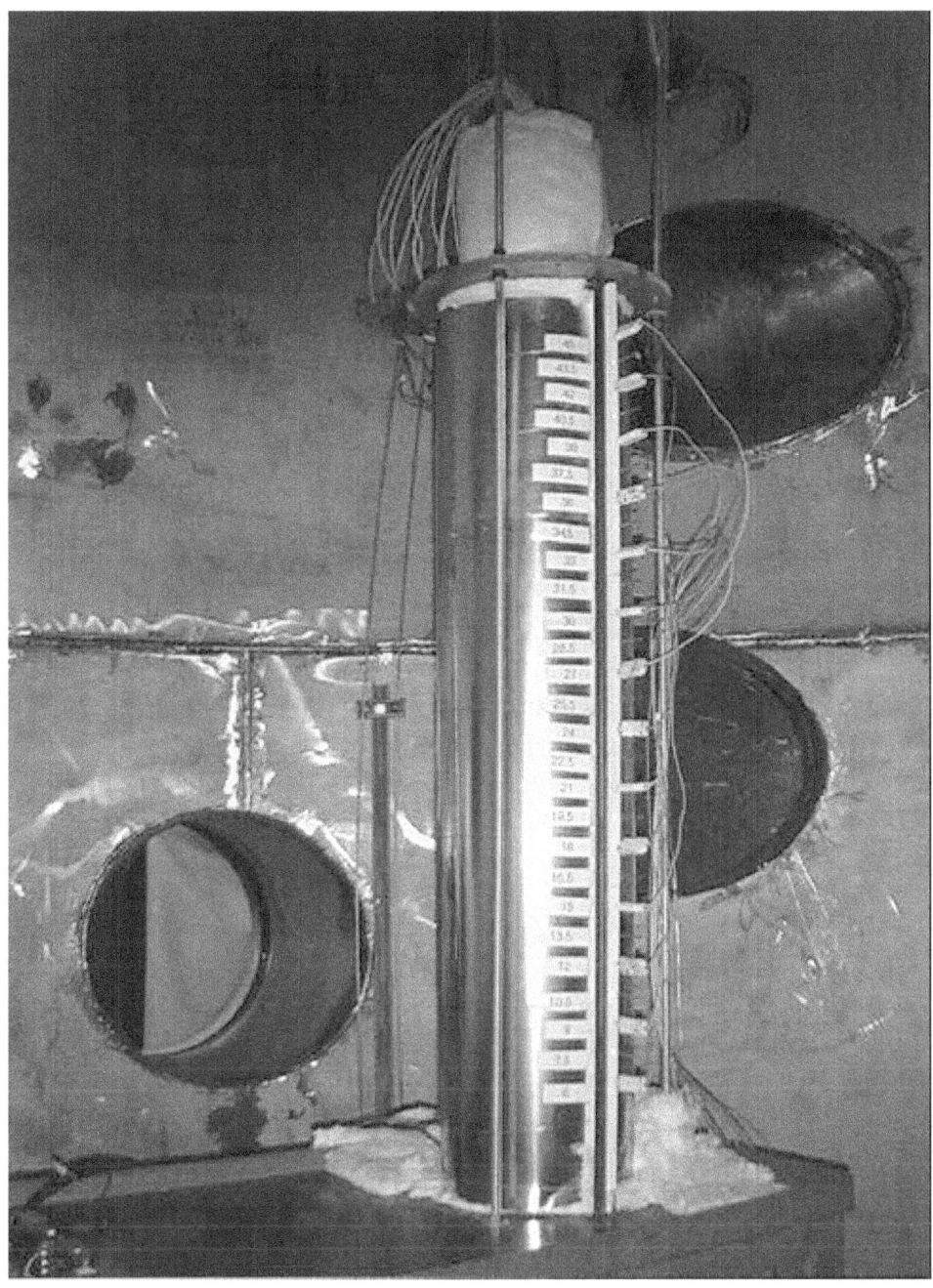

Figure 3.8   Initial test setup for Heater Design Test 2

3.4.1 Additional Sampling

In addition to the data collected by the instrumentation listed in Table 3.2, three other sampling efforts collected data for this experiment. The first effort was to monitor the oxygen concentration of the exit gas, the second effort was the sampling of smoke particulates for metals analysis, and the third was video recording of the experiment from a number of perspectives.

Oxygen: The oxygen concentration of the gas leaving the test bundle was monitored using an ITX (S/N: 0307010-286) portable atmosphere-monitoring instrument. The instrument measured both oxygen concentration and the %LEL. The sample was collected from the mid point of the chimney through a stainless steel tube that was routed through an ice bath and outside the CYBL vessel where the oxygen monitor was located. The sample was drawn using the instrument's internal sample pump. The %O_2 and %LEL data was collected internally along with the time at 5-second intervals.

Smoke: Health physics personnel installed three particulate sampling devices inside the CYBL vessel to analyze the total metal content (this provides no information on the chemical nature of the metal) of the smoke generated by the experiment. The sample filters were analyzed for the metals added to Zirconium to make the Zircaloy 2 alloy, which include tin, iron, nickel, and chromium. The samples were also analyzed for magnesium, a principal constituent of the magnesium oxide insulation used inside the heaters. Table 3.1 presents the initial mass of each of these constituents that was initially present in the test bundle.

Video: Video perspectives included a close-up looking down the chimney, a side view of the experimental setup, and a distant top view from the top opening of the CYBL vessel.

3.4.2 Test Operation

The experiment was conducted on May 24, 2005. The chronology of significant events during the experiment is listed in Table 3.3. The air preheater (and data collection system) was started at 11:58 a.m., and the air flow of 30.0 slpm was started 53.9 minutes into the test. The preheat air temperature set point was 600 K (327°C). The base plate heater was powered at 61.8 minutes and collection of oxygen data began at 132.5 minutes.

Table 3.4   Chronology of events during experiment

Clock Time	Experiment Time (min)	Event
11:58:26	0.00	Start: data on, air heat on, 327 C set point
12:52:18	53.90	Air flow on 30.0 slpm
13:00:10	61.77	Base plate heat on 327 C set point
14:10:56	132.54	O_2 data on
14:11:02	132.64	Heater power on
14:18:20	139.94	Lower smoke samplers into CYBL
14:28:34	150.17	Switch flow from air to 7.5% O_2
14:50:28	172.07	Rod E shorts
14:50:56	172.54	Rod E burns out
14:51:10	172.77	Reset O_2 valve, no sample until now
15:12:00	193.60	Brief fire, rod H shorts
15:12:22	193.97	Rod H burns out
16:22:02	263.64	D48 fails
16:23:18	264.90	D42 fails
16:24:36	266.20	W, X, and Y36 fail
16:25:08	266.74	O_2 going to zero: ignition
16:26:42	268.30	D36 fails
16:26:44	268.34	Z36 fails
16:28:32	270.14	All rod power off
17:15:50	317.44	Gas flow off
17:35:02	336.64	Data off

Figure 3.9 shows the total power history and individual heater rod current histories, while Figure 3.10 shows the applied voltage during the test. The power to the heater rods was applied at 132.6 minutes. The total initial power was 975 W, which corresponds to about 20.3 W/ft. At 150.2 minutes, the input gas was changed from air to 7.5 vol% O_2 in nitrogen. At 172.1 minutes, heater rod E shorts and fails. The constant power control increased the current to the remaining heater rods, and the total power increased slightly to about 1000 W. This is 22.7 W/ft on the remaining heater rods. At 193.6 minutes, heater rod H shorts and fails. Again, the constant power control increased the current to the remaining heater rods. The total power increased slightly again to 1050 W or 26.3 W/ft on the remaining heater rods. The total power was maintained at 1050 W until ignition was achieved at 266.7 minutes, shortly after which all heater rods failed.

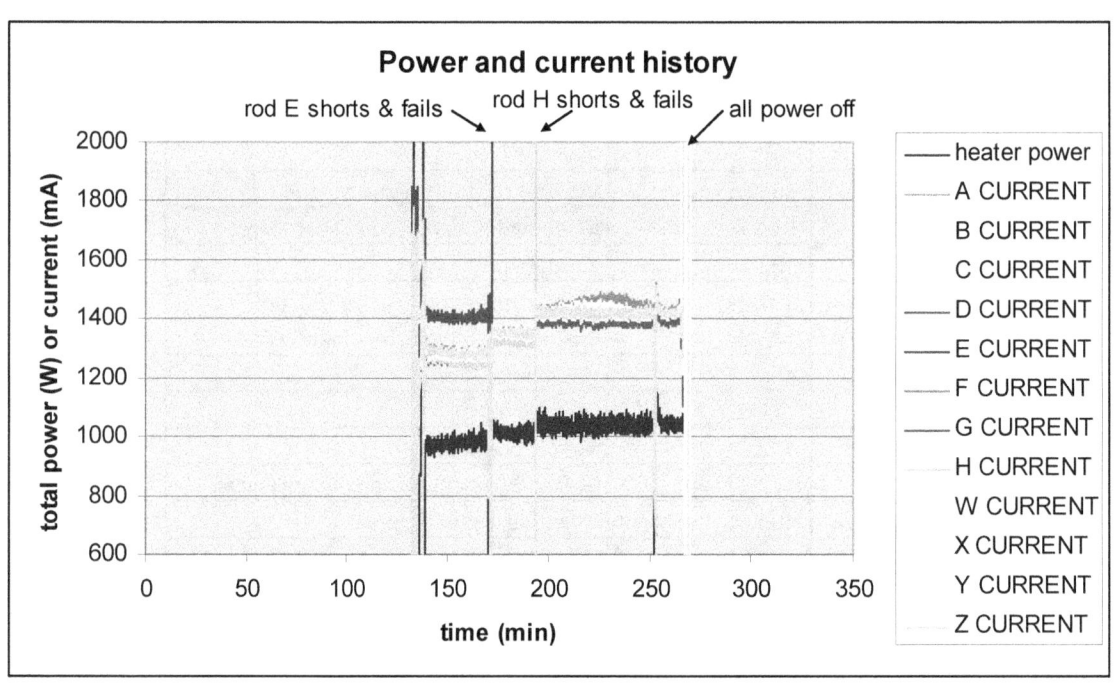

Figure 3.9   Heater currents during time of Zircaloy ignition with major experimental events noted

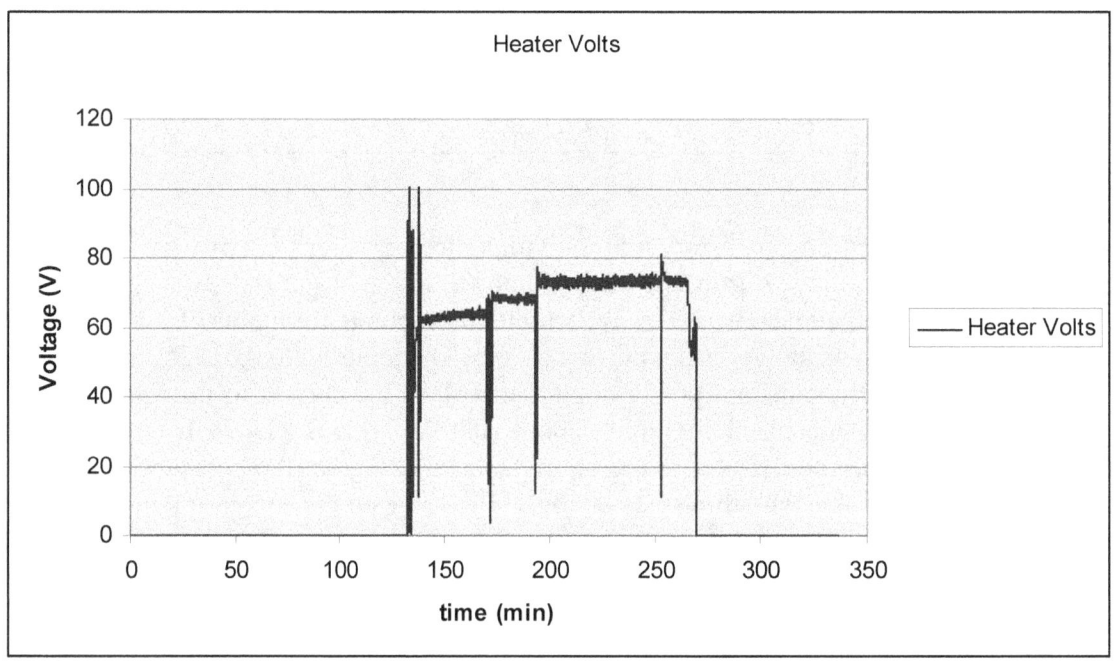

Figure 3.10   Voltage applied to heater rods

3.4.3   Ignition Details

The point of Zirconium ignition is most easily seen in the oxygen concentration data shown in Figure 3.11. At the point of ignition, the oxygen concentration drops sharply to zero. This occurred at 266.7 minutes into the test. Figure 3.12 and Figure 3.13 show the axial heater rod temperature profile and the first derivative of these temperatures, respectively. These plots show that D36 and D42 approached ignition in similar fashion, but D42 failed before D36. This indicates that the fire ignited above D36 and below D42. Once ignition occurred, the TCs leading up to D42 (and D48) failed. Twelve minutes after D36 failed, the temperature at D30 peaked at 1225°C, the highest temperature measured by a type K TC. However, this TC did not fail; therefore, the burn front did not quite reach this level. Figure 3.14 shows the radial temperature profiles at the 12-in., 24-in., and 36-in. levels. Generally, the radial profiles are even at each level, with heater rod Z being a bit lower in temperature than the other heater rods. Note that heater rod Z was adjacent to the light pipe array, which could account for the slightly lower temperature. Ignition and TC failure is evident with all the TCs at the 36-in. level and not at the lower levels. Note that as soon as all the heater rods failed, the temperatures on the heater rods at the 24-in. level and below dropped sharply. Also, note that the TCs that did fail did not exhibit any of the peculiar behavior evident in the first heater design test indicating that the full wrap installation method worked as expected.

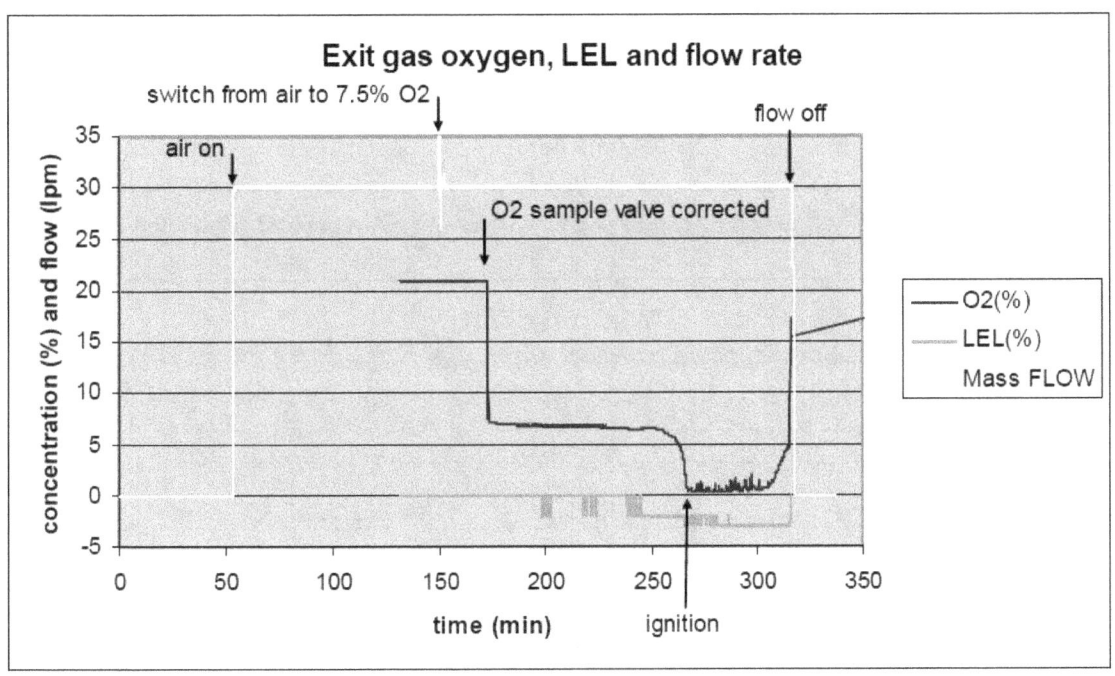

Figure 3.11   Oxygen concentration of off gas with major experimental events noted

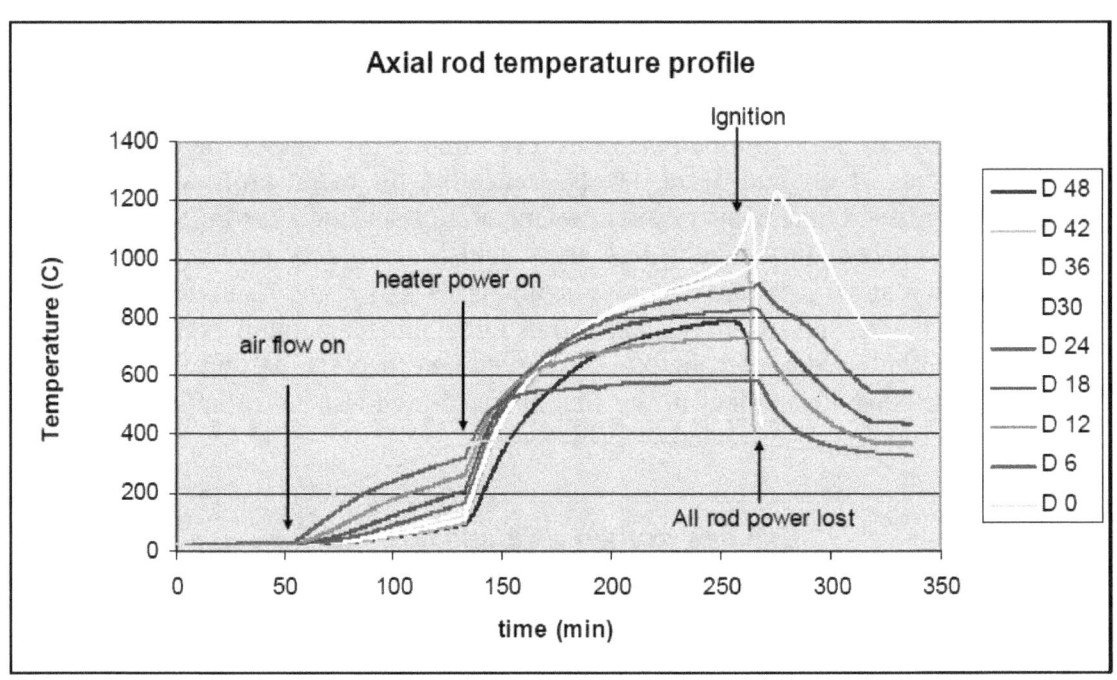

Figure 3.12   Axial temperature profile with major experimental events noted

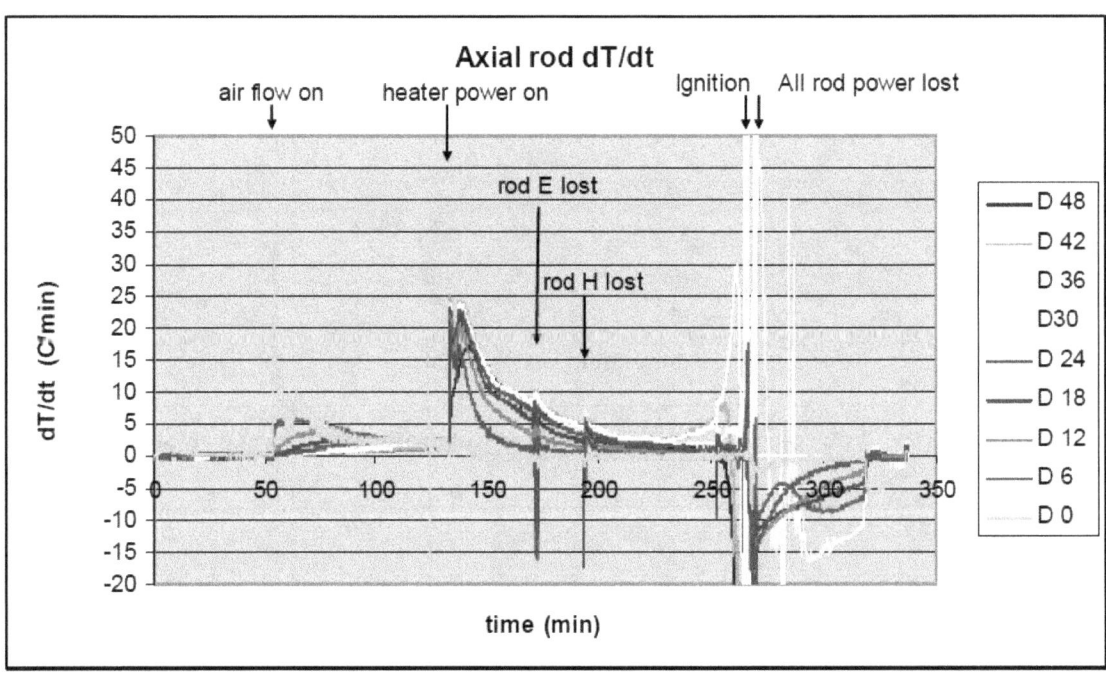

Figure 3.13   First derivative of axial temperature profile with major experimental events noted

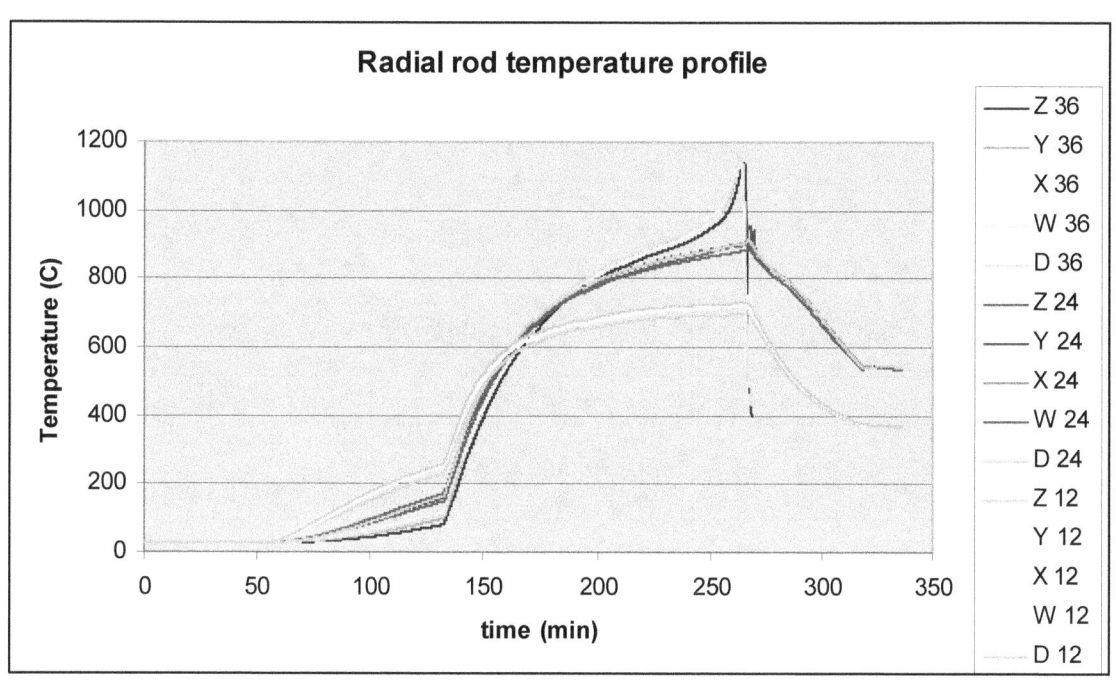

Figure 3.14   Radial temperature profile at the 12-in., 24-in., and 36-in. elevations

Figure 3.15 shows the axial temperature profile along the outside surface of the quartz tube. Figure 3.16 shows the light pipes glowing during the burn. The platinum (types S and B) TCs operate to higher temperatures (1500 to 1600°C) than the type K TCs used on the heater rods. Because these TCs are not located inside the quartz tube where the fire is located, they are much less prone to failure due to the fire. While this also means the temperature indicated does not accurately represent the temperature anywhere in the heater rod bundle and was generally lower than the temperature inside the heater rod bundle, the wall temperature measurements did provide an excellent indication of the location of the burn front. When all the heaters failed, all the temperatures at the 24-in. level and below dropped off quickly indicating that there was little oxidation energy being released at these levels. There was a 15-minute delay in the temperature drop at the 27-in. level, suggesting a small amount of oxidation energy release. The temperatures at the 30-in. level and above all increased significantly after the loss of the heater power, indicating significant energy input from Zirconium oxidation.

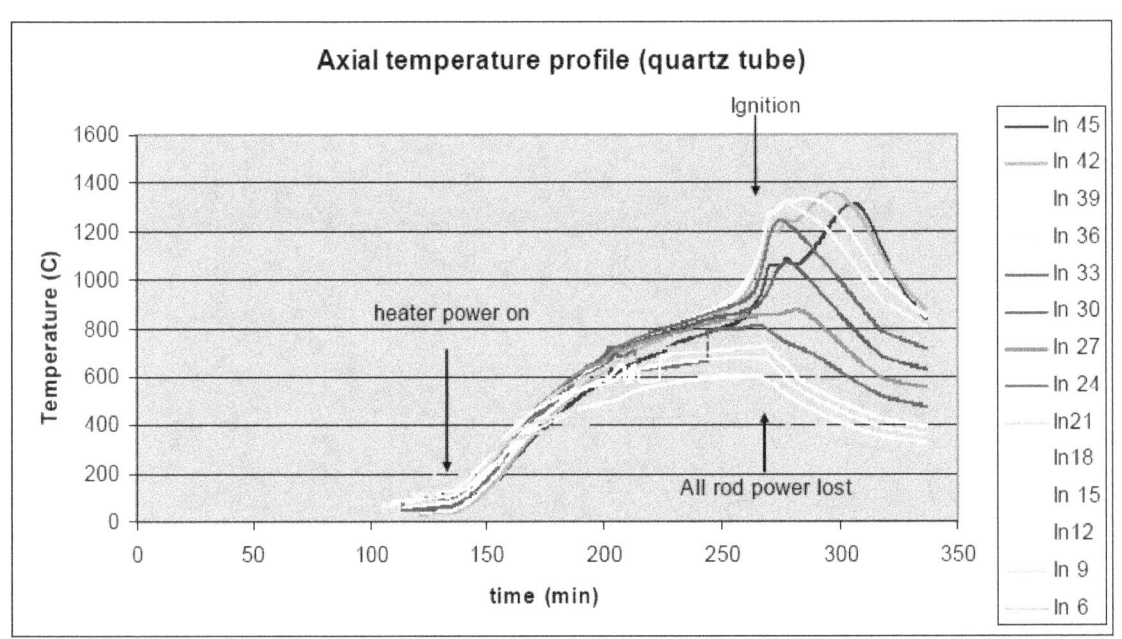

Figure 3.15   **Axial temperature profile along quartz tube**

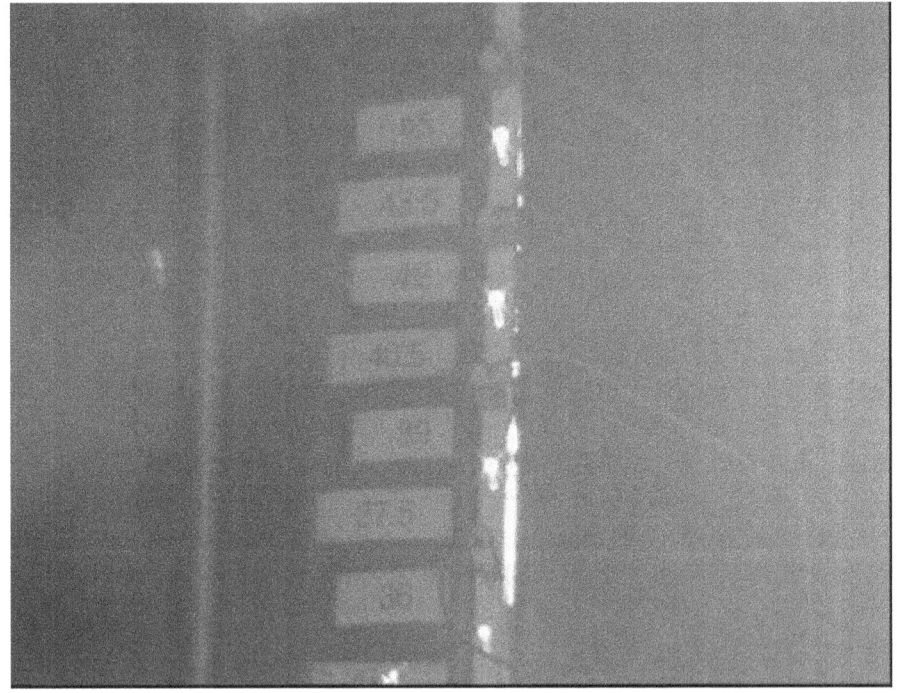

Figure 3.16   **Close-up of light pipes during burn**

From the data presented in Figure 3.15, it can be seen that TCs In 36 and In 39 heated up first and together indicating the burn initiated in this region. A few minutes later, TCs In 33 and In 42 both sharply increased in temperature, indicating that the burn front moved both upward and downward 3 in. A few minutes later, the temperature at the 30-in. and 45-in. levels spiked suggesting further upward and downward spread of the oxidation front. The peak temperatures at the 30-in., 33-in., and 36-in. levels were all reached at about the same time; however, the peak temperature was not as high at the progressively lower levels. After the temperature at these

54

levels peaked, the temperature peaked at progressively higher levels, the next being at the 39-in. level, the 42-in. level, and finally the 45-in level. The highest peak temperature measured was 1364°C at the 42-in. level.

As stated previously, the quartz wall temperature measurements were generally lower than the corresponding fuel rod temperatures. It is possible to correlate quartz wall temperature measurement with the measured fuel rod temperature measurements before the type K TC failure in order to estimate the offset of the wall temperature from the rod temperature. Figure 3.17 shows the axial heater rod temperatures at the 36-in. and 42-in. levels. The higher temperature data starting at about 160 minutes is obtained from the quartz wall temperature measurements after adjusting the wall temperature to match the rod temperature shortly before the type K TCs failed. The peak rod temperature estimated in the rod bundle was about 1475°C. This temperature is far short of the 1850°C required to melt the Zircaloy cladding. The absence of any appreciable smoke during this experiment also suggests that the cladding did not melt, and the molten Zr-MgO reaction evident in the first heater design test did not occur.

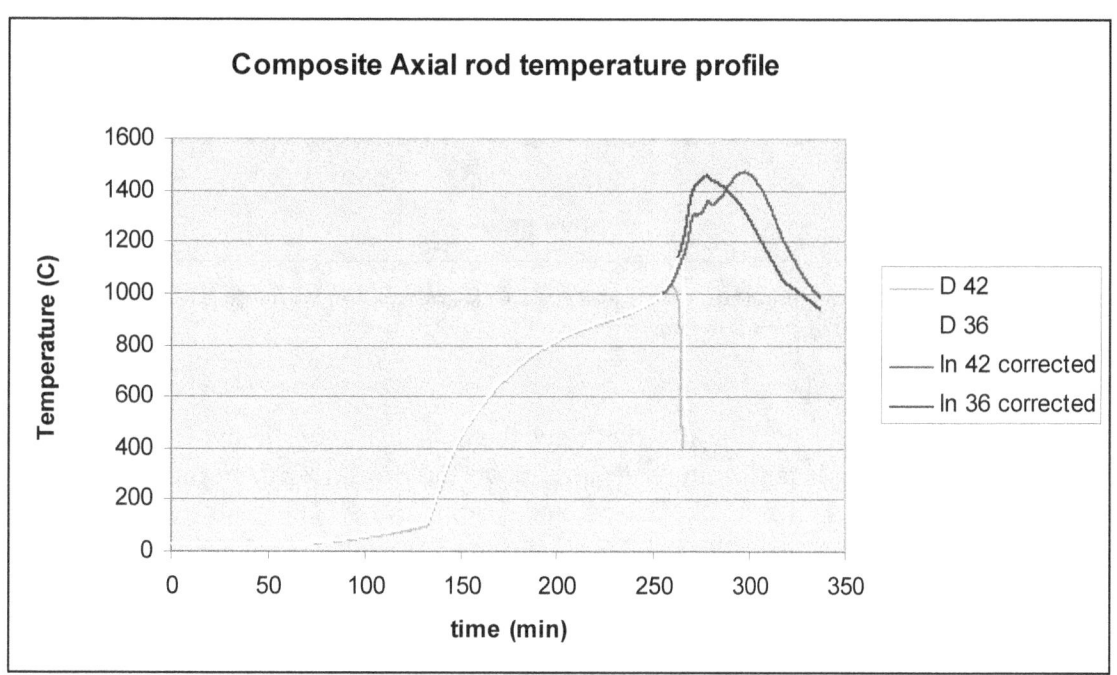

Figure 3.17   Composite axial rod temperature profile using corrected wall temperature measurements

3.4.4   Post-Test MELCOR Analysis

Incorporation of the new ANL Zircaloy 4 air oxidation kinetic information [2] into MELCOR resulted in excellent prediction of the time of ignition. Figure 3.18 shows the estimated peak cladding temperature (see Figure 3.17) measured during the experiment along with the MELCOR prediction using the new kinetics. The new ANL kinetic study identified a breakaway phenomena in the oxidation of Zircaloy after which the kinetic rate of oxidation greatly increased. Thus, the study provides two oxidation kinetic parameters, one for pre-breakaway and one for post-breakaway and an empirical time-at-temperature criterion for the occurrence of breakaway. The time-at-temperature criterion was used to transition the MELCOR calculation from pre- to post-breakaway kinetics, which occurred at 250 minutes in excellent agreement with the data. Also shown is the MELCOR prediction using only the pre-breakaway kinetics. Without the added oxidation energy, the rod temperature was predicted to drop sharply at 268 minutes when all the heater rods failed.

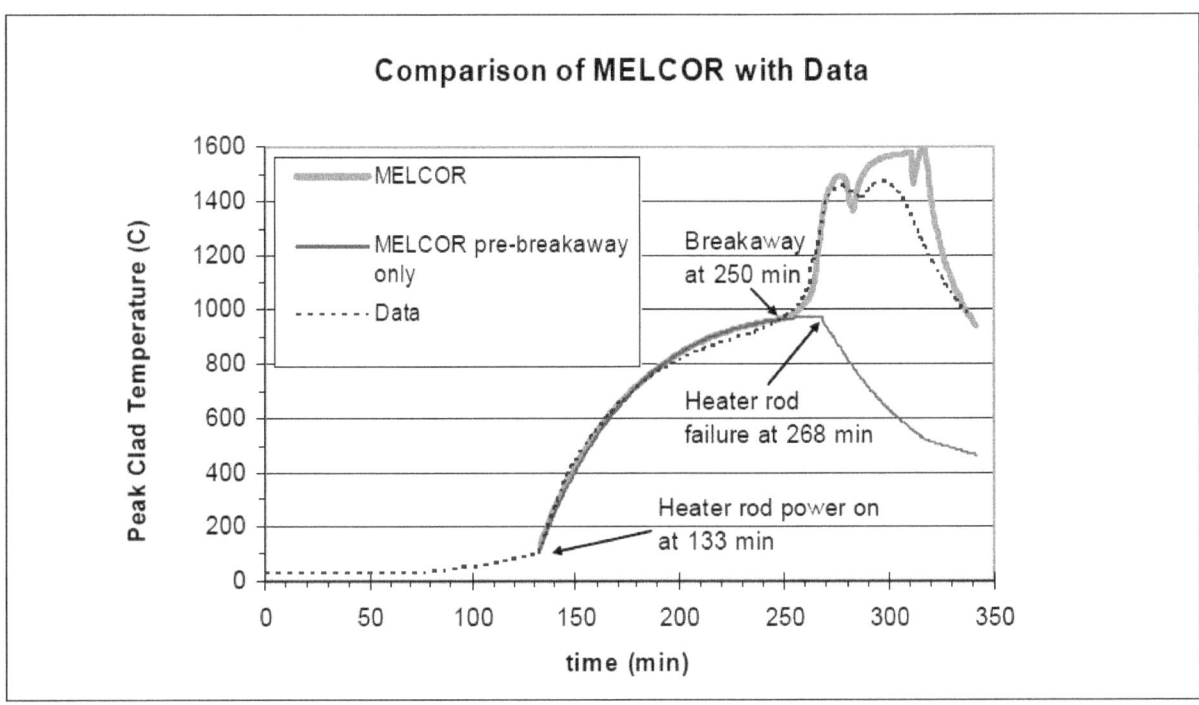

Figure 3.18   **Peak cladding temperature data and MELCOR calculation with and without post-breakaway oxidation kinetics**

3.4.5   Post-Test Inspection

Figure 3.19 shows the outer insulation jacket after the conclusion of the experiment. The labels show obvious signs of high temperature at and above the 36-in. level. Figure 3.20 shows the temperature of the outer jacket at the 12-in. and 36-in. levels. The peak jacket temperature measured at the 36-in. level was 200°C. The condition of the labels at the 42-in. level suggests the surface temperature here reached a higher temperature. Figure 3.21 shows the inner quartz tube and inner thermal radiation reflector with the insulation removed. The upper portion of the bundle shows obvious signs of the high temperature burn. Figure 3.22 shows a detail of the heater rod rubble before the bundle was removed from the test stand. Figure 3.23 shows a detail of heater rod rubble at the 36-in. to 39-in. levels after bundle removal from the test stand. Note that much of the rubble is still held within the test bundle by intact Nichrome heater elements. Nichrome melts at about 1400°C; therefore, the core of many of the heater rods did not reach this temperature. This is consistent with the estimate of a maximum clad temperature of 1475°C. Because the heater rods were no longer heated once the cladding ignited, the cores of the heater rods were cooler than the burning surface.

Figure 3.19   Outer insulation jacket after burn

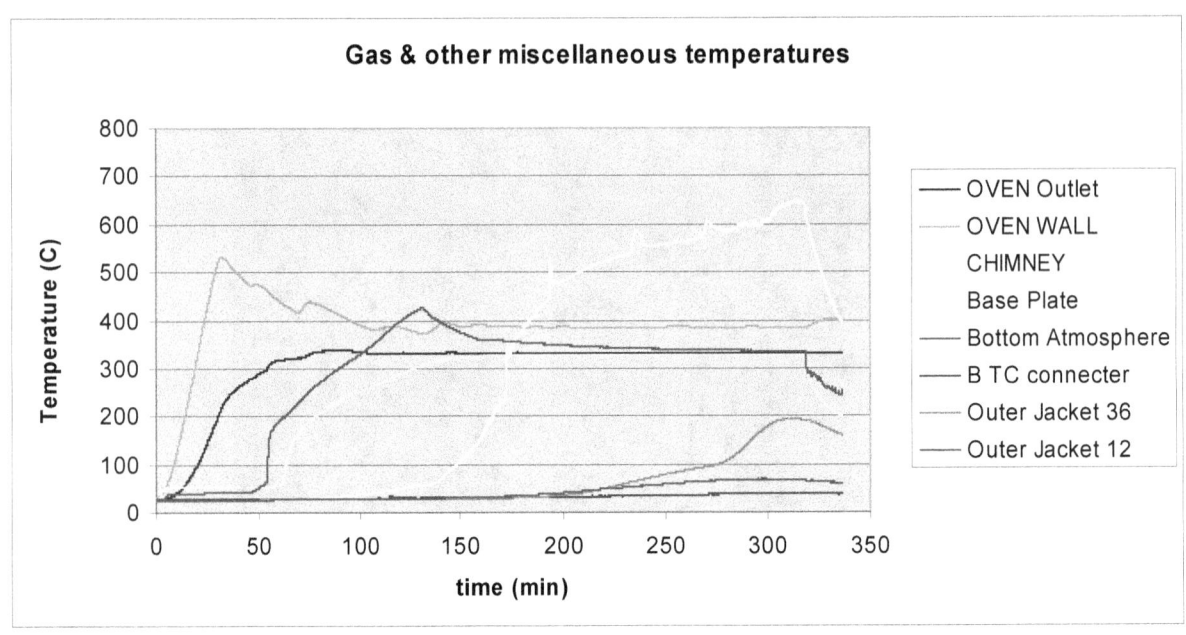

Figure 3.20 Inlet gas temperature, exit gas temperature in chimney, and other miscellaneous temperatures

Figure 3.21   Post-test quartz tube and inner thermal radiation shield after removal of insulation

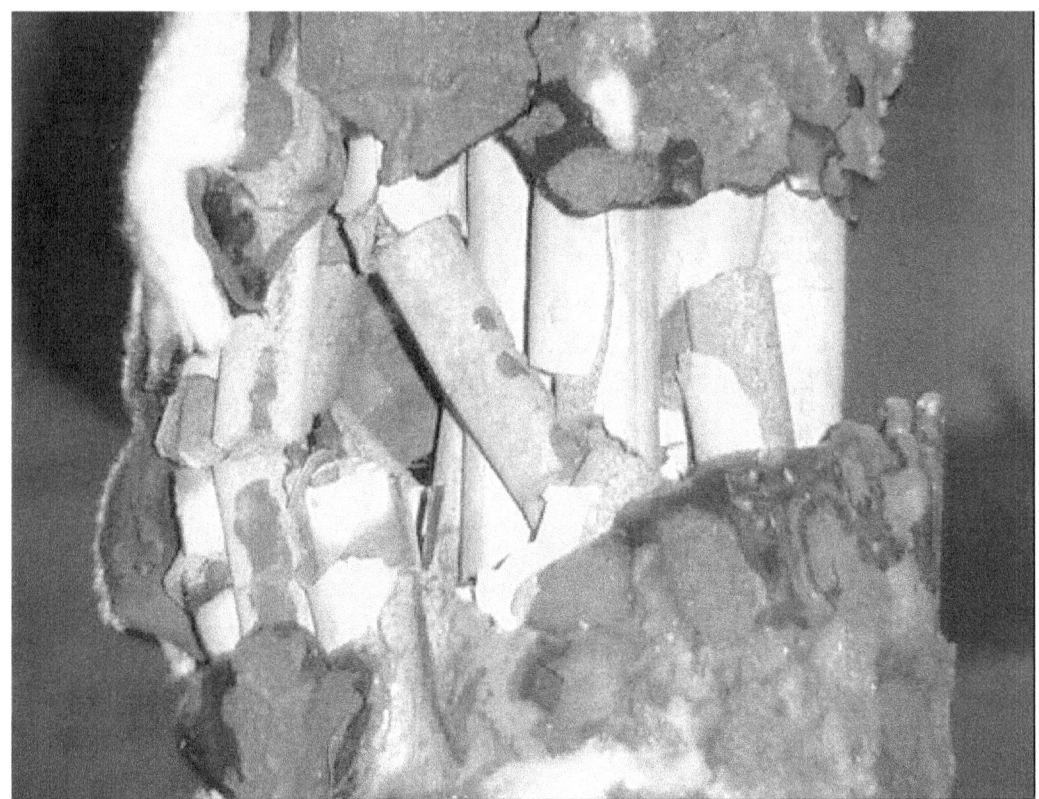

Figure 3.22   Detail of heater rod rubble

Figure 3.23   **Detail of heater rod rubble at the 36-in. level after bundle removal from test stand**

3.4.6   Smoke Sampling Results

Total metals analysis of the three smoke sampling filters yielded no trace of any metals including magnesium, tin, iron, nickel, or chromium. This is consistent with the experimental observation that no smoke was generated during the experiment.

3.5   Issues

The main experimental issue identified by this experiment was the nature of the downward burn front advance. With the reduced oxygen expected at the burn ignition point, the burn front does not advance downward substantially once the heat input from the electric heaters is lost. With prototypic decay heated fuel rods, the heat input from the fuel rods would, of course, not be lost once ignition occurs and significant downward burn front migration would be expected. This is an experimental artifact that will be taken into account in the numerical model calibration phase of this project.

3.6   Conclusion

The main objective of the test was achieved. The Zircaloy clad was successfully ignited using the Watlow heater design under reduced oxygen conditions. As expected, the resulting burn was slower and cooler than during the previous heater design test. The peak temperature of the clad was estimated to be 1475°C. This is consistent with the observations that much of the Nichrome wire was still intact (melting point of 1400°C), and no smoke was produced by the reaction of molten Zirconium (melting point of 1850°C) and the magnesium oxide core.

4 HYDRAULIC ANALYSIS OF THE SPENT FUEL POOL EXPERIMENT

This report summarizes the findings of the SFP pressure drop experiments conducted between August 25 and October 12, 2005. The stated purpose of these investigations was to determine hydraulic coefficients, namely S_{LAM} and k values, for the calibration of the MELCOR severe accident analysis code. The apparatus was operated in the laminar regime with Reynolds numbers spanning from 70 to 900, based on the bundle velocity and hydraulic diameter.

4.1 Experimental Apparatus and Procedures

Commercial components were purchased to create a prototypical BWR assembly including the top and bottom tie plates, spacers, water rods, channel box, and all related assembly hardware. Stainless steel conduit was substituted for the fuel rod pins for hydraulic testing. The diameter of the stainless steel rods was slightly smaller than prototypic pins, 1.11×10^{-2} m versus 1.12×10^{-2} m. The slightly simplified stainless steel mock fuel pins were fabricated based on drawings and physical examples supplied typical fuel type . The dimensions of the assembly components are listed in Table 4.1.

Table 4.1   Dimensions of assembly components in the 9×9 BWR

Description	Lower (Full) Section	Upper (Partial) Section
Number of Pins	74	66
Pin Diameter (m)	1.11×10^{-2}	1.11×10^{-2}
Pin Pitch (m)	1.44×10^{-2}	1.44×10^{-2}
Pin Separation (m)	3.28×10^{-3}	3.28×10^{-3}
Water Rod OD (main section) (m)	2.49×10^{-2}	2.49×10^{-2}
Water Rod ID (m)	2.34×10^{-2}	2.34×10^{-2}

Figure 4.1 shows the layout of the SFP pressure drop experimental assembly, including all available pressure port locations. Two Paroscientific Digiquartz differential pressure transducers (Model 1000-3D) were plumbed directly to the desired pressure ports. These pressure gauges use a highly sensitive quartz crystal to measure slight changes in differential pressure (resolution ~0.02 Pa). For this report the pressure drop across two ports is reported as the low side port first followed by the high side port, e.g., the overall pressure drop across the entire assembly is reported as 1–B.

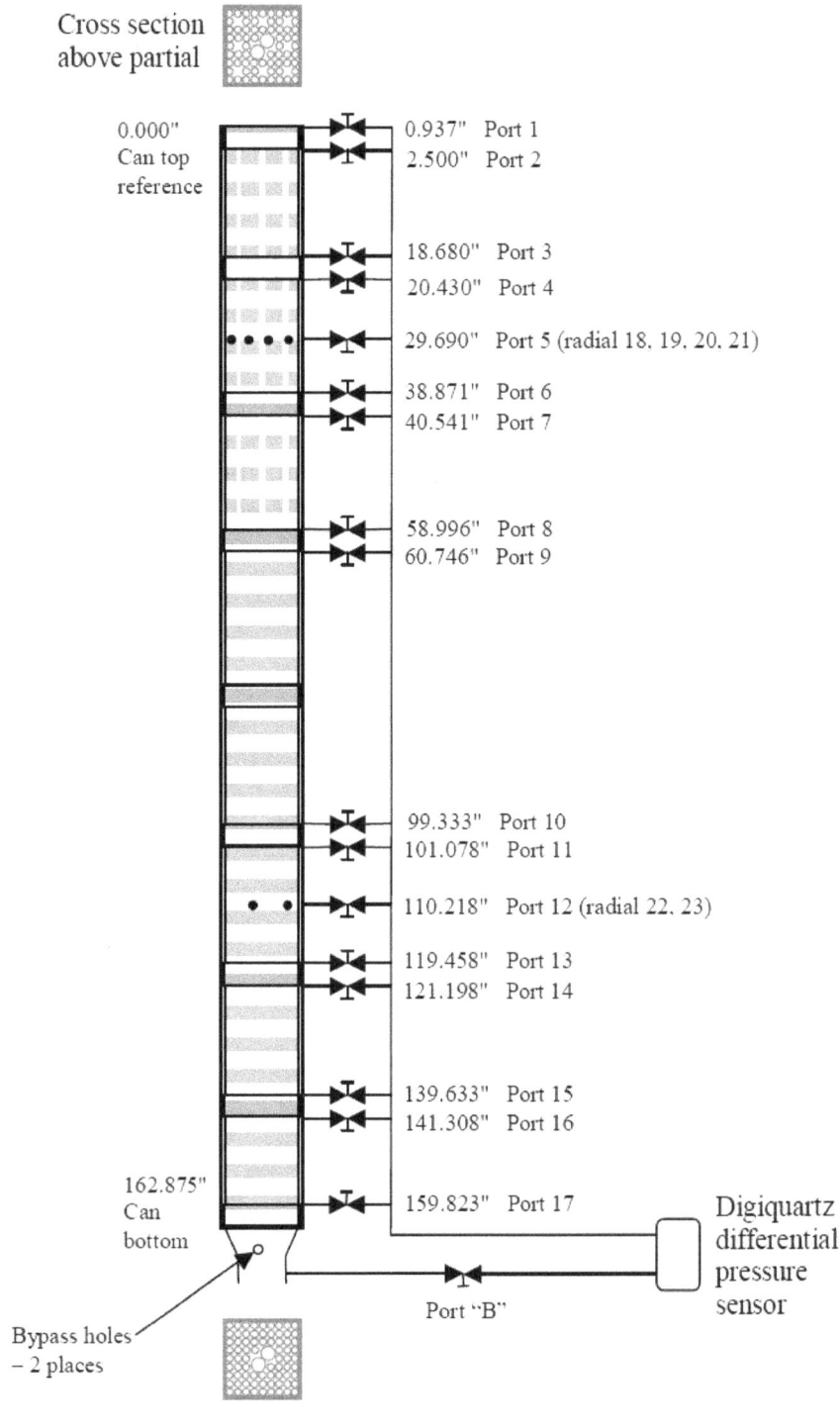

Figure 4.1   **Experimental apparatus showing as-built port locations**

Measurements were recorded directly to the hard drive of a PC-based data acquisition system every 3 seconds using a LabView 7.1 interface. These measurements included the air flow rate through the assembly, ambient air temperature, ambient air pressure, and the assembly pressure drops. The procedure for collecting the data is detailed next.

With each pressure transducer plumbed to two set port locations and with the air flow off, pressure drop measurements were recorded for a period of roughly 1 minute. These measurements were termed zero flow measurements and allowed for correction of any zero drift in the transducer. Next, the air flow was set to the desired rate with pressure drop readings subsequently acquired for 2 minutes. The air flow was then stopped, and zero flow measurements were again taken for 1 minute. This procedure was repeated for different air flow rates. The resulting pressure traces are shown in Figure 4.2. The pressure spikes evident during the reestablishment of flow are discarded before averaging for the pressure drops. Also, the slight zero drift of the transducer was corrected by subtracting the average of the zero flow measurements taken prior to and after each respective flow test. The zero corrections of the pressure drops were less than 0.15 N/m^2, which ocurred during an overall 1-B pressure drop measurement.

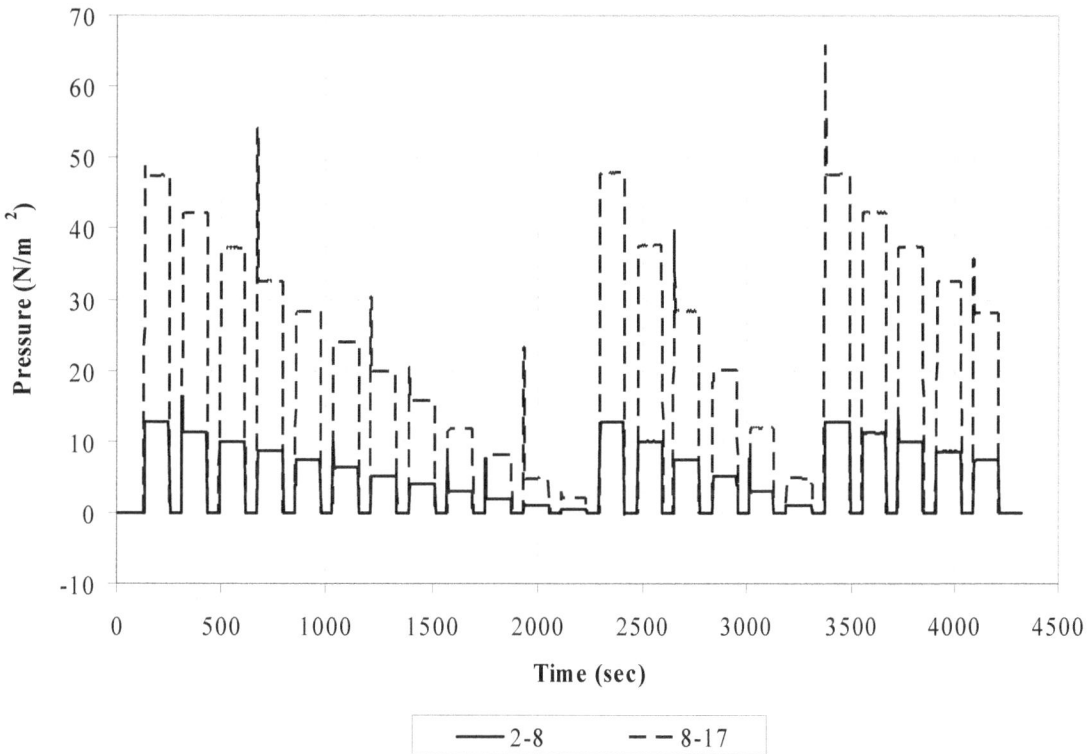

Figure 4.2   Pressure traces recorded during the blocked water rod testing for measurements across 2–8 and 8–17

Experimental runs may be divided into two distinct groups, blocked and unblocked water rods. The bypass holes were blocked for all testing. To understand the influence of flow through the water rods, acrylic sleeves were inserted over the water rod exit holes for one series of experiments to force all of the air flow through the bundle (Figure 4.3). These experimental runs were then repeated using the apparatus with unblocked water rods.

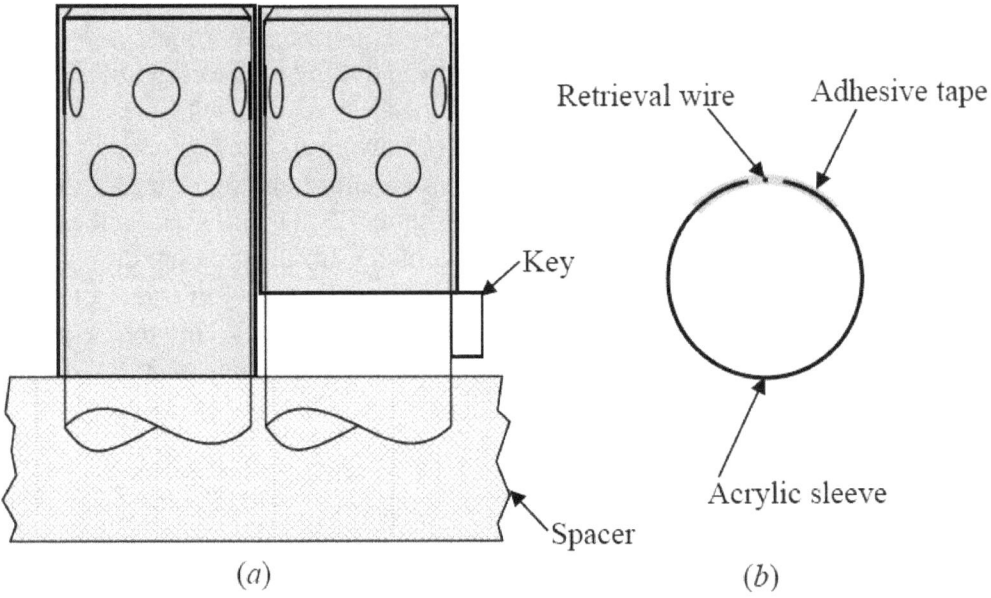

Figure 4.3 Acrylic sleeves are inserted over the top of the water rods to block the flow. Sleeve alignment on the water rods (a) and the sleeve cross-section (b) are shown above

4.2 Hydraulic Analysis

As discussed previously, the goal of this research was to determine the S_{LAM} and k coefficients for use with the MELCOR code. The cross-sectional areas and hydraulic diameters used in the hydraulic analyses to follow are presented in Table 4.2.

Table 4.2 Summary of the flow areas and hydraulic diameters of the SFP experiment.

Description	A (m^2)	D_H (m)
Upper Bundle	1.06×10^{-2}	1.41×10^{-2}
Upper Bundle plus Water Rods	1.14×10^{-2}	1.45×10^{-2}
Upper Spacer	8.35×10^{-3}	3.57×10^{-3}
Lower Bundle	9.79×10^{-3}	1.19×10^{-2}
Lower Bundle plus Water Rods	1.06×10^{-2}	1.24×10^{-2}
Lower Spacer	7.37×10^{-3}	2.89×10^{-3}
Water Rod ($\times 1$)	4.29×10^{-4}	2.49×10^{-2}

Curve fits to the pressure drop data were used to determine the S_{LAM} and k coefficients of the assembly. The determination of these coefficients is discussed next. The major, or viscous, pressure loss is expressed in Equation 1.

$$\Delta P_{major} = f \left(\frac{L}{D_H} \right) \left(\frac{\rho \cdot V_{bundle}^2}{2} \right) \qquad 1$$

The friction factor for laminar flow is written explicitly as

$$f = \frac{S_{LAM}}{Re}, \text{ where } S_{LAM} = 64 \text{ (pipe flow)}$$
$$= 100 \text{ (bundle flow)}$$

2

Substituting for the Reynolds number yields

$$\Delta P_{major} = S_{LAM}\left(\frac{L}{D_H^2}\right)\left(\frac{V_{bundle} \cdot \mu}{2}\right)$$

3

The minor, or form, pressure drops across the assembly are given by

$$\Delta P_{minor} = \sum k\left(\frac{\rho \cdot V_{bundle}^2}{2}\right)$$

4

Curve fits to pressure drop data are presented in the following format. In Equation 5, the quadratic term accounts for the minor losses and the linear term for the major losses.

$$\Delta P_{total} = a_2 \cdot V_{bundle}^2 + a_1 \cdot V_{bundle}$$

5

Because the total pressure drop is simply the sum of the major and minor pressure drops, the S_{LAM} and k coefficients may now be determined explicitly.

$$S_{LAM} = 2 \cdot a_1\left(\frac{D_H^2}{L \cdot \mu}\right)$$

$$\sum k = \frac{2 \cdot a_2}{\rho}$$

6

4.2.1 Blocked Water Rod Results

The following is an example S_{LAM}-k analysis of the curve fit to pressure drop data. Please refer to Figure 4.1 for the location of the pressure ports described next. The data in Figure 4.4 refer to the pressure drops across pressure ports 2–8, 8–17, and 2–17 for the assembly with blocked water rods (compiled from data file, "blocked3 full 8-17 partial 2-8.csv," created October 4, 2005).

Note: The pressure drops are plotted versus the corresponding bundle velocity for each section. The hydraulic diameter and flow area of the lower portion of the assembly were used to calculate S_{LAM} and k coefficients for the 2–17 data. Also, the pressure drop across 2–17 was not independently measured but is the summation of the pressure drops across 2–8 and 8–17. See Appendix B, Error Analysis, for details concerning the 95% confidence error bars on the data.

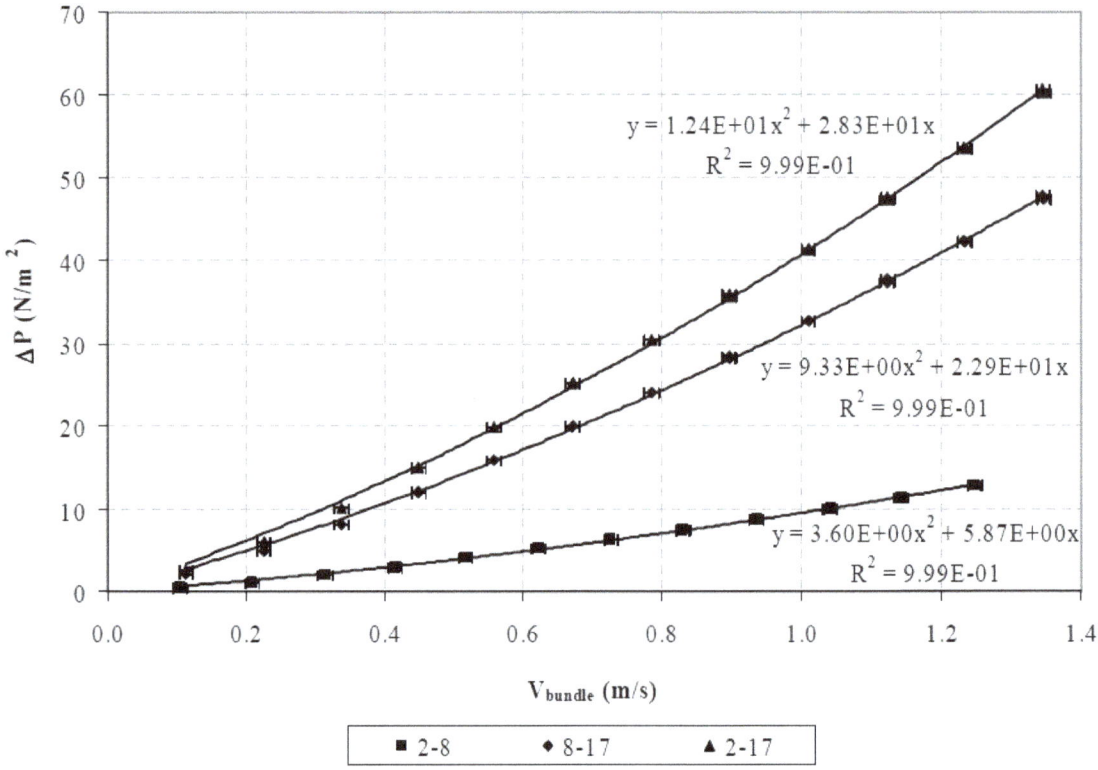

Figure 4.4   Pressure drop as a function of bundle velocity for the SFP assembly with <u>blocked</u> water rods

The relevant information for the calculation of the S_{LAM} and k coefficients is given in Table 4.3. Uncertainties in the S_{LAM} and k coefficients listed in this report are taken as ±5 and ±1, respectively. See Appendix B, Error Analysis, for details concerning this uncertainty estimate.

Table 4.3 S_{LAM} and k coefficient analysis data for pressure drops between 2–8, 8–17, and 2–17 for blocked water rods

Pressure Drop	L (m)	A (m^2)	D$_H$ (m)	a_1 (N·s/m^3)	a_2 (N·s^2/m^4)	S_{LAM}	Σ k	# of spacers	k per spacer
2–8	1.43	0.0106	0.0141	5.87	3.60	88	7.3	2	3.7
8–17	2.56	0.0098	0.0119	22.9	9.33	138	19	5	3.8
2–17	4.00	0.0098	0.0119	28.3	12.4	109	25	7	3.6

This analysis assumes air properties at local ambient conditions, typically $\rho = 0.98$ kg/m^3 and $\mu = 1.85 \times 10^{-5}$ N·s/m^2. Changes in air temperature and pressure are taken into account for measurements collected during different experimental runs.

The influence of the experimental flow rate range is examined in Table 4.4. For this analysis, subsets of the entire data set were used to determine the curve fit coefficients. These results are also presented graphically in Figure 4.5. The hydraulic loss coefficients tend towards constant values as the experimental flow range is increased. The errors associated with these measurements are also more significant in the lower flow rate ranges. Therefore, the S_{LAM} and k values determined from the largest data range are considered to be more applicable.

Table 4.4   S_{LAM} and k values for different ranges of experimental air flow rate

Flow Rate Range (slpm)	2–8		8–17		2–17	
	S_{LAM}	$\Sigma\,k$	S_{LAM}	$\Sigma\,k$	S_{LAM}	$\Sigma\,k$
50–200	46	20	99	47	75	64
50–300	60	14	113	34	86	46
50–400	75	10	126	25	99	33
50–500	84	8.1	134	20	106	27
50–600	88	7.3	138	19	109	25

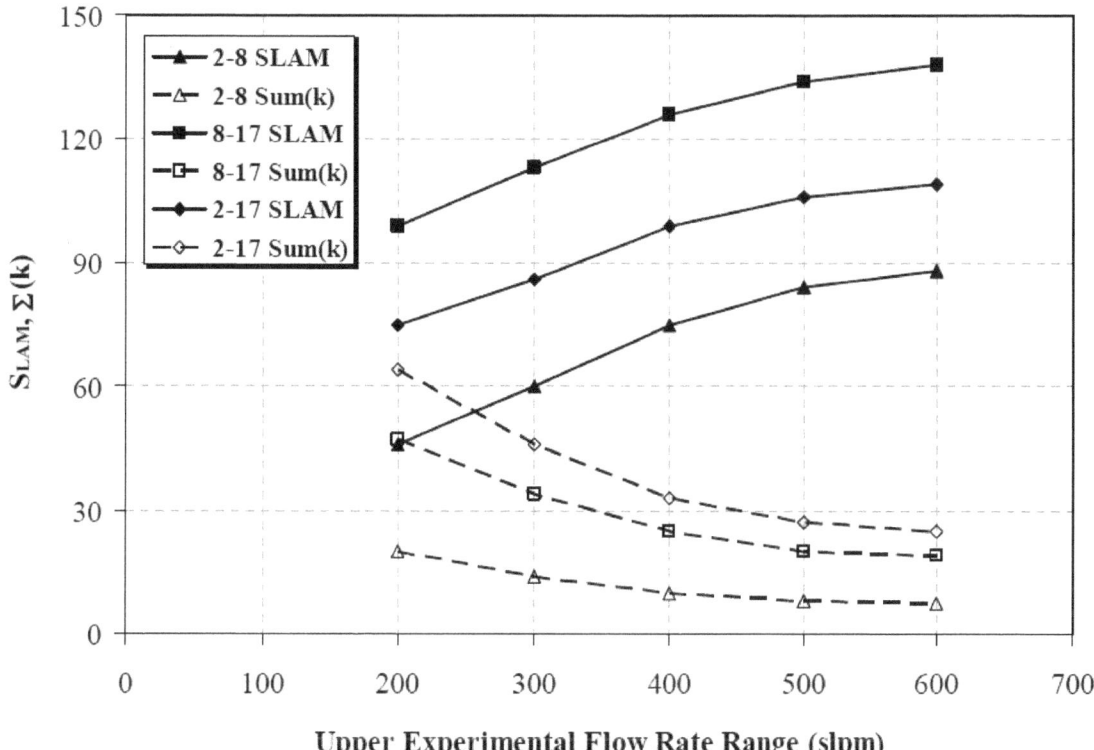

Figure 4.5   Hydraulic loss coefficients as a function of the upper, measured experimental flow range

A summary of the S_{LAM} and k coefficients for blocked water rod testing is shown in Table 4.5. These values were determined from the full experimental flow rate range of 50 to 600 slpm, or Reynolds numbers of 70 to 900 in the fully populated segment, respectively. The hydraulic diameter and flow area of the fully populated section was used to calculate the S_{LAM} and k values for any span including the lower section, e.g., 1–B. Values of S_{LAM} and k for segments 1–2, 4–6, and 6–7 are calculated using the characteristic hydraulics of both the partially and the fully populated sections.

Table 4.5 S_{LAM} and k coefficients for the blocked water rod assembly

Description	Segment	A (m²)	L (m)	D_H (m)	S_{LAM}	Σ k	# of spacers	k per spacer
Long Spans	1–B	0.0098	4.19	0.0119	106	37	–	–
	2–8	0.0106	1.44	0.0141	88	7.3	2	3.6
	8–17	0.0098	2.56	0.0119	138	19	5	3.8
	2–17	0.0098	4.00	0.0119	109	25	7	3.6
	4–8	0.0106	0.98	0.0141	85	4.0	1	4.0
	8–15	0.0098	2.05	0.0119	137	15	4	3.8
	4–15	0.0098	3.03	0.0119	111	18	5	3.6
Individual Bundle Runs	4–5	0.0106	0.24	0.0141	69	0.88	–	–
	5–6	0.0106	0.23	0.0141	51	0.00	–	–
	4–6	0.0106	0.47	0.0141	60	0.88	–	–
	4–6	0.0098	0.47	0.0119	40	0.71	–	–
	11–12	0.0098	0.23	0.0119	89	0.70	–	–
	12–13	0.0098	0.23	0.0119	65	0.13	–	–
	11–13	0.0098	0.47	0.0119	77	0.84	–	–
Spacers/Tie Plates	1–2	0.0106	0.040	0.0141	49	0.42	–	–
	1–2	0.0098	0.040	0.0119	33	0.36	–	–
	6–7	0.0106	0.045	0.0141	408	3.1	1	3.1
	6–7	0.0098	0.045	0.0119	272	2.6	1	2.6
	13–14	0.0098	0.044	0.0119	729	3.2	1	3.2
	17–B	0.0098	0.13	0.0119	119	11	–	–

Next, the S_{LAM} and k coefficients across 6–7 and 13–14 are recalculated using the appropriate spacer hydraulic diameter and flow area. As shown in Table 4.6, the values determined from this analysis are significantly reduced.

Table 4.6   Spacer S_{LAM} and k coefficients calculated with the appropriate spacer hydraulic diameter and flow area

Segment	A (m²)	L (m)	D_H (m)	S_{LAM}	Σ k
6–7	0.00835	0.045	0.00357	21	1.9
13–14	0.00737	0.044	0.00289	32	1.8

The values of S_{LAM} through the spacers 6–7 and 13–14 are rather large compared to the other values in Table 4.5. These large values are due to the use of the bundle hydraulic diameter when calculating the S_{LAM} coefficients. As a result, the S_{LAM} values for segments including spacers, such as 1-B, have inflated values of S_{LAM} when compared to a single section without spacers (11–13).

The single section/spacer S_{LAM} and k values can be manipulated to recreate the values observed in the multiple section/spacer data. To accomplish this, the values of S_{LAM} must be weight averaged based on flow length. Equation 7 shows the general format for calculating the effective S_{LAM} coefficient of an assembly span with an overall flow length of L_{tot}, "I" number of spacers, and "J" number of sections.

$$S_{LAM\,eff} = \frac{\left(\sum_{i=1}^{I} L_{sp,i} \cdot S_{LAM\,sp,i} + \sum_{j=1}^{J} L_{sect,j} \cdot S_{LAM\,sect,j} \right)}{L_{tot}} \qquad 7$$

The loss coefficients will simply be the sum of the k values for the individual segments. The values for 4–6 and 6–7 are considered to be typical for a segment and spacer in the upper section. Similarly, 11–13 and 13–14 were used for the lower section. The top and bottom tie plate S_{LAM} and k coefficients are measured separately and are used to calculate the effective values for the 1–B span. Two values are listed for segments 1–2, 4–6, and 6–7 in Table 4.5. These values denote the loss coefficients determined from the upper and lower assembly hydraulic areas and diameters. The first values listed in the table are used for spans including the upper portion of the assembly only. For spans including the lower assembly (e.g., 2–17), the values calculated with the lower assembly area and hydraulic diameter are used for segments in the partially populated region (for 6–7 S_{LAM} = 272, k = 2.6). Table 4.7 gives the calculated effective S_{LAM} and k coefficients based on the values from the individual sections/spacers. The S_{LAM} values are within about 10% of each other. However, the effective k values are somewhat higher than the measured values. Discrepancies in these two sets are most likely due to the assumption that all segments and spacers have the same loss coefficient values throughout the assembly.

Table 4.7   Calculation of effective S_{LAM} and k values for different span lengths

Pressure Drop	Measured		Effective	
	S_{LAM}	Σ k	S_{LAM}	Σ k
1–B	106	37	104	39
2–8	88	7.3	81	8.8
8–17	138	19	133	20
2–17	109	25	105	28
4–8	85	4.0	75	4.9
8–15	137	15	133	16
4–15	111	18	106	20

Finally, Figure 4.6 shows the cumulative pressure drop in the assembly as a function of axial location for various flow rates. The lines in the plot represent the pressure drop determined from Equations 3 and 4. The form loss and laminar friction coefficients were taken from Table 4.5. The dashed line at z = 2.7 m indicates the demarcation of the partially and fully populated sections.

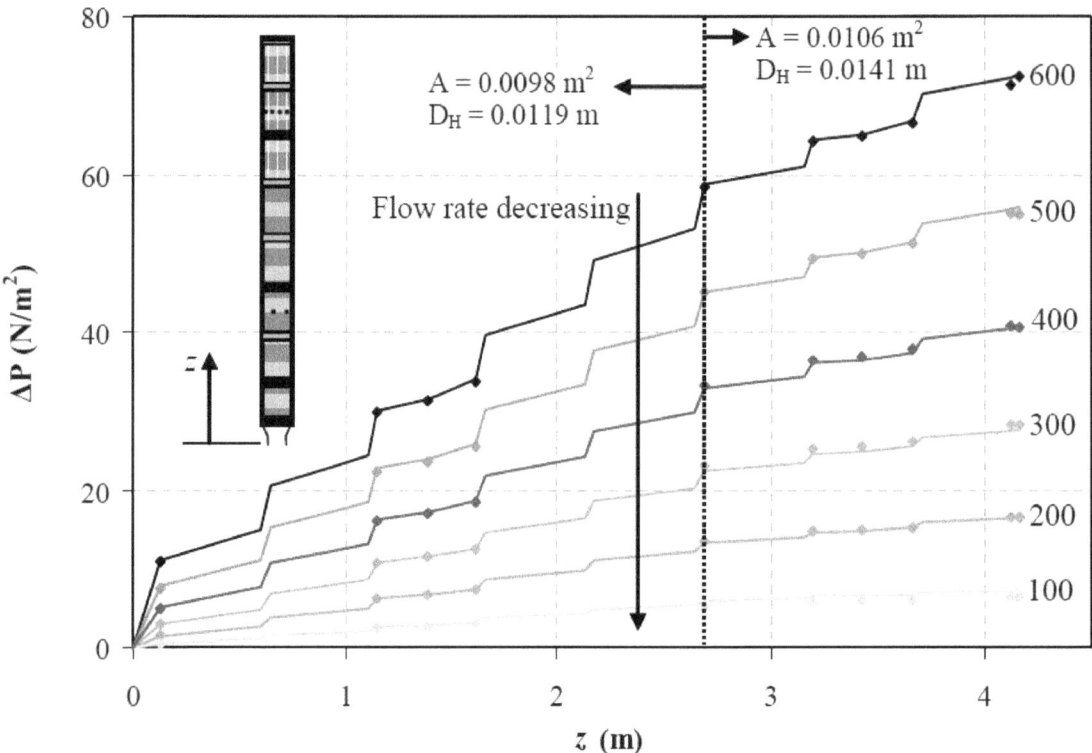

Figure 4.6   Cumulative pressure drop as a function of axial location for flow rates of 100, 200, 300, 400, 500, and 600 slpm

The lines represent the analytically determined pressure drops using Equations 3 and 4 along with the values of S_{LAM} and k listed in Table 4.5.

4.2.2   Unblocked Water Rod Results

The pressure drops across pressure ports 2–8, 8–17, and 2–17 are shown in Figure 4.7 in the SFP assembly with unblocked water rods. The pressure drop across 2–17 is again the summation of the measurements across 2–8 and 8–17. The pressure drops are slightly less than those presented in Figure 4.4 for a given velocity. (compiled from data files, "unblocked2 full 17-8 partial 8-2.csv," "unblocked3 full 8-17 partial 2-8.csv," and "unblocked4 full 8-17 partial 2-8.csv," created September 28, October 3, and October 12, 2005, respectively). Again, the pressure drops are plotted versus the corresponding section bundle velocity. The bundle velocities shown in this plot were calculated using the bundle area only and are discussed next. The hydraulic diameter and flow area of the lower portion of the assembly were used to calculate S_{LAM} and k coefficients for the 2–17 data.

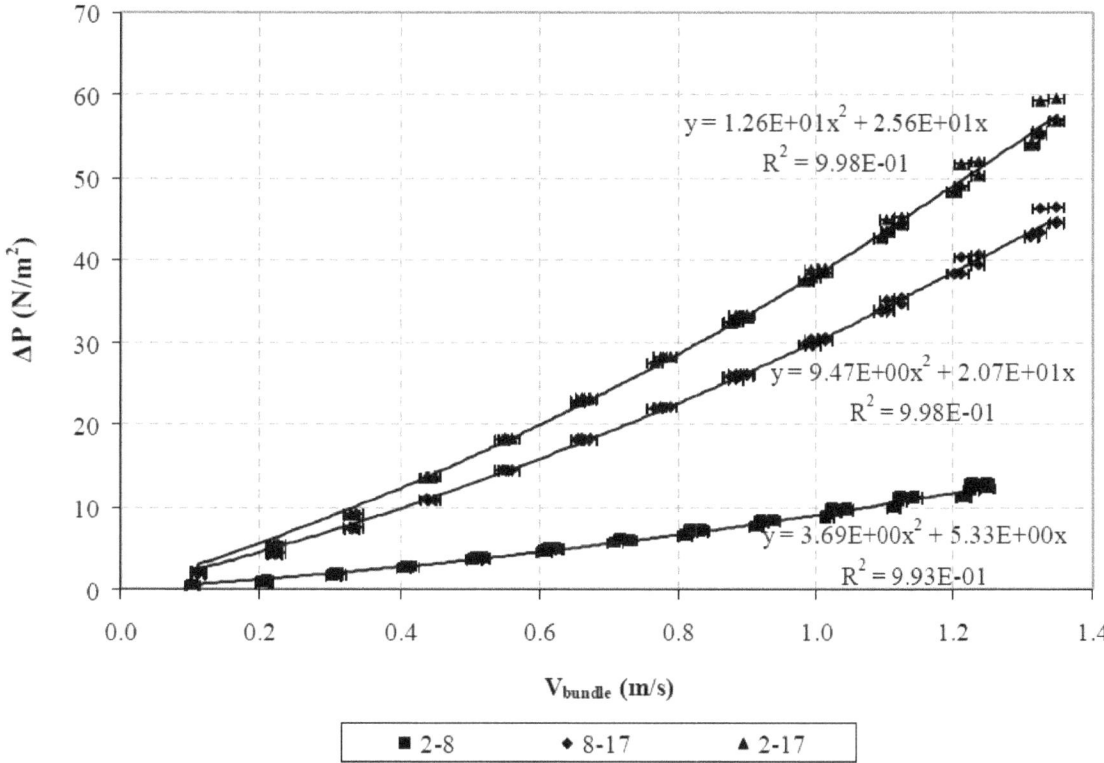

Figure 4.7 **Pressure drop as a function of bundle velocity for the SFP assembly with unblocked water rods**

Note: The bundle velocities shown in the plot were calculated for all the air flowing through the bundle.

This analysis was performed to directly compare with the previous blocked runs. The values used for this analysis are shown in Table 4.8. The inertial losses appear to be similar, but the S_{LAM} coefficient decreased somewhat from the blocked water rod values in Table 4.3. Again, the hydraulic diameter and flow area of the lower portion of the assembly were used to calculate S_{LAM} and k coefficients for the 2–17 data.

Table 4.8 **S_{LAM} and k coefficient analysis data for pressure drops between 2–8, 8–17, and 2–17 for unblocked water rods**

All flow is assumed to pass through the bundle for this analysis.

Pressure Drop	L (m)	A (m²)	D_H (m)	a_1 (N·s/m³)	a_2 (N·s²/m⁴)	S_{LAM}	Σk	# of spacers	k per spacer
2–8	1.43	0.0106	0.0141	5.33	3.69	80	7.3	2	3.7
8–17	2.56	0.0098	0.0119	20.7	9.47	125	19	5	3.8
2–17	4.00	0.0098	0.0119	25.6	12.6	99	25	7	3.6

A summary of the S_{LAM} and k coefficients for unblocked water rod testing is shown in Table 4.9. Again, this analysis assumes all flow passes through the bundle. The hydraulic diameter and flow area of the fully populated section was used to calculate the S_{LAM} and k values for any span including the lower section, e.g., 1–B

73

Table 4.9 S_{LAM} and k coefficients for the unblocked water rod assembly assuming all flow passes through the bundle

Description	Segment	A (m^2)	L (m)	D$_H$ (m)	S$_{LAM}$	Σ k	# of spacers	k per spacer
Long Spans	1–B	0.0098	4.19	0.0119	96	37	–	–
	2–8	0.0106	1.44	0.0141	80	7.3	2	3.7
	8–17	0.0098	2.56	0.0119	125	19	5	3.8
	2–17	0.0098	4.00	0.0119	99	25	7	3.6
	4–8	0.0106	0.98	0.0141	75	3.4	1	3.4
	8–15	0.0098	2.05	0.0119	130	14	4	3.5
	4–15	0.0098	3.03	0.0119	104	17	5	3.4
Tie Plates	1–2	0.0106	0.04	0.0141	64	0.39	–	–
	17–B	0.0098	0.13	0.0119	172	12	–	–

Next, the above analysis was repeated using the sum of the bundle and water rod areas to determine the bundle velocity. This approach sets the velocity of the air inside the water rods and the bundle to be equal. For this case, the fraction of flow through the water rods is 7.5 to 8.1% of the total flow rate for the upper and lower assembly sections, respectively. These values are simply the ratio of the water rod area to the total flow area (bundle plus water rod area) in the two sections. Table 4.10 lists a summary of the S_{LAM} and k coefficients calculated from this analysis. The tie plates are omitted since the water rods have no flow area through these sections. The hydraulic diameter and flow area of the fully populated section was used to calculate the S_{LAM} and k values for any span including the lower section.

Table 4.10 S_{LAM} and k coefficients for the unblocked water rod assembly assuming equal air velocities in the bundle and water rods

Description	Segment	A (m^2)	L (m)	D$_H$ (m)	S$_{LAM}$	Σ k	# of spacers	k per spacer
Long Spans	1–B	0.0106	4.19	0.0124	113	43	–	–
	2–8	0.0114	1.44	0.0145	92	8.6	2	4.3
	8–17	0.0106	2.56	0.0124	147	22	5	4.4
	2–17	0.0106	4.00	0.0124	117	30	7	4.3
	4–8	0.0114	0.98	0.0145	86	4.0	1	4.0
	8–15	0.0106	2.05	0.0124	154	17	4	4.3
	4–15	0.0106	3.03	0.0124	123	20	5	4.0

4.2.3 Water Rod Flow Rate

In the previous section, the air flow rate through the water rods was assumed to be based on flow area fractions. The water rod flow rate may be explicitly determined by comparing pressure drop data for both blocked and unblocked water rod assemblies. Figure 4.8 shows the total air flow rate through the assembly as a function of pressure drop across 2–17. This particular pressure drop was chosen because of the abundance of existing data across 2–17 and the fact that it spans the entire water rod. Power law fits were chosen to continuously represent the data based on a goodness of fit criteria. The flow rate through the bundle must be the same to create the same pressure drop. Therefore, the difference in the flow rate represented by the two curves is the flow rate through the water rods.

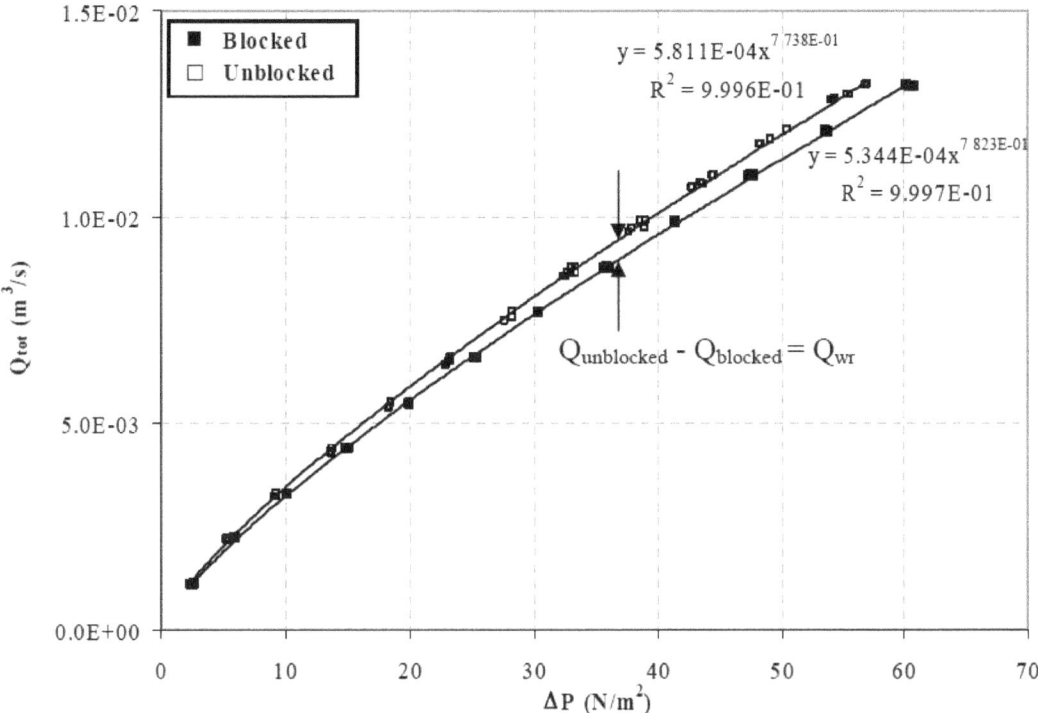

Figure 4.8   Total air flow rate through the assembly as a function of pressure drop across 2–17 for blocked (closed) and unblocked (open) water rods

Figure 4.9 shows the fraction of flow rate through the water rods as a function of total flow rate through the assembly. This curve was determined from the pressure drop data across 2–17. The experimental data span a range of 1.1×10^{-3} to 1.3×10^{-2} m³/s. Initially, the flow rate through the water rods decreases sharply with increasing total flow rate. The flow fraction then approaches a value of approximately 5% at the highest total flow rate.

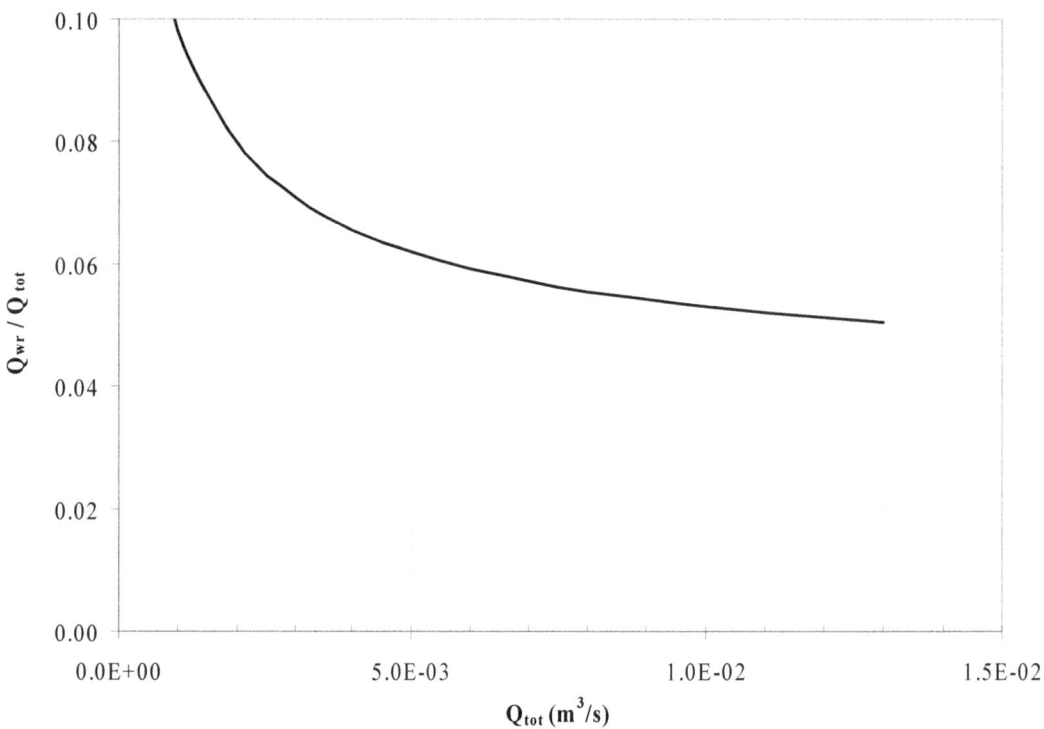

Figure 4.9   Fraction of flow rate through the water rods as a function of total flow rate as determined from the 2–17 pressure drop data

Now, the S_{LAM} and k analysis shown in Section 4.2.2 may be repeated with experimentally determined water rod flow rates. Figure 4.10 shows the bundle pressure drop as a function of the bundle velocity calculated using the experimentally determined flow rate through the bundle $(Q_{tot} - Q_{wr})$. As expected, the bundle velocities are slightly less than those shown in Figure 4.7.

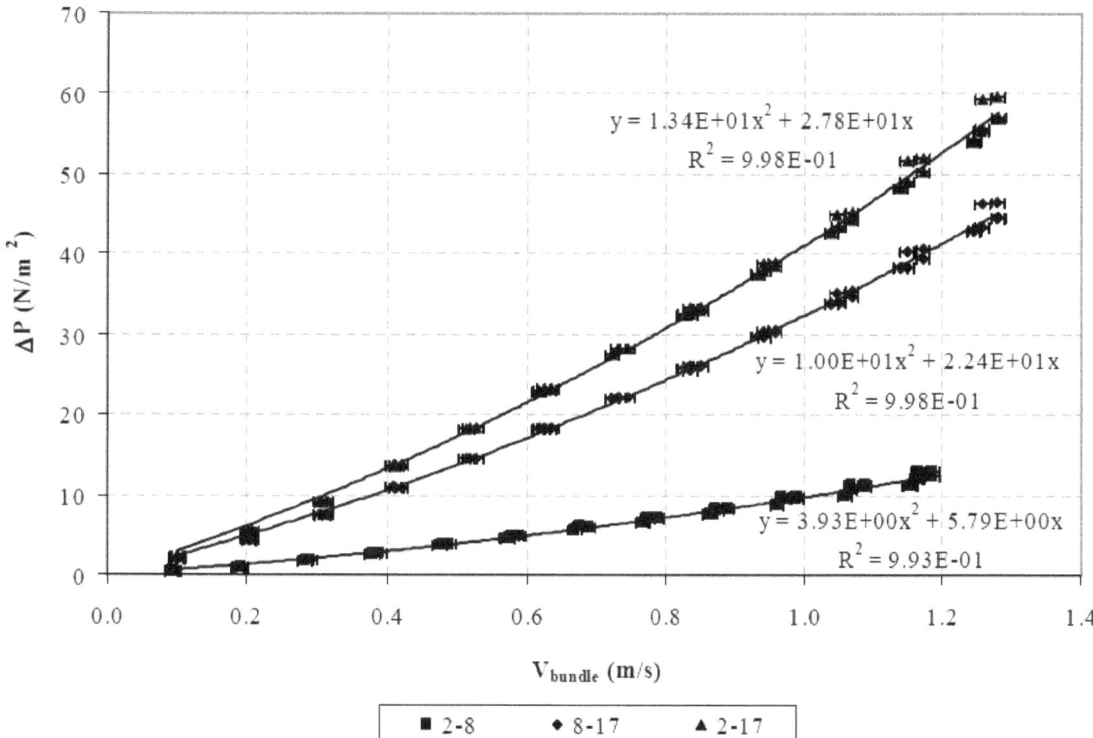

Figure 4.10   **Pressure drop as a function of bundle velocity for the SFP assembly with <u>unblocked</u> water rods**

The bundle velocities shown in the plot were calculated using the experimentally determined bundle flow rate ($Q_{tot} - Q_{wr}$).

The values used for the S_{LAM} and k analysis of this pressure drop data are presented next in Table 4.11. These values are quite similar to the blocked water rod values in Table 4.3. The hydraulic diameter and flow area of the lower portion of the assembly were used to calculate S_{LAM} and k coefficients for the 2–17 data.

Table 4.11   **S_{LAM} and k coefficient analysis data for pressure drops between 2–8, 8–17, and 2–17 for unblocked water rods**

Note: Flow through the bundle ($Q_{tot} - Q_{wr}$) was determined experimentally for this analysis.

Pressure Drop	L (m)	Λ (m^2)	D_H (m)	a_1 (N·s/m^3)	a_2 (N·s^2/m^4)	S_{LAM}	Σ k	# of spacers	k per spacer
2–8	1.43	0.0106	0.0141	5.79	3.93	87	7.8	2	3.9
8–17	2.56	0.0098	0.0119	22.4	10.0	135	20	5	4.0
2–17	4.00	0.0098	0.0119	27.8	13.4	107	27	7	3.9

Table 4.12 presents the results of the S_{LAM} and k analysis using the experimentally determined bundle flow rate. These S_{LAM} and k values are remarkably similar to the values obtained during the blocked water rod tests (Table 4.5). This agreement indicates that the flow rates through the water rods reported here are accurately portrayed over the experimental range of flow rates. The hydraulic diameter and flow area of the fully populated section was used to calculate the S_{LAM} and k values for any span including the lower section.

Table 4.12 S_{LAM} and k coefficients for the unblocked water rod assembly using the experimentally determined bundle flow rate ($Q_{tot} - Q_{wr}$)

Description	Segment	A (m^2)	L (m)	D$_H$ (m)	S_{LAM}	Σ k	# of spacers	k per spacer
Long Spans	1–B	0.0098	4.19	0.0119	105	38	–	–
	2–8	0.0106	1.44	0.0141	87	7.8	2	3.9
	8–17	0.0098	2.56	0.0119	135	20	5	4.0
	2–17	0.0098	4.00	0.0119	107	27	7	3.9
	4–8	0.0106	0.98	0.0141	81	3.6	1	3.6
	8–15	0.0098	2.05	0.0119	141	15	4	3.8
	4–15	0.0098	3.03	0.0119	113	18	5	3.6

4.2.4 Bypass Hole Flow Rate

Two bypass holes are located in the bottom tie plate (Figure 4.10). With the pool cell in place, these holes supply flow to the annular region between the canister and the pool cell. These holes were blocked for all testing referred to earlier in this report. To measure the flow rate through the bypass holes, the blockages were removed one at a time from these holes. The flow rate was then varied until the pressure drop across 1–B matched the blocked case. The difference in the flow rates was taken to be the flow rate through the bypass holes. These tests were conducted with the unblocked water rod configuration.

Figure 4.11 shows the fraction of flow rate through the bypass holes as a function of total flow rate. The experimental range of flow rates was 2.2×10^{-3} to 5.4×10^{-3} m^3/s. For the prototypic situation of two open bypass holes, the fraction of total flow rate through these holes is considerable (up to 6.8% at the highest flow rate measured). These values represent a preliminary attempt to characterize the bypass flow rate in an isothermal assembly unbounded by a pool cell. Further investigations provided better insight into the flow across the bypass holes (See Figure 5.9).

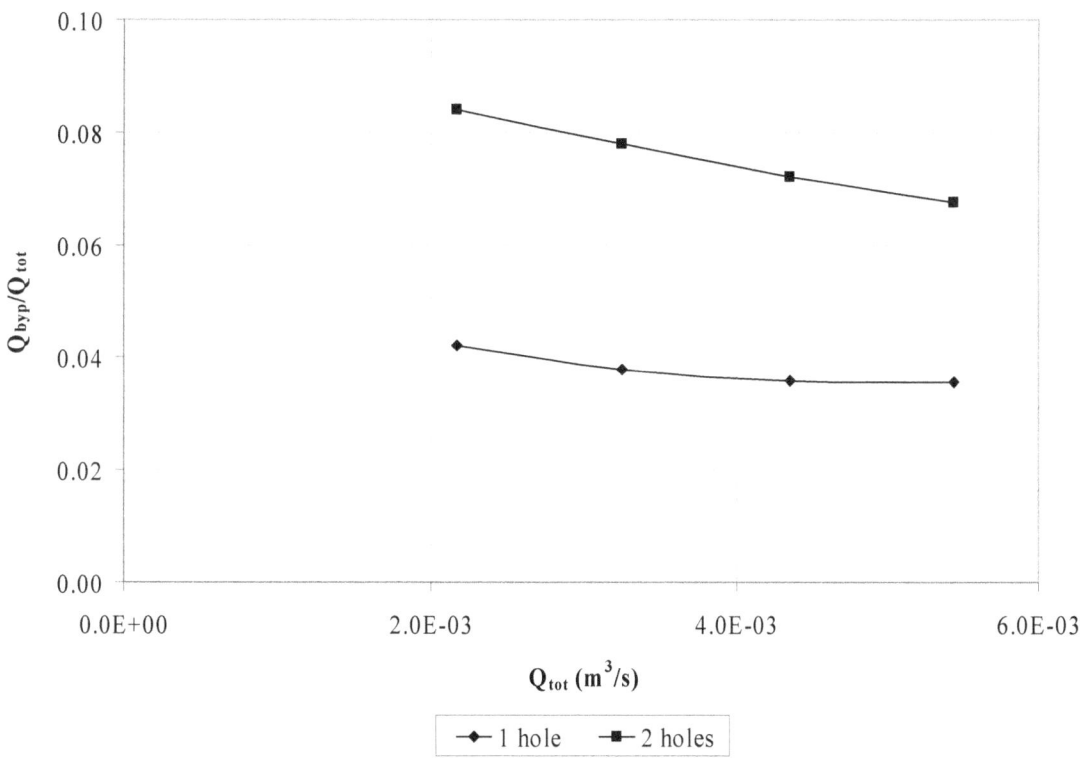

Figure 4.11   Fraction of flow rate through the bypass holes as a function of total flow rate as determined from 1–B pressure drop data

4.3   Summary

Extensive hydraulic experiments have been conducted for a prototypic spent fuel assembly modeled after the prototypic 9×9 BWR to determine the appropriate S_{LAM} and k coefficients for use with the MELCOR severe accident analysis code. The experimental range spanned the laminar region with volumetric flow rates of 1.1×10^{-3} to 1.3×10^{-2} m^3/s. The resulting Reynolds numbers based on the bundle velocity and hydraulic diameter were 70 to 900, respectively.

These experiments included testing with the water rods in the assembly, both blocked and unblocked. The values best suited for use with MELCOR modeling the 9×9 prototypical BWR geometry are given in Table 4.9.

4.3.1   Application of Experimental Results to a MELCOR Model

MELCOR, like other control volume codes, includes constitutive relationships to specify form losses (i.e., minor losses) and wall friction losses (i.e., major or viscous) along a flow path as a hydraulic flow loss term to the momentum equation. Because a single MELCOR flow path may be used to represent a rather complicated flow path, the wall friction terms may be computed for one or more segments that are connected in series. The format of the user-specified input follows Equations 3 and 4. In the context of recent BWR SFP analysis applications, a BWR bundle may be divided into five control volumes with six connecting flow paths that represent the major and minor flow losses. Similarly, the interstitial bypass region may be analogously subdivided.

As shown in Figure 4.1, the BWR fuel assembly contains seven grid spacers, upper and lower tie plates, full and partial rod regions, two water rods, and an inlet nozzle. It is generally not practical to include a detailed representation of all the geometry changes in the MELCOR model. In addition, the MELCOR code includes some hard-wired geometry models that further limit the modeling of the two large water rods inside the assembly and their associated flow. Consequently, modeling choices are required to represent the geometry of the BWR assembly. Relative to the hydraulic modeling in the MELCOR model, the experimental data are used in the following manner:

- Based on the hydraulic impact of the fully populated versus the partial regions, the control volume boundaries were specified to span uniform geometry regions. Control volume boundaries were placed at the bottom of the lower tie plate, the transition from the fully populated rod region to the partial region, and at the top of the upper tie plate. By spatially dividing the two regions, the distinct flow loss effects can be extended to heated conditions, where the flow will accelerate along the length of the assembly. This division of hydraulic segments has been implemented into whole pool analyses.

- The flow resistance in a flow path spans the region from cell-center of the lower control volume to cell-center of the upper control volume. Hence, the pressure in a given control volume represents the pressure in the center of the control volume. A single flow path may span more than one grid spacer and perhaps a tie plate. Segment data on a particular flow path defines the form and wall friction losses along the geometric regions encompassed in the flow path length. Several flow paths exhibited multiple flow segments due to geometry changes with the flow path and these net effects are calculated by the code.

- For the initial application for BWR SFP analysis, the results from Table 4.9 were used, which include prototypical water rod flow effects. For the flow segments in the fully populated tube region, a S_{LAM} of 125 was used and k losses of 3.8 were used for each spacer included in the range of the flow path. The flow area and hydraulic diameters were preserved. The total length across all flow paths in the fully populated region was 2.56 m and the total k was 19. Therefore, the total flow losses from 8–17 were exactly preserved, including the effect of flow within the water rods. Similarly, the partially populated tube region used a S_{LAM} of 80 and k losses of 3.7 per spacer in the range of the flow path. The total length across all flow paths in the fully populated region was 1.44 m and the total k was 7.3. The flow path segments in the partial rod region used the larger flow area and hydraulic diameter as specified in 2–8 from Table 4.9.

- Table 4.9 was used to specify the pressure drop across the upper tie plate (1–2), lower tie plates, and the inlet region (i.e., 17–B). Similar to above, the appropriate flow areas and hydraulic for these regions were used.

- Finally, the results from Table 4.10 were used to specify the flow loss terms for the inlet nozzle leakage to the interstitial bypass region. The form losses in the MELCOR model were adjusted to match the measured total to bypass flow split.

Until this hydraulic data became available, previous BWR SFP MELCOR analyses did not include the sophistication of separate flow resistances for the partial and fully populated regions of the BWR bundle.

Other S_{LAM} and k formulations found in this report are expected to be used for other analysis efforts. The following lists some on-going and expected future applications.

- The results from the blocked water rods (Table 4.1 and Table 4.5) were used to estimate the flow resistance for a pressurized water reactor (PWR) assembly in a SFP rack. The PWR assembly has guide tubes, which may have some flow depending upon the internal storage configuration (i.e., various types of control rods and/or plugs may be used). However, in most circumstances, the guide tubes are not expected to have a significant flow.

- The results in Table 4.6 may be used for more detailed scaling of BWR grid spacers to other types of grid spacers. They also help support the unexpected experimental finding of high flow losses in the grid spacer.

- Using new modeling features in MELCOR, a water rod model was developed. The model uses explicit coupling to the assembly model and runs too slowly for "production" BWR SFP analysis. However, the water rod model and the water rod hydraulic data (Figure 4.8) will be used in the experimental analysis to assess the impact of heating on the water rod hydraulics versus the faster running "production" model approach. The results from the blocked water rods (Table 4.3 and Table 4.5) were used to estimate the flow resistance in conjunction with the water rod model.

Finally, it is worth noting that while the experimental analysis could not separately discern the contribution of wall friction distinct from form losses, the results analysis in the previous tables precisely satisfies the model input requirements for MELCOR, as well as other control volume codes, where the overall hydraulic losses are averaged across larger regions. The linear (i.e., S_{LAM}) and quadratic (i.e., the k term) hydraulic loss coefficients are easily put into the MELCOR input format and will replicate the measured flow losses across the Reynolds number range of 70 to 900, including the overall nonlinear hydraulic effects from:

- Flow development regions,

- the grid spacer entrance, internal, and exit effects,

- the inlet and exit flows to the water rods, and

- the transition from the fully to partially populated tubes regions.

5 THERMAL−HYDRAULIC ANALYSIS OF THE SPENT FUEL POOL EXPERIMENT

This report summarizes the findings of the SFP thermal-hydraulic experiments conducted between January 16 and February 16, 2006. The stated purpose of these investigations was to determine the thermal and hydraulic response for the validation of both the MELCOR severe accident analysis code and the COBRA spent fuel storage code. Two annular flow configurations were considered, namely closed bypass and open drain holes and open bypass and closed drain holes. The power applied to the apparatus spanned 200 to 2500 W with inlet natural draft rates for the open bypass/closed drain configuration of 83 to 122 slpm, or Reynolds numbers of 100 to 150 in the bundle.

5.1 Experimental Apparatus and Procedures

5.1.1 Hardware Components

Commercial components were purchased to create the prototypic 9×9 BWR assembly including the top and bottom tie plates, spacers, water rods, channel box, and all related assembly hardware. Incoloy heater rods were substituted for the fuel rod pins for heated testing. The diameter of the Incoloy heaters was slightly smaller than prototypic pins, 1.09×10^{-2} m versus 1.12×10^{-2} m. The slightly simplified Incoloy mock fuel pins were fabricated based on drawings and physical examples of a prototypic 9x9 BWR assembly. The dimensions of the assembly components are listed in Table 5.1.

The entire fuel assembly was housed inside a stainless steel enclosure, simulating the rack walls of a SFP. This enclosure was then insulated with 0.15 m (6 in.) of Fiberfrax Durablanket® type S insulation. Two radiation barriers, consisting of stainless steel shim stock, were placed at 0.075-m (3-in.) intervals.

Table 5.1   Dimensions of assembly components in the 9×9 BWR

Description	Lower (Full) Section	Upper (Partial) Section
Number of Pins	74	66
Pin Diameter (m)	1.09×10^{-2}	1.09×10^{-2}
Pin Pitch (m)	1.44×10^{-2}	1.44×10^{-2}
Pin Separation (m)	3.48×10^{-3}	3.48×10^{-3}
Water Rod OD (main section) (m)	2.49×10^{-2}	2.49×10^{-2}
Water Rod ID (m)	2.34×10^{-2}	2.34×10^{-2}

5.1.2 Thermocouple Layout

A total of 135 TCs were used to characterize the thermal response of the apparatus. TCs were placed both internal and external to the bundle. Figure 5.1 shows schematically the location of all internal TCs. The TCs were named according to the two-axis alpha scheme followed by the axial height, as referenced from the top of the bottom tie plate. For example, the TC at CS119 was the third rod from the left and third rod from the top at the 119-in. axial level (Figure 5.1).

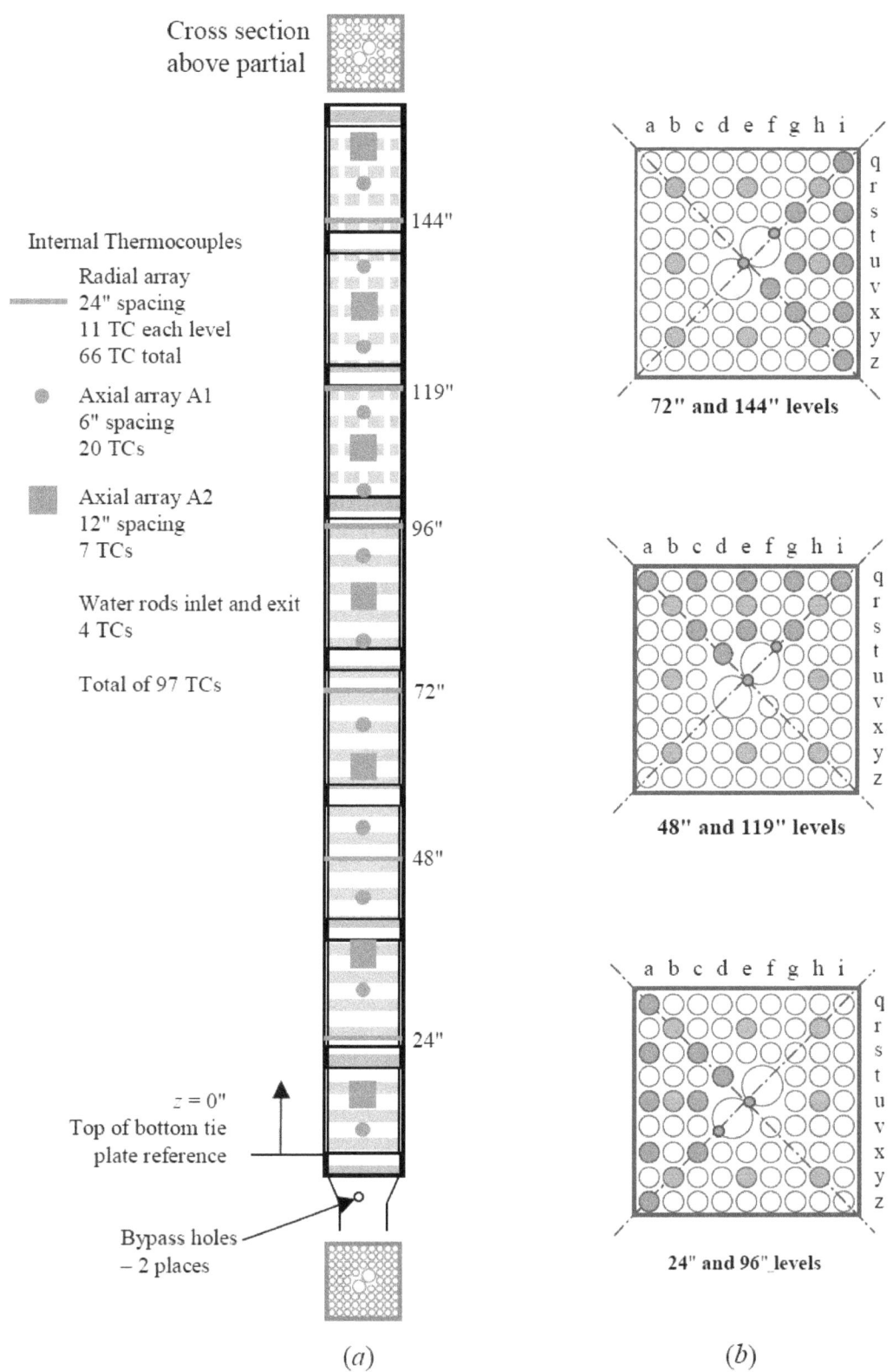

Figure 5.1. Experimental mock fuel assembly showing internal, as-built (a) axial and (b) lateral thermocouple (TC) locations

External TCs were placed on the channel box, pool cell, and outer radiation barrier as shown in Figure 5.2. The arrangement of the insulation and radiation barriers is also illustrated.

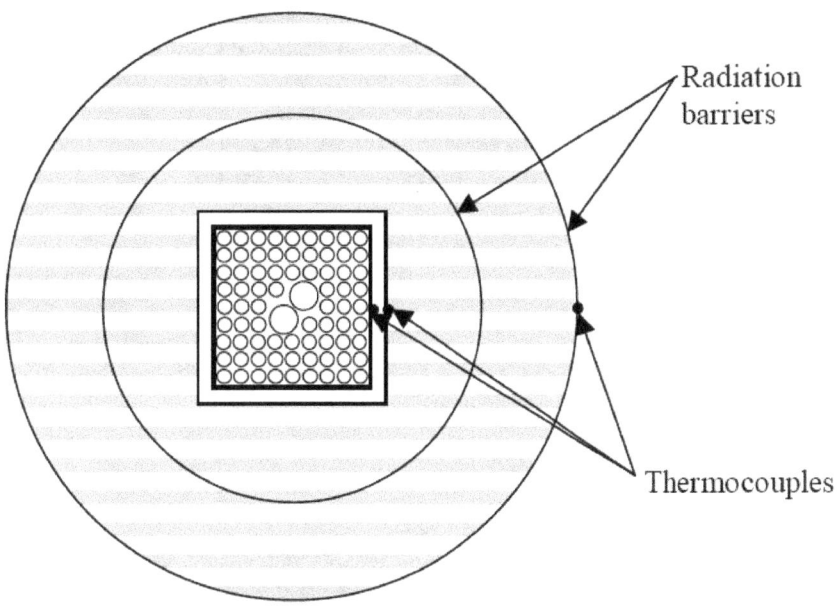

Figure 5.2   Arrangement of insulation, radiation barriers, and external TCs

5.1.3   Flow Measurements and Calibrations

Prior to heated testing, pressure drop measurements were acquired across the bottom tie plate (17–B) and the entire assembly (1–B). All hydraulic loss coefficients, S_{LAM} and k, were within experimental uncertainty of previous hydraulic testing (stainless steel assembly) with the exception of form loss of the entire assembly (Table 5.2). The uncertainties in these data are $u_{SLAM} = \pm 5$ and $u_k = \pm 1$.

Table 5.2 S_{LAM} and k analysis across 1–B and 17–B for Incoloy and stainless steel assemblies.

Note: All flow is assumed to pass through the bundle.

Description	Segment	A (m²)	L (m)	D_H (m)	S_{LAM}	Σ k
Incoloy	1–B	0.0100	4.19	0.0124	98	34
	17–B	0.0100	0.13	0.0124	171	11
Stainless Steel	1–B	0.0098	4.19	0.0119	96	37
	17–B	0.0098	0.13	0.0119	172	12

Two flow configurations were studied during the course of testing, referred to as open bypass/closed drains and closed bypass/open drains. By restricting the annular flow inlet to one pathway, the bundle and annular flow rates could be determined with much greater certainty.

Five hot wire anemometers were used to characterize the flow at various locations in the apparatus. The assembly inlet was instrumented with two Omega FMA-900-V-R constant temperature hot wires as shown in Figure 5.3. For all heated testing, these probes were inserted a distance of 25.4 mm as measured from the pipe wall.

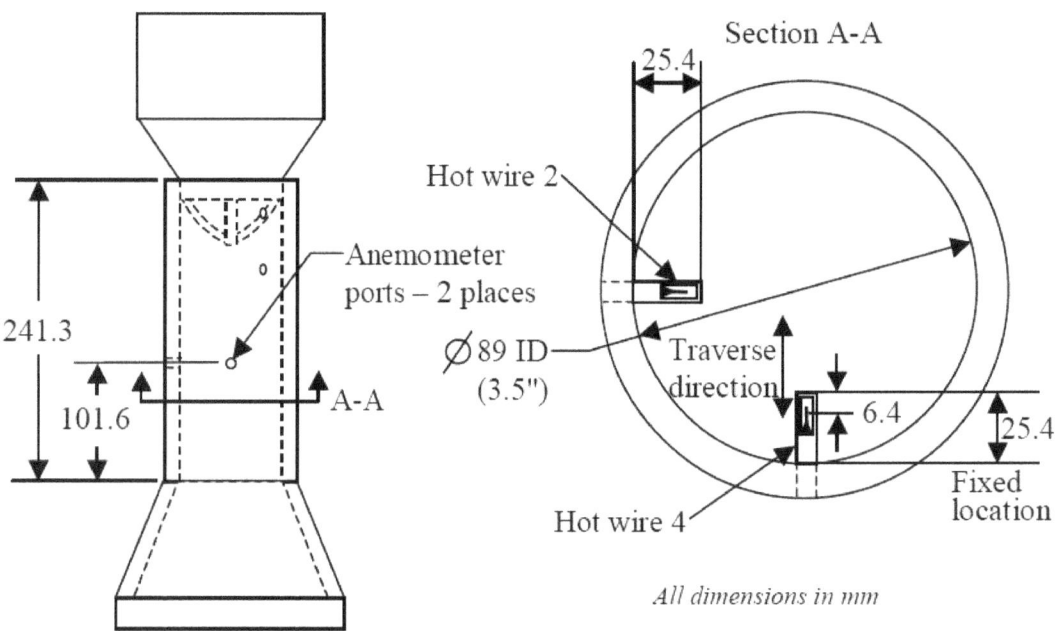

Section A-A

All dimensions in mm

Figure 5.3   Schematic showing the layout of the assembly inlet hot wires (hot wires 2 and 4)

Figure 5.4 shows the velocity inside the assembly inlet pipe as a function of distance from the inner pipe wall and input flow rates of 50 and 125 slpm. The figure also shows the fixed location of the sensor head during heated testing, which is approximately 6.4 mm inside the protective sheath. These probes were positioned at this location to measure inside the bulk of the flow while minimizing blockages due to the insertion of the sensors. The integrated average values were used to verify the output of the hot wires with the input of the mass flow controllers. Table 5.3 shows the comparison of the input flow rates versus the integrated average values. Typical errors were less than 7% with a maximum error of 9.39%. The volumetric flow rate measured with hot wire 4 was consistently high, perhaps due to the blockage introduced into the pipe during the insertion of the instrument. The hot wire represents a 4.5% area blockage when inserted to the center line of the pipe.

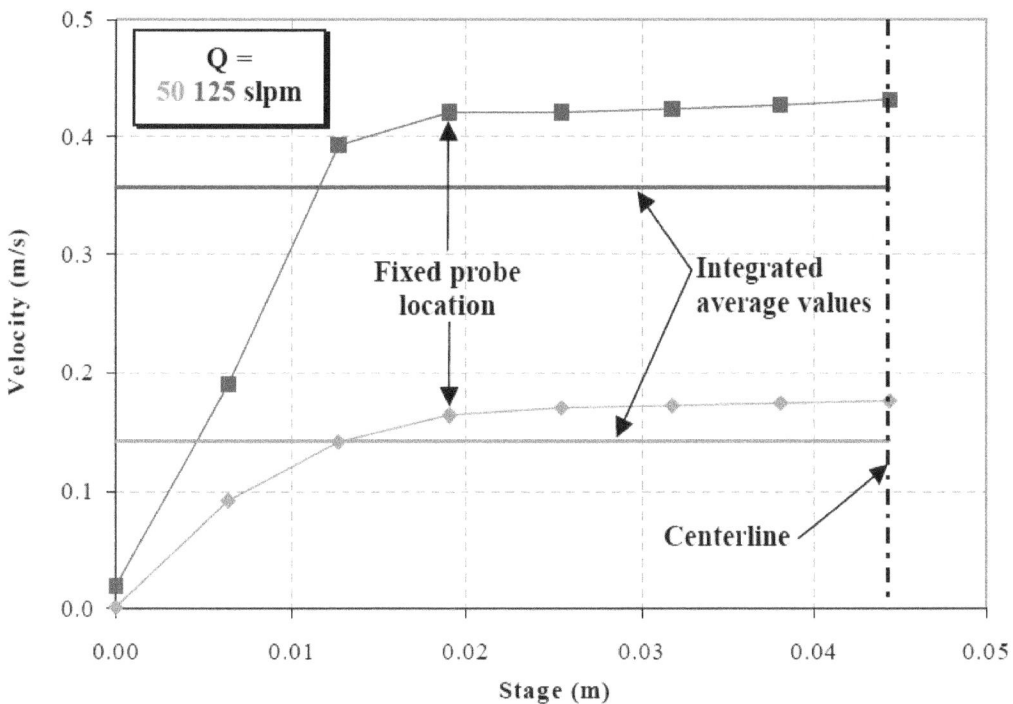

Figure 5.4   Hot wire 4 velocity traverse across the assembly inlet pipe for input volumetric flow rates of 50 (red) and 125 slpm (blue)

Table 5.3   Integrated flow rates from the hot wire velocity profiles

Input flow rate (slpm)	Integrated average velocity (m/s)	Integrated flow rate (slpm)	$\frac{\text{Integrated} - \text{Input}}{\text{Input}}$ (%)
50	0.143	53.3	6.58
75	0.220	82.0	9.39
100	0.285	106	6.18
125	0.358	133	6.56

Figure 5.5 shows the calibrations for the inlet hot wires performed on January 20 and January 31, 2006. These calibrations were used to determine the volumetric flow rate into the assembly inlet for the heated tests. The difference in outputs in the two hot wires is attributed to slight variations in positioning and instrument gains. The electronics of hot wire 2 were replaced prior to re-calibration on January 31, 2006, due to failure of the original unit on January 26, 2006. This exchange caused an overall decrease in the instrument gain reflected by the change in the sensor output.

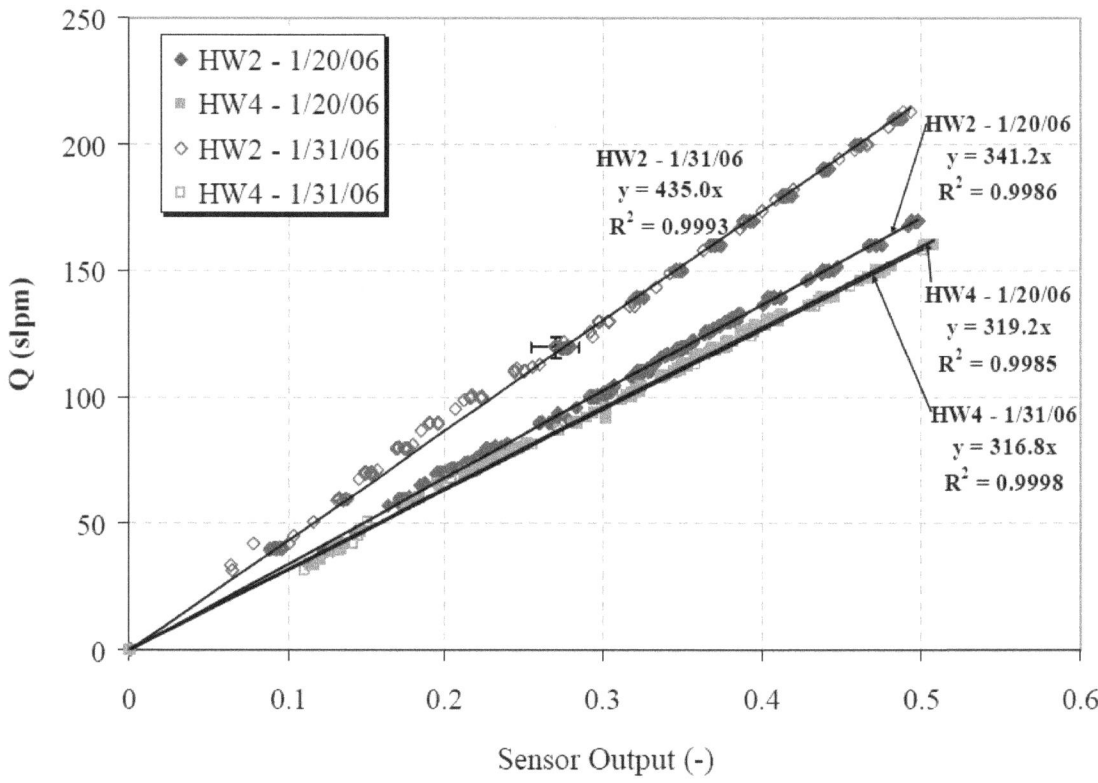

Figure 5.5   Inlet hot wire calibrations for hot wire 2 (HW2) (diamonds) and hot wire 4 (HW4) (squares) on 1/20/06 (closed) and 1/31/06 (open)

Figure 5.6 gives the location of the remaining two Omega FMA-900-V-R hot wires, hot wires 1 and 3, in the drain hole entry pipes. Again, the protective sheath was inserted to a depth of 25.4 mm from the inner pipe wall. The flow path through the drain holes into the annulus began through the drain entry pipes with an inner diameter of 40.6 mm (1.6 in.) followed by an abrupt contraction to a hole of 19.1-mm diameter.

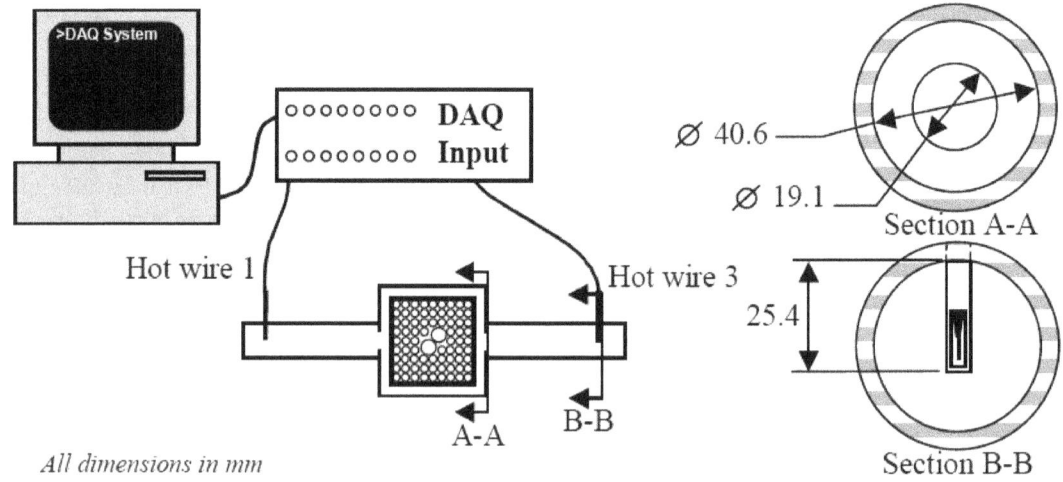

Figure 5.6   Schematic showing layout of the drain hole hot wire (HW1 and HW3)

Calibrations similar to before were performed for hot wires 1 and 3. Figure 5.7 shows the calibration curves for the drain hole hot wires conducted on January 20 and January 31, 2006. The electronics for hot wire 1 were damaged by the same electrical failure noted with hot wire 2 on January 26, 2006. A substitute component was used for the calibration performed on January 31, 2006, which resulted in an output signal of approximately half the original values for a given flow rate.

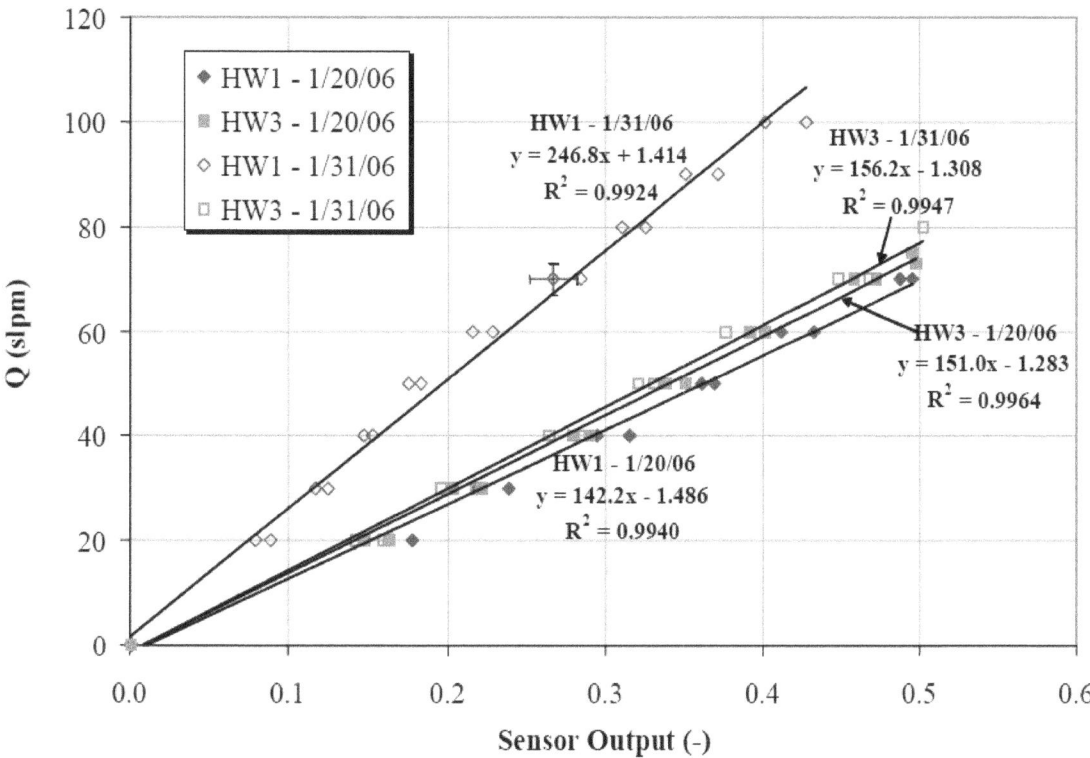

Figure 5.7   **Drain hole hot wire calibrations for hot wire 1 (HW1) (diamonds) and hot wire 3 (HW3) (squares) on 1/20/06 (closed) and 1/31/06 (open)**

Additional hot wire measurements were conducted to characterize the flow in the annulus. See Appendix C for further details. The flow in the annulus was shown to be nearly laminar and confirmed other measurements of the annular flow rate within instrument capabilities.

The pressure drop across the bypass holes was measured to determine the flow rate in the annulus for testing with open bypass/closed drain holes (Figure 5.8). First, a bottom tie plate was externally calibrated to measure the pressure drop for different flow rates with known flows passing only through the bypass holes. For heated testing, the instrument was again configured to measure the pressure drop across the bypass.

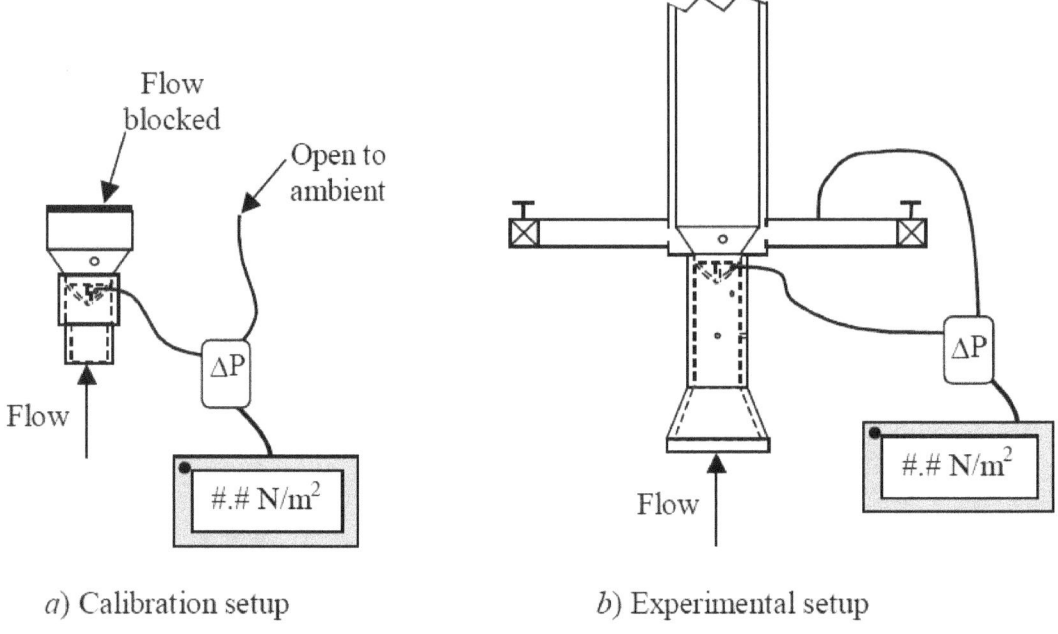

a) Calibration setup   *b*) Experimental setup

Figure 5.8.　　**Schematic of the measurement of the pressure drop across the bypass holes for (a) calibration and (b) heated testing**

Figure 5.9 shows the results of the bypass pressure drop calibrations from the external tie plate and previous hydraulic testing efforts. Also, the analytic solution for an infinite contraction and expansion is shown ($k_c = 0.5$, $k_e = 1$). The actual pressure drop indicates that the infinite limits of loss coefficients are slightly too high. A power law curve fit provides a reasonable estimate of the bypass volumetric flow rate as a function of pressure drop.

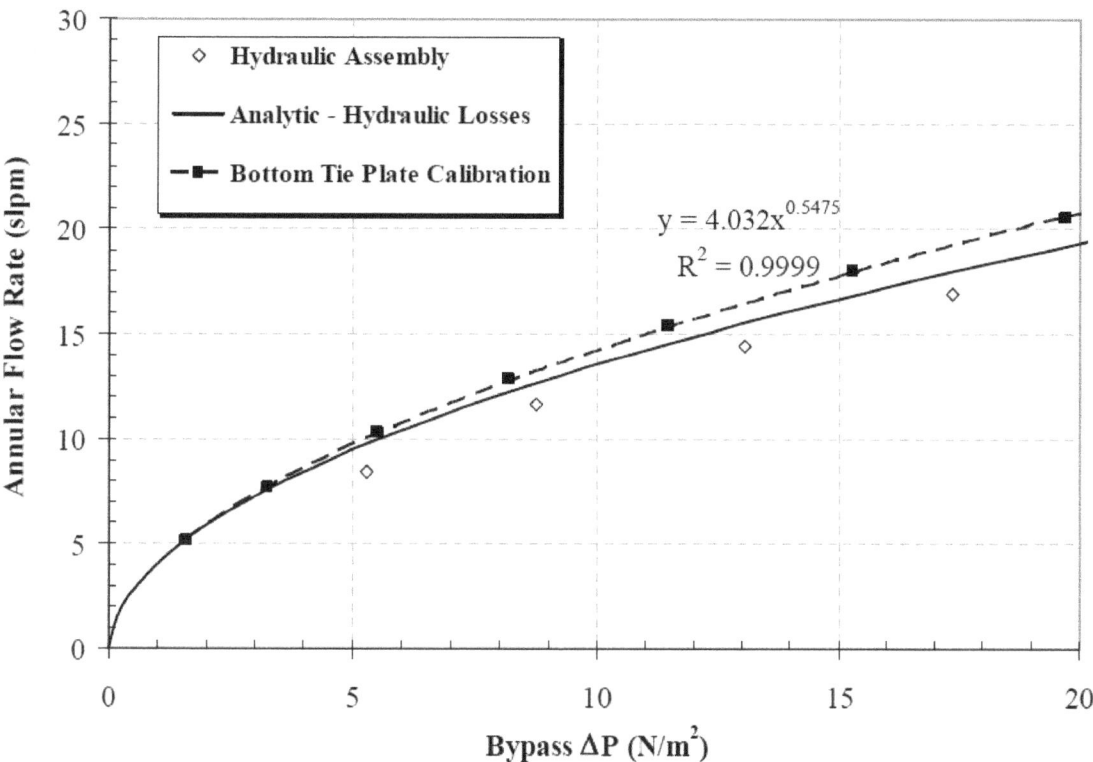

Figure 5.9. Annular flow rate as a function of the bypass pressure drop for the hydraulic assembly (diamonds), analytic solution (line), and the external bottom tie plate calibration (squares)

5.2 Modeling Efforts

5.2.1 COBRA-SFS (Coolant Boiling in Rod Arrays-Spent Fuel Storage)

COBRA-SFS is a lumped parameter finite difference code, written in FORTRAN, used to conduct thermal-hydraulic analysis of multi-assembly spent fuel storage. It is a fully implicit code which utilizes the Newton-Raphson technique to solve the equations of mass, momentum, and energy conservation. The code solves three-dimensional convective and conductive heat transfer and two-dimensional radiation heat transfer. The flow solution is limited to single-phase flow with or without buoyancy driven natural circulation. It can also determine flow and pressure fields in which the net flow is zero, simulating spent fuel stored in casks. COBRA-SFS has been validated and verified with experimental data for both steady state and transient calculations.

The code is broken down into fluid channels, rods, solid nodes, radiation heat transfer, and boundary conditions. Both the fluid and solid thermal properties are temperature dependant. The fluid channels describe the flow geometry between the rods and other solid structures. The rods are solid nodes containing an inner core and outer cladding. The solid nodes define every other structure in the assembly such as the canister, pool cell, and storage cask. Radiation heat transfer is assumed to be gray and diffuse. The view factors between the rods and canister are generated from a supplementary program called RADGEN. This program generates view factors for a variety of rod arrangements. The view factors between other components in the assembly are manually calculated using Hottel's crossed-string method. The boundary conditions are broken down into five groups: heat transfer, flow drag, plenum and outer boundary conditions, power and loading, and inlet/outlet conditions. All the information is entered into an input file in which an executable file of the FORTRAN code solves the model.

5.2.1.1 Full Length Heated Assembly Model

The COBRA model included relatively high lateral detail for the assembly components and rod array as shown in Figure 5.10 and Figure 5.11. Model details are summarized below.

- Various powers with two flow arrangements: two open bypass holes or two open drain holes.

- Specified steady state flow rate through both the annulus and bundle based on experimental values at 12 hours.

- Eight partial rods included (102 in.): power across the entire axial length.

- Internal flow assumed to be laminar (Nu = 4.36 bundle, Nu = 5.385 annulus).

- Temperature dependant properties for materials and air.

- Full length rods: 26 axial levels, 156-in. long, axial power from 0–144 in.

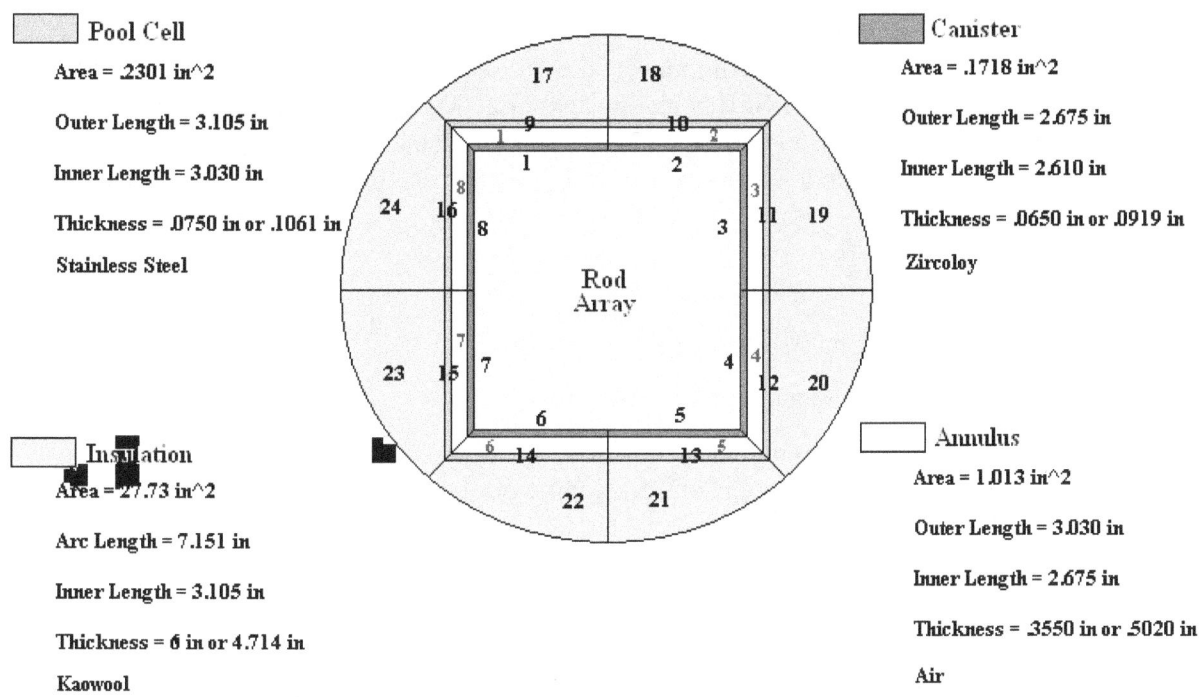

Pool Cell
Area = .2301 in^2
Outer Length = 3.105 in
Inner Length = 3.030 in
Thickness = .0750 in or .1061 in
Stainless Steel

Canister
Area = .1718 in^2
Outer Length = 2.675 in
Inner Length = 2.610 in
Thickness = .0650 in or .0919 in
Zircoloy

Insulation
Area = 27.73 in^2
Arc Length = 7.151 in
Inner Length = 3.105 in
Thickness = 6 in or 4.714 in
Kaowool

Annulus
Area = 1.013 in^2
Outer Length = 3.030 in
Inner Length = 2.675 in
Thickness = .3550 in or .5020 in
Air

Figure 5.10.   Schematic of the COBRA model layout

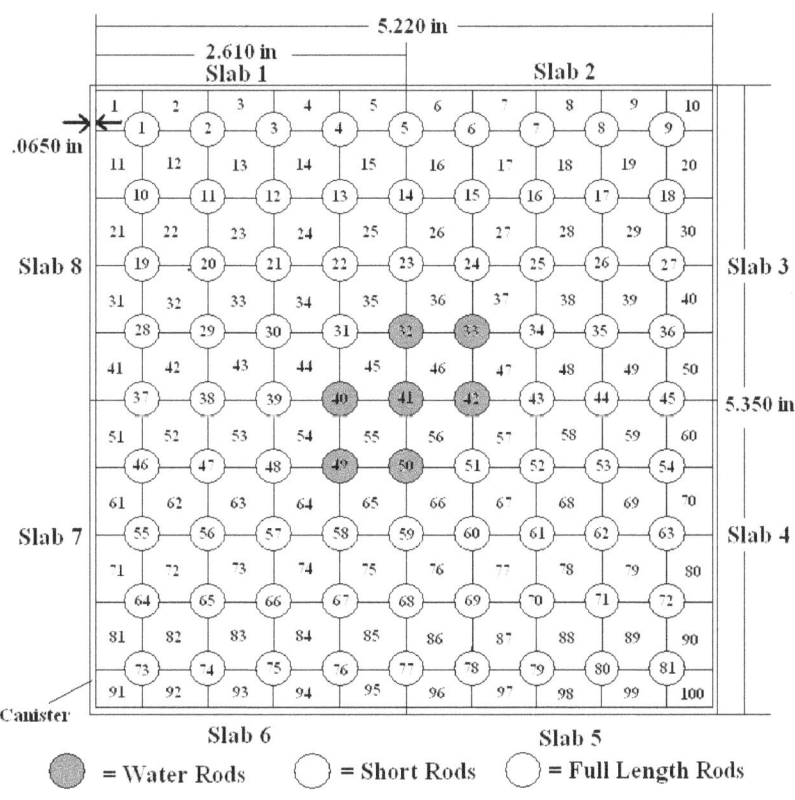

= Water Rods   = Short Rods   = Full Length Rods

Figure 5.11.   Detail view of the COBRA bundle layout

93

5.2.2 MELCOR – Severe Accident Analysis

The MELCOR 1.8.5 severe accident computer code with enhancements through Version RP was used to simulate the experimental SFP accident response. MELCOR is a full non-equilibrium, two-phase thermal-hydraulics code for the simulation of nuclear reactor severe accidents. It includes a full spectrum of models for simulation of reactor accidents progressing from design conditions to severe accidents with:

- Comprehensive fuel degradation,

- Steam and air oxidation with hydrogen production,

- Fuel and structural component degradation,

- Fission product release, transport, deposition, chemisorption, vapor condensation and evaporation, aerosol hygroscopic effects, iodine pool evolution, and gamma heating,

- Core-concrete interactions, and

- Simulation of containment engineering safety features (e.g., sprays, fan coolers, filters, and recombiners).

Of particular relevance for application to the SNL experimental SFP tests, MELCOR has new fuel geometry models for boiling and pressurized water reactor SFP rack geometry. The specialized SFP geometry models include basic heat transfer models for convection, conduction, and radiation as well as a new breakaway oxidation kinetics model and fuel and rack degradation models.

Version RP includes three recent modeling enhancements applicable to BWR SFP modeling:

3) A new rack component, which permits better modeling of a SFP rack,

4) A new oxidation kinetics model, and

5) A simplified flow regime model.

The new BWR SFP rack component permits proper radiative modeling of the SFP rack between groups of different assemblies. The new oxidation kinetics predicts the transition to breakaway oxidation in air environments on a node-by-node basis. The simplified flow regime model permitted simulation of liquid films draining down the BWR fuel assemblies during spray operation.

5.2.2.1 Full Length Heated Assembly Model

A MELCOR full-length assembly model was developed to analyze the temperature response to the SNL experimental SFP testing program. The model simulates a single assembly in a uniform pattern (Figure 5.12). For implementation into MELCOR, the uniform pattern model was represented by a single, high-powered assembly with a heavily insulated radial boundary (i.e., simulated with 6 in. of Kaowool™ insulation).

94

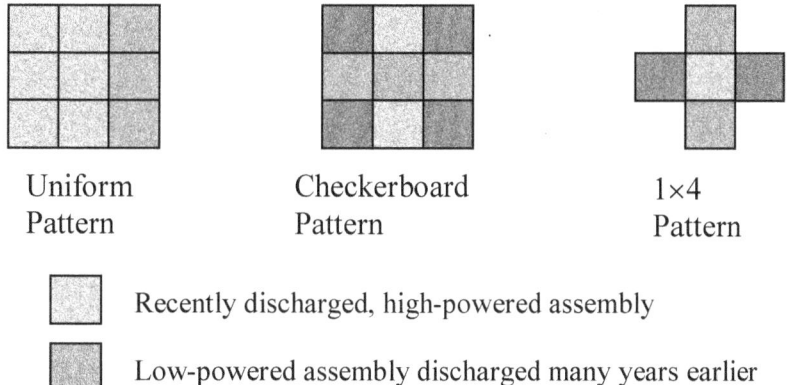

Uniform
Pattern

Checkerboard
Pattern

1×4
Pattern

Recently discharged, high-powered assembly

Low-powered assembly discharged many years earlier

Figure 5.12.   Repeating patterns used in the MELCOR separate effects models

The interstitial area between the canister wall and the rack wall is modeled as the BWR bypass region. The bypass uses the same 9-control volume axial nodalization as the bundle region. A control volume represents the inlet tube connected to the bottom of the assembly. A flow path was included that represented the two tubes connected to the rack drain holes. The inlet tube, drain lines, and assembly exit were connected to a time-independent volume that represented the atmospheric conditions during the test (i.e., pressure and temperature).

The hydraulic resistance was specified using the results from the experimental test program. The drain line and inlet tubes included standard losses for flow contractions, expansions, and wall friction.

The BWR assembly canister is modeled with the MELCOR canister component. The rack walls are modeled using the new SFP rack component with stainless steel and aluminum (i.e., to represent Boral). MELCOR does not include an option to model the two large water rods in the assembly. Consequently, the water rod mass and surface area were included in the canister wall.

The BWR assembly is represented by 14 axial levels in the COR package:

- Level 1 is the pipe leading to the inlet of the assembly.

- Level 2 is the base plate.

- Level 3 is the inlet region between the inlet nozzle and the lower tie plate.

- Level 4 is the start of the active fuel region.

- Level 8 is the top of the partial rod active fuel region.

- Level 9 represents the region of the plenum of partial rods.

- Level 12 is the top of the full-length rod active fuel region.

- Level 13 represents the plenum region of the full-length rods.

- Level 14 represents the region between the upper tie plate and the top of the racks.

The upper and lower tie plates are modeled as supporting plate structures, made of stainless steel. The 0.5-in. rack support plate at Level 2 is also modeled as a stainless steel supporting structure.

MELCOR requires that the canister, water rods, and fuel cladding be specified as the same material. The default Zircaloy properties for these components were replaced with inconel. In reality, the experimental apparatus had a Zircaloy canister, Zircaloy water rods, and Incoloy heater rods. MELCOR's Zircaloy oxidation kinetics model was disabled to prevent oxidation of the heater rods. The interior of the heater rods was modeled as compacted magnesium oxide. MELCOR's gamma heating model was disabled to deposit all the power from the heater rods in the *simulated* fuel (i.e., the magnesium oxide).

A constant power profile was used in each heater rod. Because eight partial heater rods did not extend the full length of the assembly, there was a step change in the overall, axial power profile at the top of the heated zone of the partial rods.

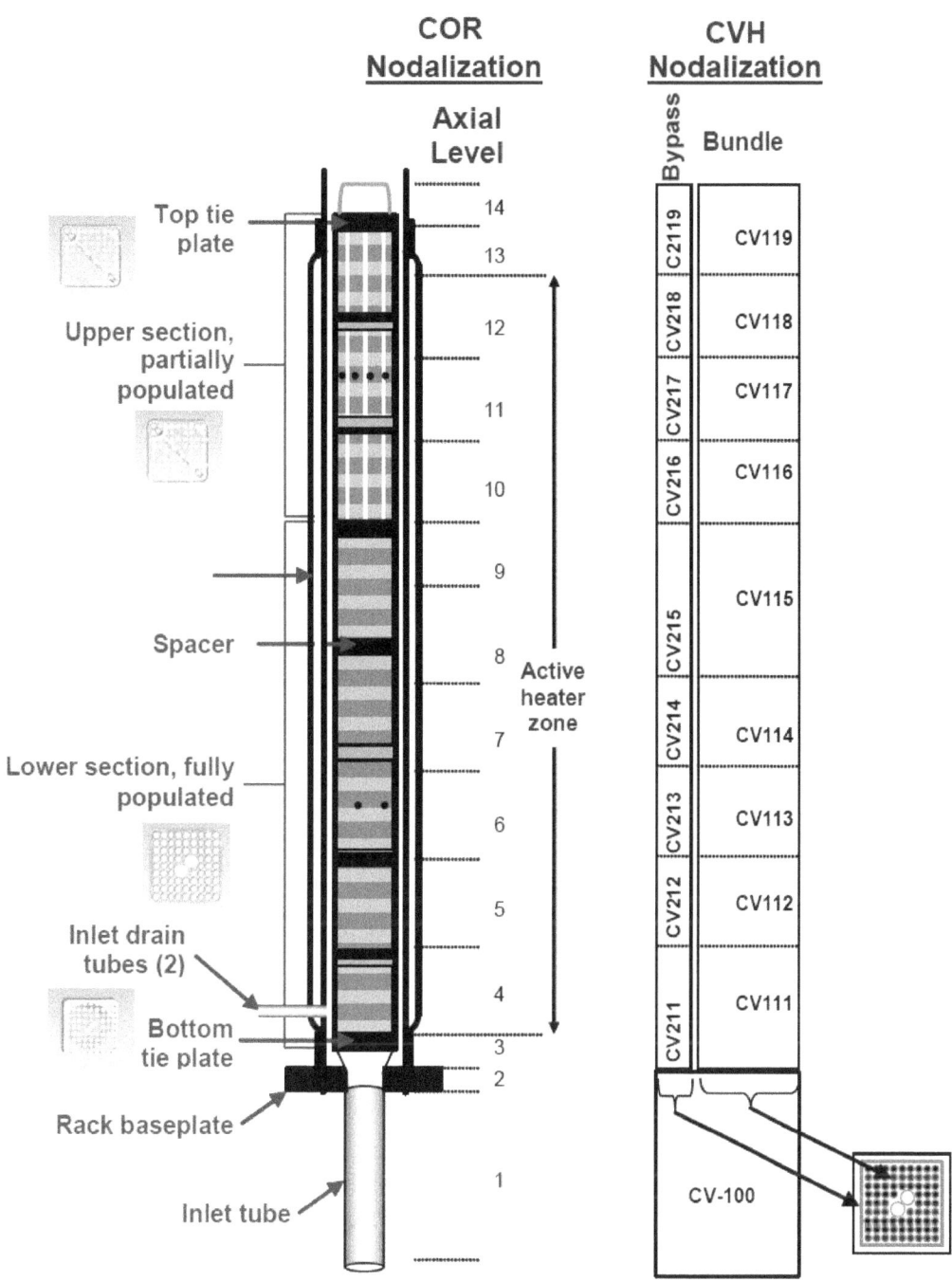

Figure 5.13.   MELCOR nodalization of the full-length assembly experiment

5.3   Results

5.3.1   Transient Thermal Response

The transient thermal response of the apparatus at 0.610-m (24-in.) increments for an input power of 1370 W and closed bypass/open drains flow configuration is shown in Figure 5.14. The location of the peak temperature in the assembly increases upward with time until the top of the stack is the hottest at around 7 hours.

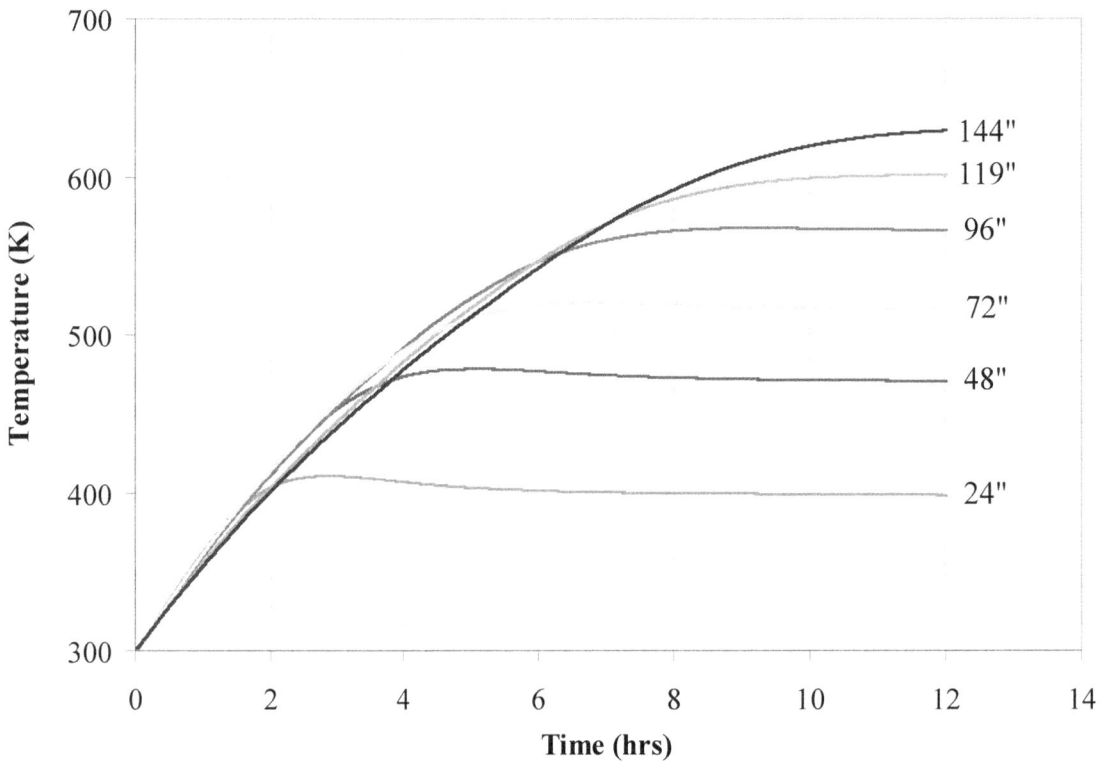

Figure 5.14   **Transient thermal response at different axial locations for a power input of 1370 W and closed bypass/open drains flow configuration**

The transient thermal response for an input power of 1370 W and open bypass/closed drains flow configuration in 0.610-m (24-in.) increments is shown Figure 5.15. This configuration is considered the more prototypic flow arrangement in an actual SFP. The location of the peak temperature in the assembly increases upward with time until the top is hottest at around 10 hours. The peak temperature at 12 hours was 36 K hotter due to lower overall induced air flow rates and, thus, less heat convected out of the assembly.

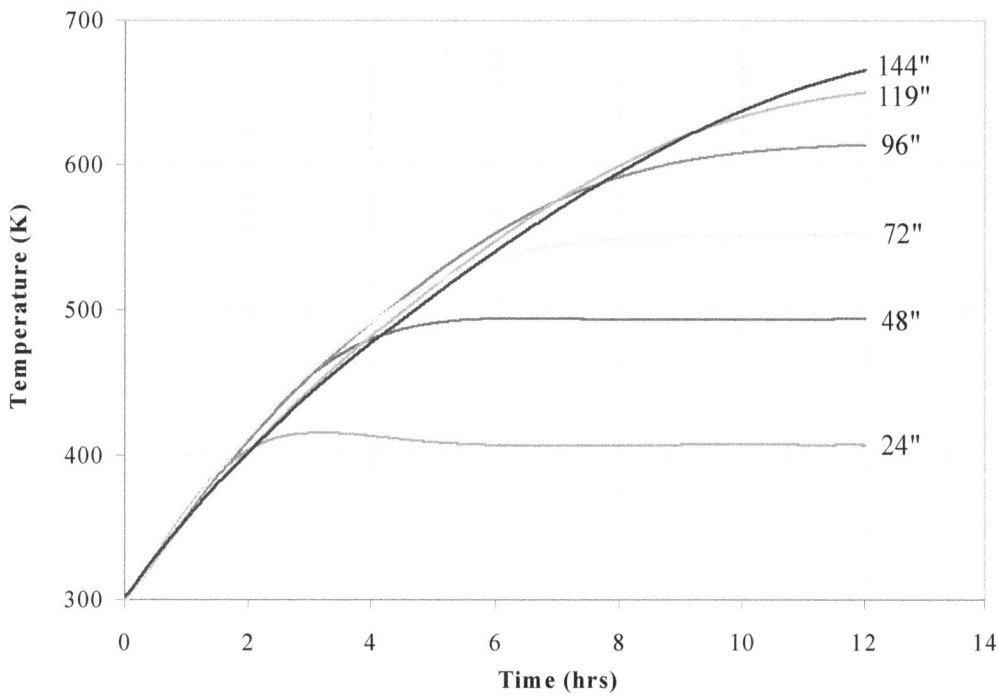

Figure 5.15 **Transient thermal response at different axial locations for a power input of 1370 W and open bypass/closed drains flow configuration**

Figure 5.16 shows the comparisons of the MELCOR and COBRA codes to the 144-in. level temperature response for the test conducted with an input power of 1370 W and open bypass/closed drains configuration. The codes are within 3% error of the experimental values at all times. MELCOR predictions of temperature are somewhat better, although under predictive.

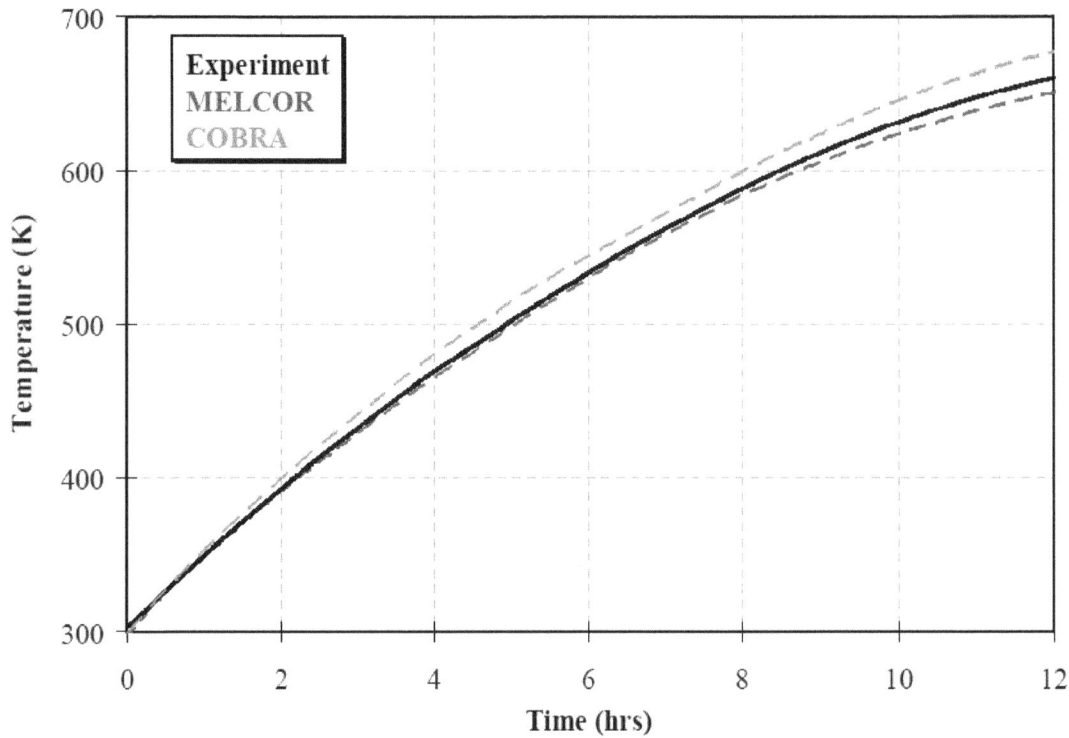

Figure 5.16. **Comparison of the thermal transient response with the MELCOR and COBRA codes for a power input of 1370 W and open bypass/closed drains flow configuration**

Figure 5.17 shows a comparison of the experimental (symbols) and code results (lines) for an input power of 1370 W and a flow configuration of closed bypass/open drains. These temperature results are presented as a function of vertical location in the assembly at 3 to 12 hours in increments of 3 hours. The code values are within 10% error over the entire assembly. However, these errors are within 6% for the top half of the assembly.

Figure 5.18 is similar to the previous graph for the open bypass/closed drains configuration. Here, the codes predict temperatures within 8% error of the experimental results. The top half of the assembly is again more accurately portrayed with errors of less than 5%.

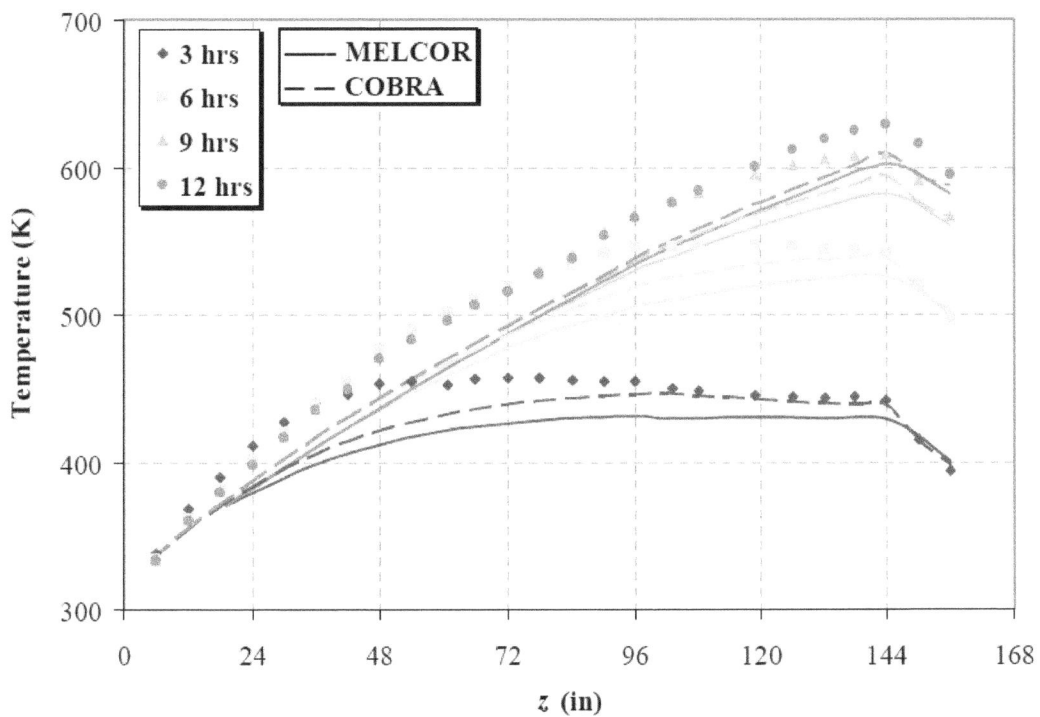

Figure 5.17   **Comparison of experimental (symbols) to MELCOR (solid lines) and COBRA (dashed lines) temperatures as a function of axial location at times of 3 (◆), 6 ( ), 9 ( ), and 12 hours (●)**

Note: The apparatus was configured with closed bypass/open drains and an input power of 1370 W.

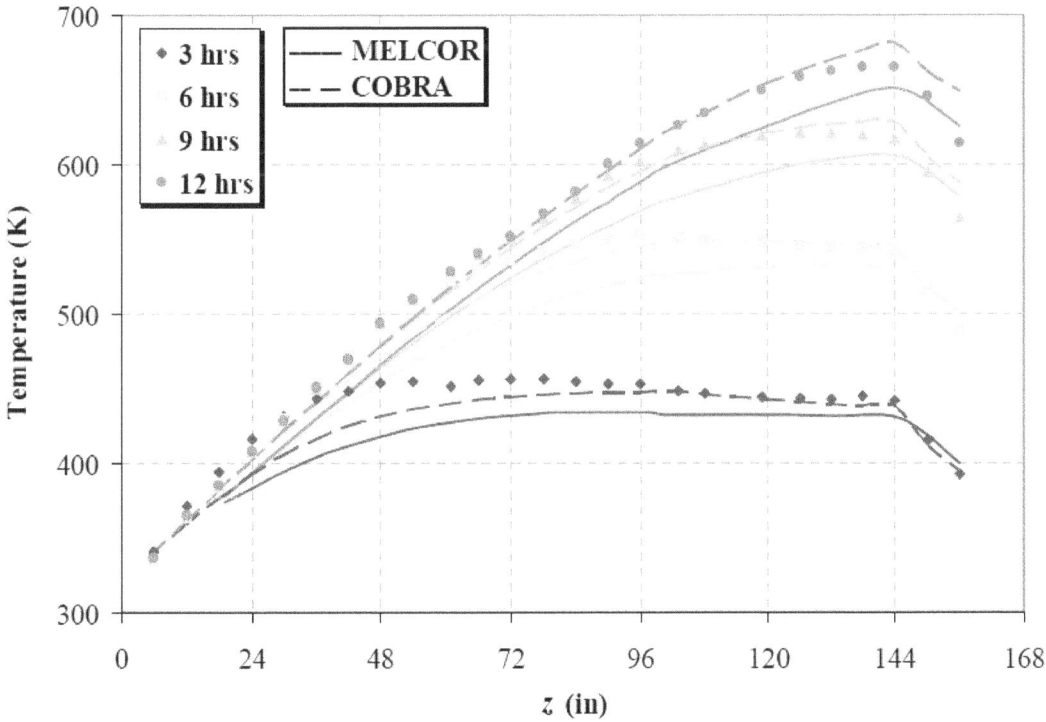

Figure 5.18   **Similar to the figure above for the open bypass/closed drains configuration**

5.3.2 "Steady State" Peak Cladding Temperature

Figure 5.19 shows the PCT of the assembly as a function of input power for test results (symbols), MELCOR (blue lines), and COBRA (red lines) in both flow configurations. MELCOR calculations were within 5% error for all cases. COBRA results for the closed bypass/open drains configuration was also within 5% error. However, errors of up to 15% were incurred while modeling the open bypass/closed drains cases.

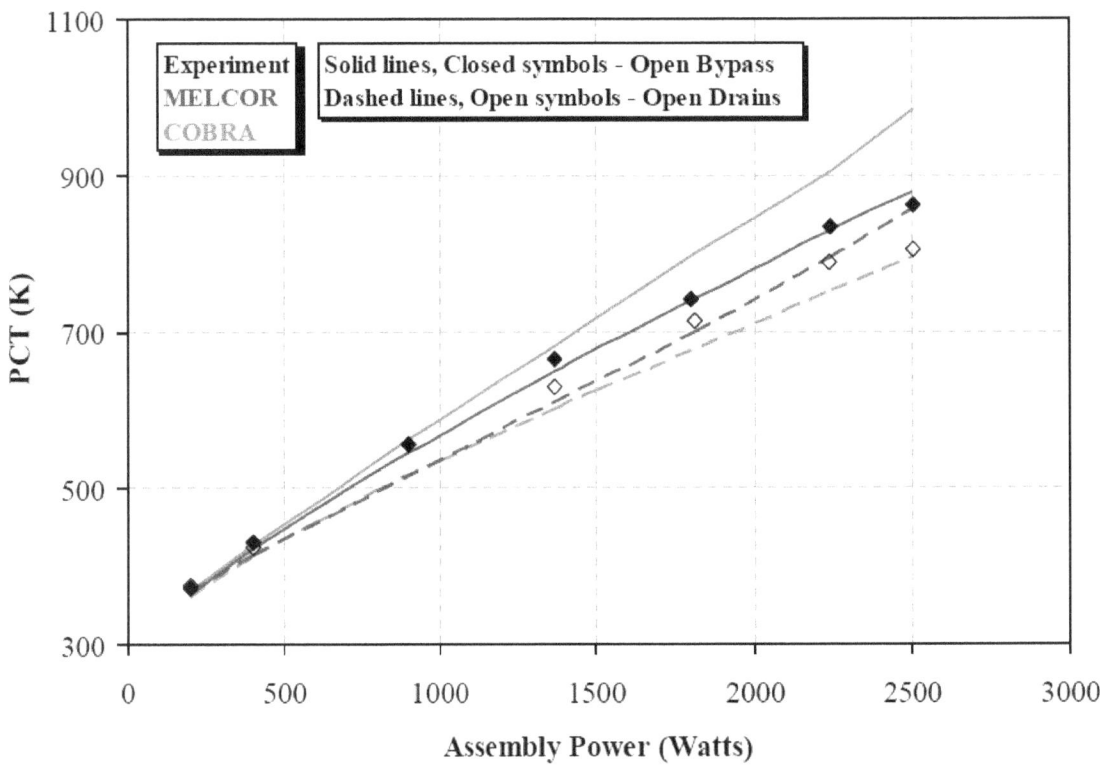

Figure 5.19 PCT as a function of assembly input power for experimental (symbols), MELCOR (blue lines), and COBRA (red lines)

Note: Data for both flow configurations are shown, open bypass/closed drains (solid lines, closed symbols) and closed bypass/open drains (dashed lines, open symbols).

5.3.3 Steady State Natural Induction Flow Rates

Figure 5.20 gives the naturally-induced volumetric flow rates in the assembly inlet and annulus as a function of input power. The red symbols and lines represent the open bypass/closed drains configuration; the blue entries represent the closed bypass/open drains configuration. Only predictions from MELCOR are shown, as flows were specified as inputs into COBRA. MELCOR results for the open bypass configuration are within experimental uncertainty for all but the lowest two input powers. The code tended to under predict the inlet flow rates for the open drains configuration but calculated annular flow rates to within uncertainty for both flow configurations. All code predictions were within 10% error.

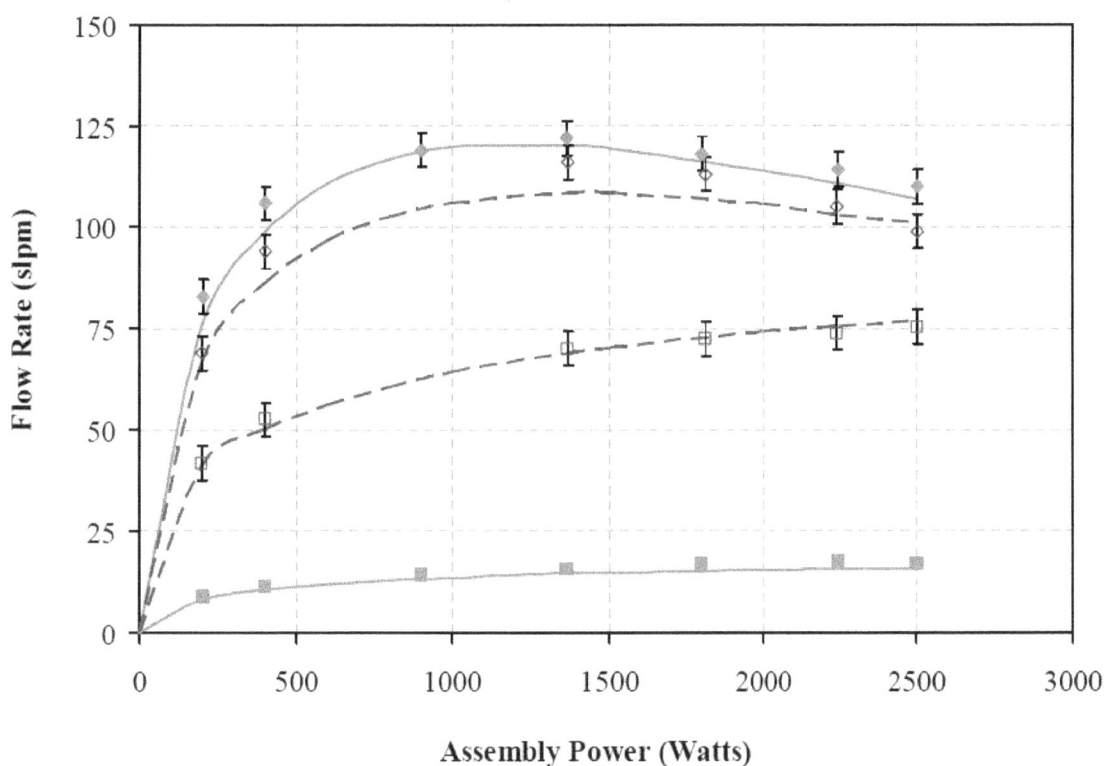

Figure 5.20   **Volumetric flow rates as a function of assembly input power forexperimental (symbols) and MELCOR (lines)**

Note: The data represent the bundle (diamonds) and the annulus (squares) flow rates. Both flow configurations are shown, open bypass/closed drains (solid red lines, closed symbols) and closed bypass/open drains (dashed blue lines, open symbols).

5.4   Summary

Benchmark experiments were conducted with a highly prototypic mock BWR spent fuel assembly to measure thermal and hydraulic response. The two flow configurations studied were open bypass/closed drains and closed bypass/open drains. The open bypass configuration is considered the more prototypic flow setup. Peak cladding temperatures for both experiments and MELCOR were within 5% error at all assembly power inputs and flow configurations. Predicted temperatures from COBRA were within 5% and 15% error for the closed bypass/open drains and open bypass/closed drains configurations, respectively. Naturally-inducted flow rates were within 10% error over the entire power input range between tests and MELCOR.

The overall agreement of the MELCOR calculations and the experimental data is attributed to the direct application of prototypic viscous and form loss coefficients determined in previous, unheated testing.

Table 5.4 gives a summary of all heated testing of the apparatus. Measurements of note are listed for values at 12 hours of elapsed test time.

Table 5.4   Summary of all thermal-hydraulic tests for the SFP experiment

Test Date	Test Configuration			Assembly Temperatures (°C)			Flow Rates (slpm)	
	True RMS Power (W)	# of drain holes	# of bypass holes	PCT	Max. Channel Temp.	Max. Pool Cell Temp.	Inlet	Annulus
01/24/06	202	0	2	99	91	84	83	9
01/26/06	402	0	2	158	144	133	106	11
02/01/06	902	0	2	282	266	250	119	14
01/17/06	1368	0	2	392	377	359	122	16
02/02/06	1801	0	2	468	441	435	118	17
01/19/06	2240	0	2	562	542	527	119	18
02/15/06	2505	0	2	589	568	557	110	17
01/23/06	200	2	0	101	92	85	69	42
01/25/06	403	2	0	150	138	125	94	53
01/16/06	1369	2	0	356	339	320	116	70
02/08/06	1812	2	0	440	423	407	113	72
01/18/06	2236	2	0	515	498	482	105	74
02/16/06	2505	2	0	532	516	502	99	75
01/27/06	1362	2	2	346	328	312	116	70

*Flow rates and temperatures reported at 12 hrs elapsed test time

6 SUMMARY OF THE 1×4 INCOLOY SHORT STACK THERMAL RADIATION EXPERIMENTS OF THE SFP PROJECT

This report summarizes the findings of the SFP thermal radiation experiments conducted between February 27 and March 23, 2006. The stated purpose of these investigations was to determine the thermal response of a spent fuel in a 1×4 arrangement for the validation of both the MELCOR severe accident analysis code and the COBRA spent fuel storage code. The radiative heat transfer between assemblies was of particular interest. To this purpose, the experimental apparatus was specifically constructed to maximize the radiative heat transfer between a single, hot center assembly and four, cooler neighbors.

6.1 Assembly Design

The 1×4 Incoloy assembly was designed to study the radiative heat transfer of a centrally located hot bundle to its cooler neighbors (Figure 6.1). Bulk air flow was prevented during testing by blocking all flow paths into and out of the assembly. However, the apparatus was not leak tight and did allow volumetric expansion of the air as it heated. The entire assembly height was approximately 52 in., modeling about one-third of a full-length assembly.

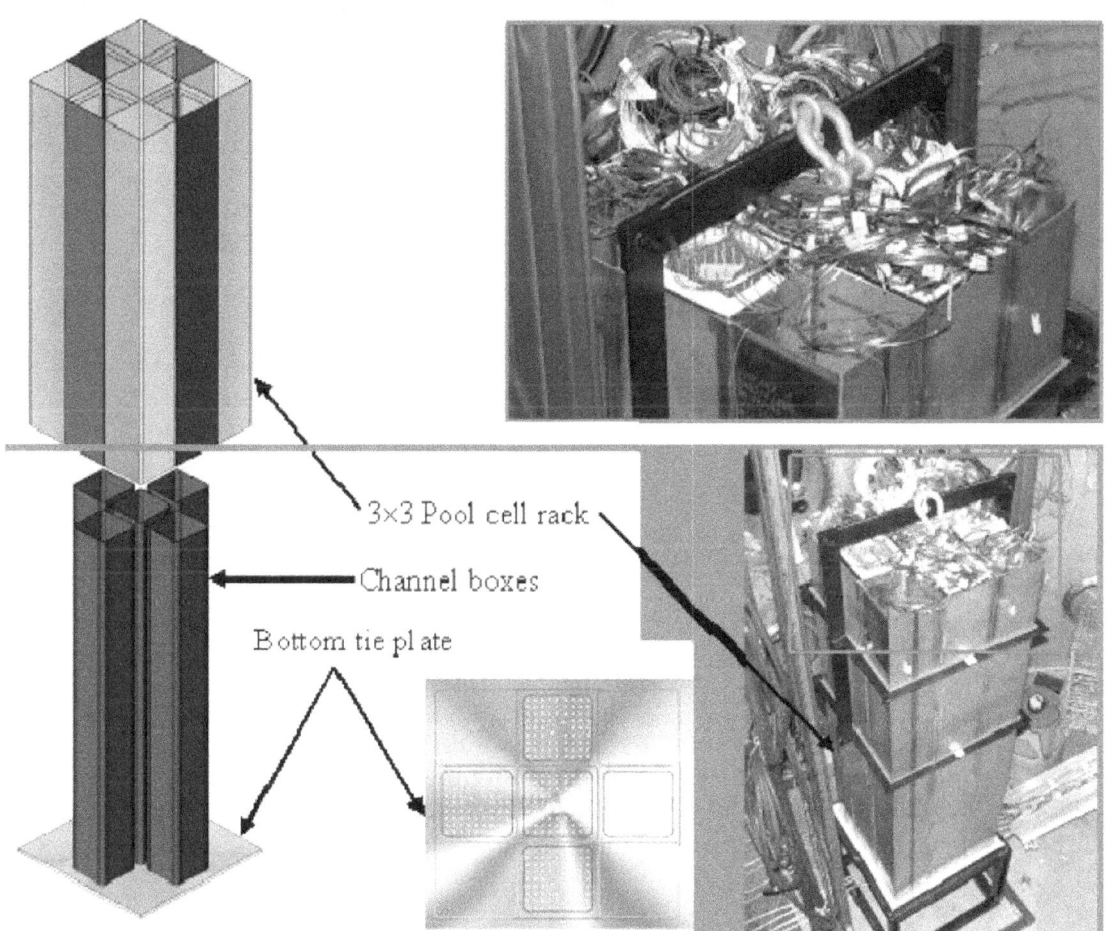

Figure 6.1   **Schematic and pictures of the assembly design**

Note: These depictions do not show the insulation applied to the outside of the pool cell rack or to the corner cells.

Commercial components were purchased to create the assembly, including two rod spacers, water rods, channel boxes, and the pool cell rack. The water rods and channel boxes were cut to size (52 in. high) from full-length prototypic components. Incoloy heater rods were substituted for the fuel rod pins for heated testing. The diameter of the Incoloy heaters was slightly smaller than prototypic pins, 1.09×10^{-2} m versus 1.12×10^{-2} m. The heater rods had unheated regions of 2 in. at the top and bottom, resulting in a 48-in heated zone. The spacers were placed at 16 in. and 36 in. as measured from the top of the bottom tie plate.

Figure 6.2 shows the assembly after 6 in. of Kaowool™ insulation was applied to the exterior of the apparatus. The cables visible in the top and side views are the electrical feeds to the five mock fuel assemblies. The alumina pipes extending through the insulation in the top view connect to the water rods in each bundle. Cooling air was fed through the bottom of the apparatus into the water rods and out the alumina tubes at the conclusion of heated testing.

Top view

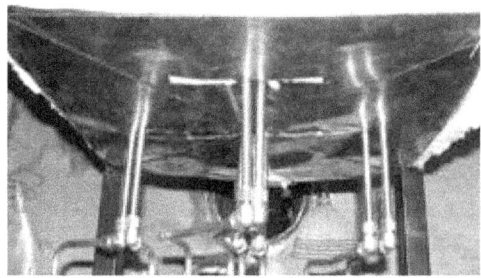

Bottom view   Side view

Figure 6.2   Photographs showing the final insulated assembly

6.1.1   TC Layout

Figure 6.3 depicts the general TC layout of the 1×4 Incoloy assembly. Three main TC levels were placed at heights of 11, 26, and 41 in. as measured from the top of the bottom tie plate. The solid circles indicate heater rods with a TC attached.

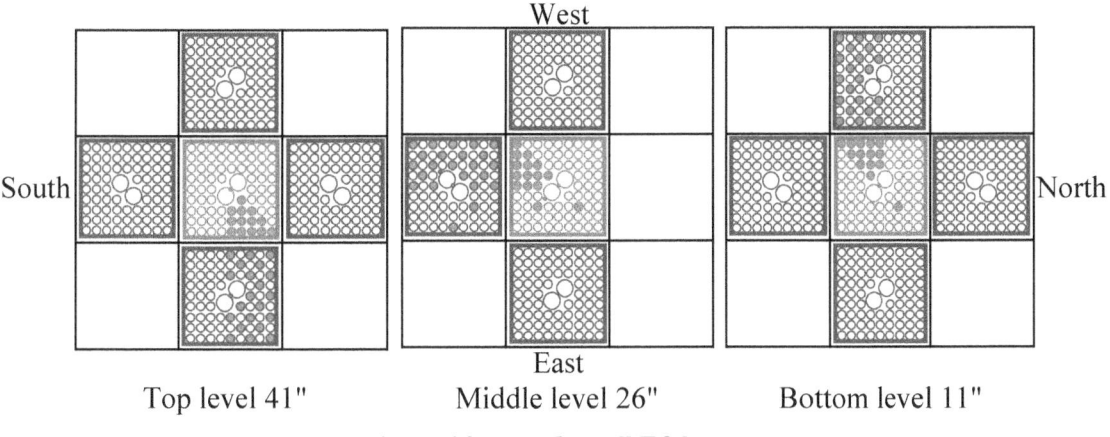

West

South

North

East

Top level 41"   Middle level 26"   Bottom level 11"

Figure 6.3   Overall TC layout

Figure 6.4 provides the location and naming convention of the TCs in the center assembly. A list of all installed TCs is presented below the figure. The naming convention first identifies the bundle, followed by the rod and the axial height. An extra identifier (W) is included for TCs attached to water rods and appears after the bundle designation.

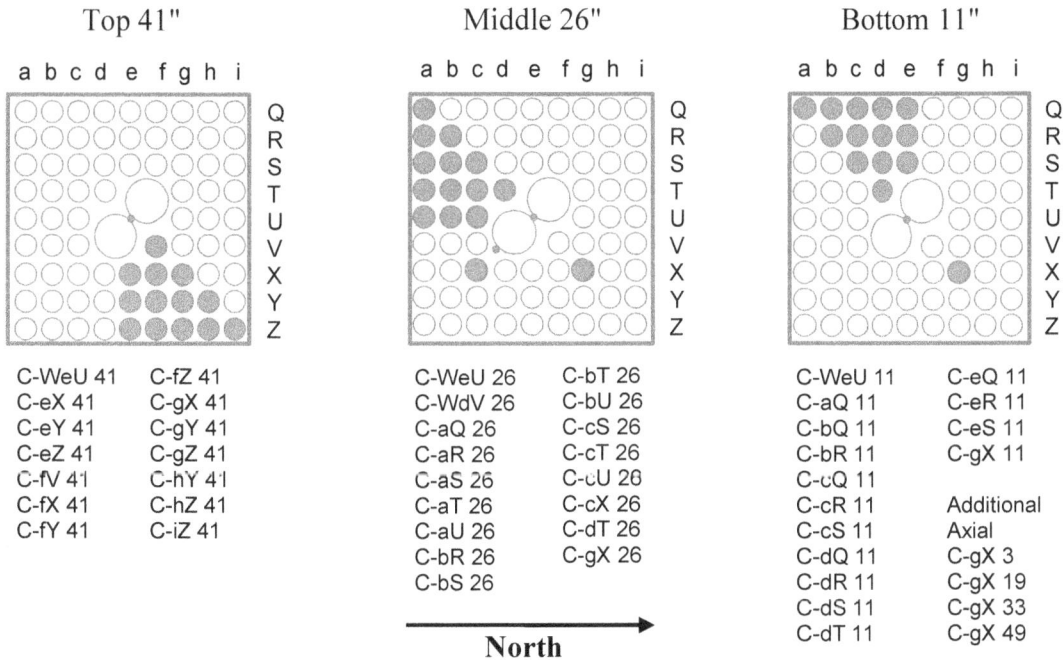

Top 41"   Middle 26"   Bottom 11"

C-WeU 41	C-fZ 41
C-eX 41	C-gX 41
C-eY 41	C-gY 41
C-eZ 41	C-gZ 41
C-fV 41	C-hY 41
C-fX 41	C-hZ 41
C-fY 41	C-iZ 41

C-WeU 26	C-bT 26
C-WdV 26	C-bU 26
C-aQ 26	C-cS 26
C-aR 26	C-cT 26
C-aS 26	C-cU 26
C-aT 26	C-cX 26
C-aU 26	C-dT 26
C-bR 26	C-gX 26
C-bS 26	

C-WeU 11	C-eQ 11
C-aQ 11	C-eR 11
C-bQ 11	C-eS 11
C-bR 11	C-gX 11
C-cQ 11	
C-cR 11	Additional
C-cS 11	Axial
C-dQ 11	C-gX 3
C-dR 11	C-gX 19
C-dS 11	C-gX 33
C-dT 11	C-gX 49

North

Figure 6.4   Center assembly TC locations and naming conventions

Similar to Figure 6.4, Figure 6.5 presents the TCs located in the peripheral assemblies. Again, TCs attached to water rods are given the extra identifier (W) after the bundle identification.

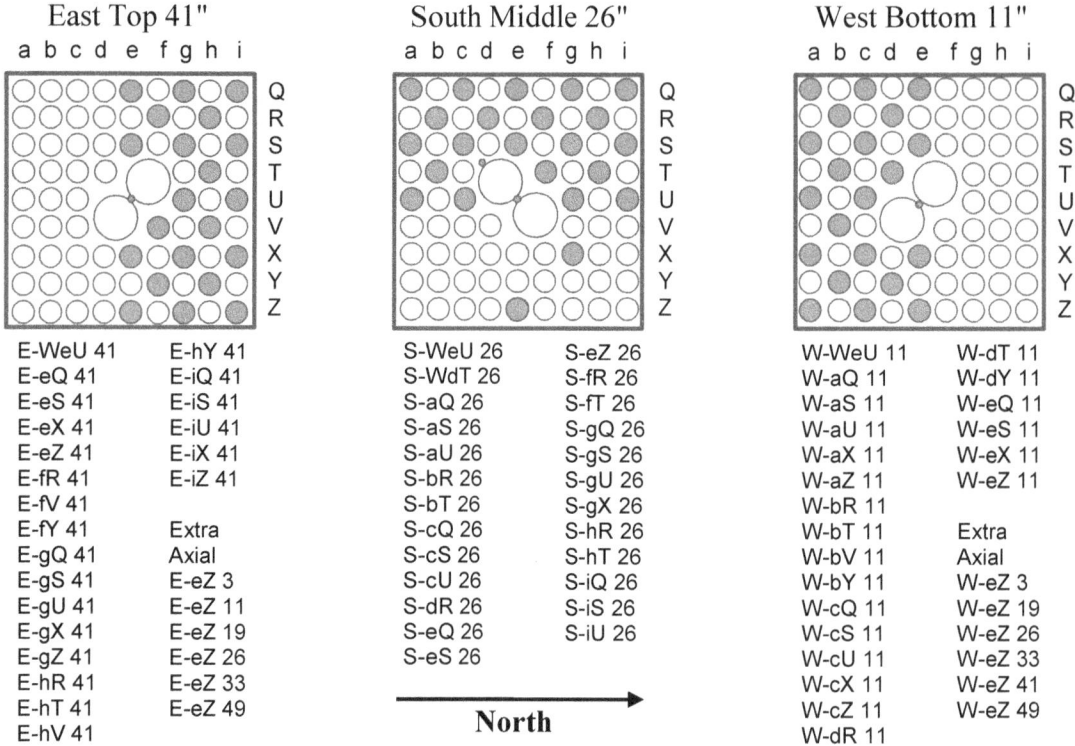

East Top 41"		South Middle 26"		West Bottom 11"	
E-WeU 41	E-hY 41	S-WeU 26	S-eZ 26	W-WeU 11	W-dT 11
E-eQ 41	E-iQ 41	S-WdT 26	S-fR 26	W-aQ 11	W-dY 11
E-eS 41	E-iS 41	S-aQ 26	S-fT 26	W-aS 11	W-eQ 11
E-eX 41	E-iU 41	S-aS 26	S-gQ 26	W-aU 11	W-eS 11
E-eZ 41	E-iX 41	S-aU 26	S-gS 26	W-aX 11	W-eX 11
E-fR 41	E-iZ 41	S-bR 26	S-gU 26	W-aZ 11	W-eZ 11
E-fV 41		S-bT 26	S-gX 26	W-bR 11	
E-fY 41	Extra	S-cQ 26	S-hR 26	W-bT 11	Extra
E-gQ 41	Axial	S-cS 26	S-hT 26	W-bV 11	Axial
E-gS 41	E-eZ 3	S-cU 26	S-iQ 26	W-bY 11	W-eZ 3
E-gU 41	E-eZ 11	S-dR 26	S-iS 26	W-cQ 11	W-eZ 19
E-gX 41	E-eZ 19	S-eQ 26	S-iU 26	W-cS 11	W-eZ 26
E-gZ 41	E-eZ 26	S-eS 26		W-cU 11	W-eZ 33
E-hR 41	E-eZ 33			W-cX 11	W-eZ 41
E-hT 41	E-eZ 49			W-cZ 11	W-eZ 49
E-hV 41				W-dR 11	

North →

Figure 6.5   Peripheral assemblies TC locations and naming conventions

6.2   MELCOR Analysis Methodology

6.2.1   MELCOR Code

The MELCOR 1.8.5 severe accident computer code [3], with enhancements through Version RP, was used to simulate the experimental SFP accident response. MELCOR is a full non-equilibrium, two-phase thermal-hydraulics code for the simulation of nuclear reactor severe accidents. It includes a full spectrum of models for simulation of reactor accidents progressing from design conditions to severe accidents with

1) Comprehensive fuel degradation,

2) Steam and air oxidation with hydrogen production,

3) Fuel and structural component degradation,

4) Fission product release, transport, deposition, chemisorption, vapor condensation and evaporation, aerosol hygroscopic effects, iodine pool evolution, and gamma heating,

5) Core-concrete interactions, and

6) Simulation of containment engineering safety features (e.g., sprays, fan coolers, filters, and recombiners).

Of particular relevance for application to the SNL experimental SFP tests, MELCOR has new fuel geometry models for boiling and PWR SFP rack geometry. The specialized SFP geometry models include basic heat transfer models for convection, conduction, and radiation, as well as a new breakaway oxidation kinetics model and fuel and rack degradation models.

Version RP includes three recent modeling enhancements applicable to BWR SFP modeling:

1) A new rack component, which permits better modeling of a SFP rack,

2) A new oxidation kinetics model, and

3) A simplified flow regime model.

The new BWR SFP rack component permits proper radiative modeling of the SFP rack between groups of different assemblies. The new oxidation kinetics predicts the transition to breakaway oxidation in air environments on a node-by-node basis. The simplified flow regime model permitted simulation of liquid films draining down the BWR fuel assemblies during spray operation.

6.2.2 Short 1×4 Assembly Model

A MELCOR short-stack assembly model was developed to analyze the temperature response to the SNL experimental SFP testing program. The model simulates a portion of five assemblies in a 1×4 pattern (Figure 6.6). The fully populated rod region at the bottom of the assembly was represented in the radiation tests. For implementation into MELCOR, the 1×4 pattern was represented by a single, high-powered assembly in the first ring and four peripheral assemblies in a second ring. The second ring was surrounded with 6 in. of Kaowool™ insulation.

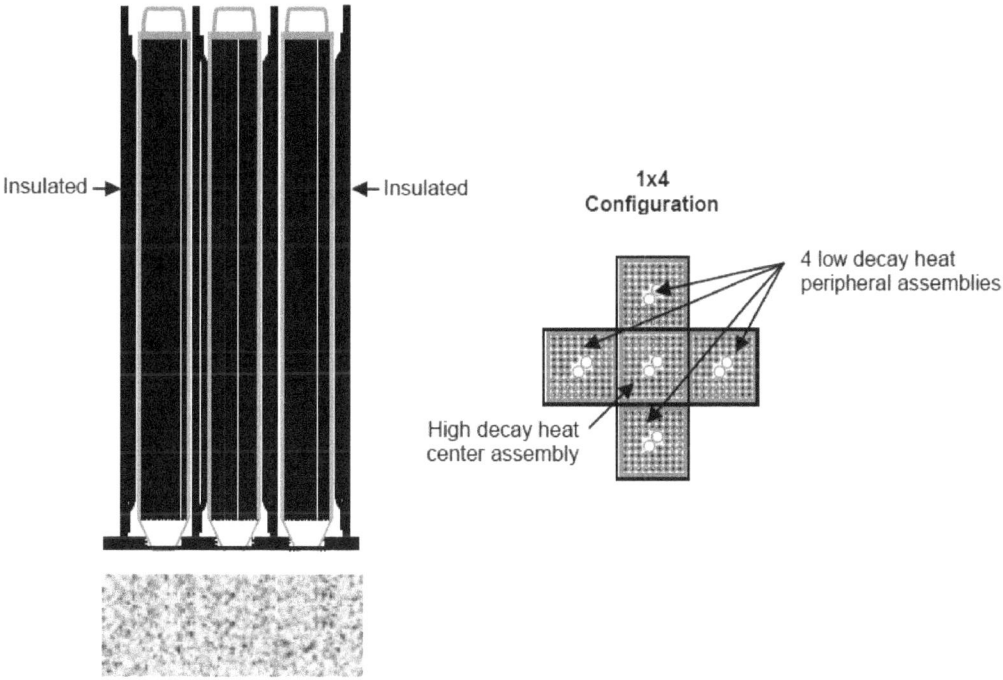

Figure 6.6   Schematic of the 1×4 configuration

Figure 6.7 shows the corresponding COR and control volume hydrodynamics (CVH) nodalizations. The BWR assembly is represented by 11 axial levels in the COR Package:

- Level 1 is the tube leading to the inlet of the assembly (blocked).

- Level 2 is the lower tie plate.

- Level 3 is the start of the active fuel region.

- Level 10 is the top of the full-length rod active fuel region.
- Level 11 is the upper tie plate.

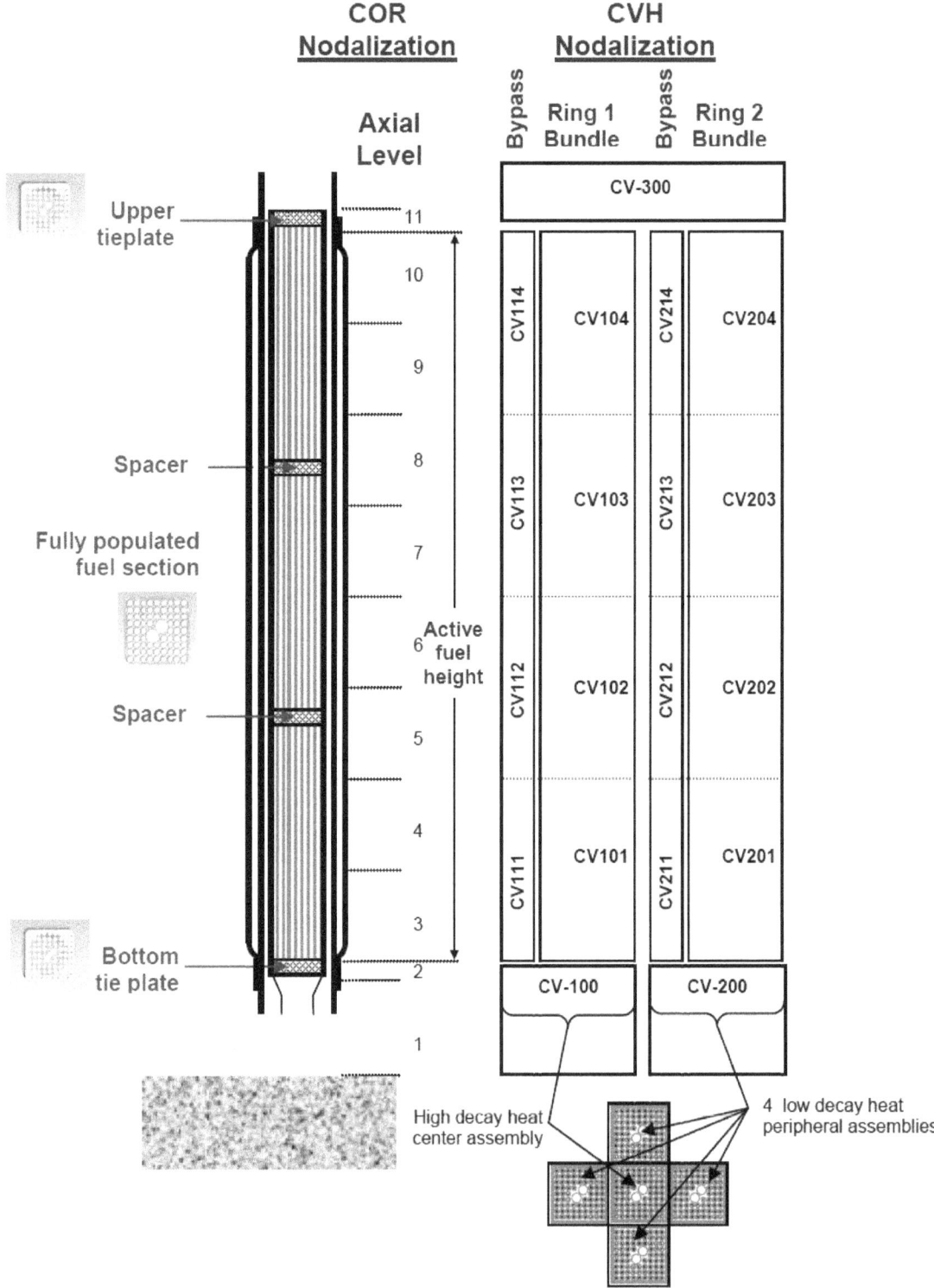

Figure 6.7   **MELCOR nodalization of the full-length assembly experiment**

The upper and lower tie plates are modeled as supporting plate structures made of stainless steel. The 0.5-in. rack support plate at Level 2 is also modeled as a stainless steel supporting structure. While the model accounts for the thermal capacitance of these support structures, MELCOR treats the top and bottom surfaces as adiabatic. The actual apparatus was insulated on the top and bottom with 6 in. of Kaowool™ insulation. This difference may account for some end effects reflected in the experimental data but not in the model.

In the heated region two COR cells were placed in each CVH volume. The interstitial area between the canister wall and the rack wall is modeled as the BWR bypass region. The bypass uses the same four-control volume axial nodalization as the channel region. At the bottom of each ring is a control volume that represents the inlet tube connected to the bottom of the assembly. However, the flow path representing the inlet tube was closed to represent the plugged inlet used in the radiation tests. The exits of the assembly were connected to a time-independent volume that represented the atmospheric conditions during the test (i.e., constant pressure and temperature). The gas in the assemblies was allowed to enter into CV-300 (an expansion plenum) as the assembly heated (i.e. the system did not pressurize just as in the actual apparatus).

The hydraulic resistance was specified using the results from the SNL hydraulic test (Section 4.2.2). However, the gas was stagnant in the MELCOR model, except for the volumetric expansion as the gas heated.

The BWR assembly canister was modeled with the MELCOR canister component. The center rack walls were modeled using the new SFP rack component with stainless steel and aluminum (i.e., to represent Boral). However, due to limitations in MELCOR to simultaneously represent the rack and the insulation, the 12-rack panels and surrounding insulation on the outside of the four peripheral assemblies were represented in MELCOR's Heat Structure (HS) Package. The primary limitations of the heat structure rack/insulation model only occur in high temperature conditions (i.e., the HS Package does not model oxidation, melting, or relocation). However, the peripheral canisters will properly radiate to the heat structure rack/insulation model. The peak temperatures in the radiation tests were well below the rack melting temperature.

MELCOR does not include an option to model the two large water rods in the assembly. Consequently, the water rod mass and surface area was included in the canister wall.

The interiors of the heater rods were modeled as compacted magnesium oxide. MELCOR's gamma heating model was disabled to deposit all the power from the heater rods in the *simulated* fuel (i.e., the magnesium oxide). A constant, linear power profile was used in each heater rod.

The emissivity of the stainless steel rack walls was modeled using the default model in MELCOR except where noted. The emissivity was a function of temperature,

$$\varepsilon = 0.042 + 0.000347 \, T$$

The Zircaloy rod and canister emissivities were also modeled using MELCOR's default model. The emissivity was a function of the oxide layer thickness,

$$\varepsilon = 0.325 + 0.1246 \text{x} 10^6 \, \Delta r_{ox} \quad (\text{for } \Delta r_{ox} < 3.88 \times 10^{-6} \text{ m})$$
$$\varepsilon = 0.808642 - 50.0 \, \Delta r_{ox} \quad (\text{for } \Delta r_{ox} > 3.88 \times 10^{-6} \text{ m})$$

MELCOR's steam emissivity model was effectively disabled because the assembly was filled with dry air.

6.3   COBRA 1×4 Model

- No-flow conditions with adiabatic surfaces at the top and bottom of model.

- Radial heat loss through insulation only.

- Density allowed to change with temperature. System does not pressurize.

- Center assembly power 1500 W, peripheral assemblies have no power.

- Internal flow assumed to be laminar (Nu = 4.36 Bundle, Nu = 5.383).

- Eight axial levels, 48-in. long

- Includes 3×3 rack arrangement with Kaowool™ insulation in the four corner cells and surrounding the entire assembly.

- Temperature dependant properties for material and air.

Figure 6.8 shows the overall slab layout of the COBRA model. The slab partitions were based on the geometry and composition of the actual hardware.

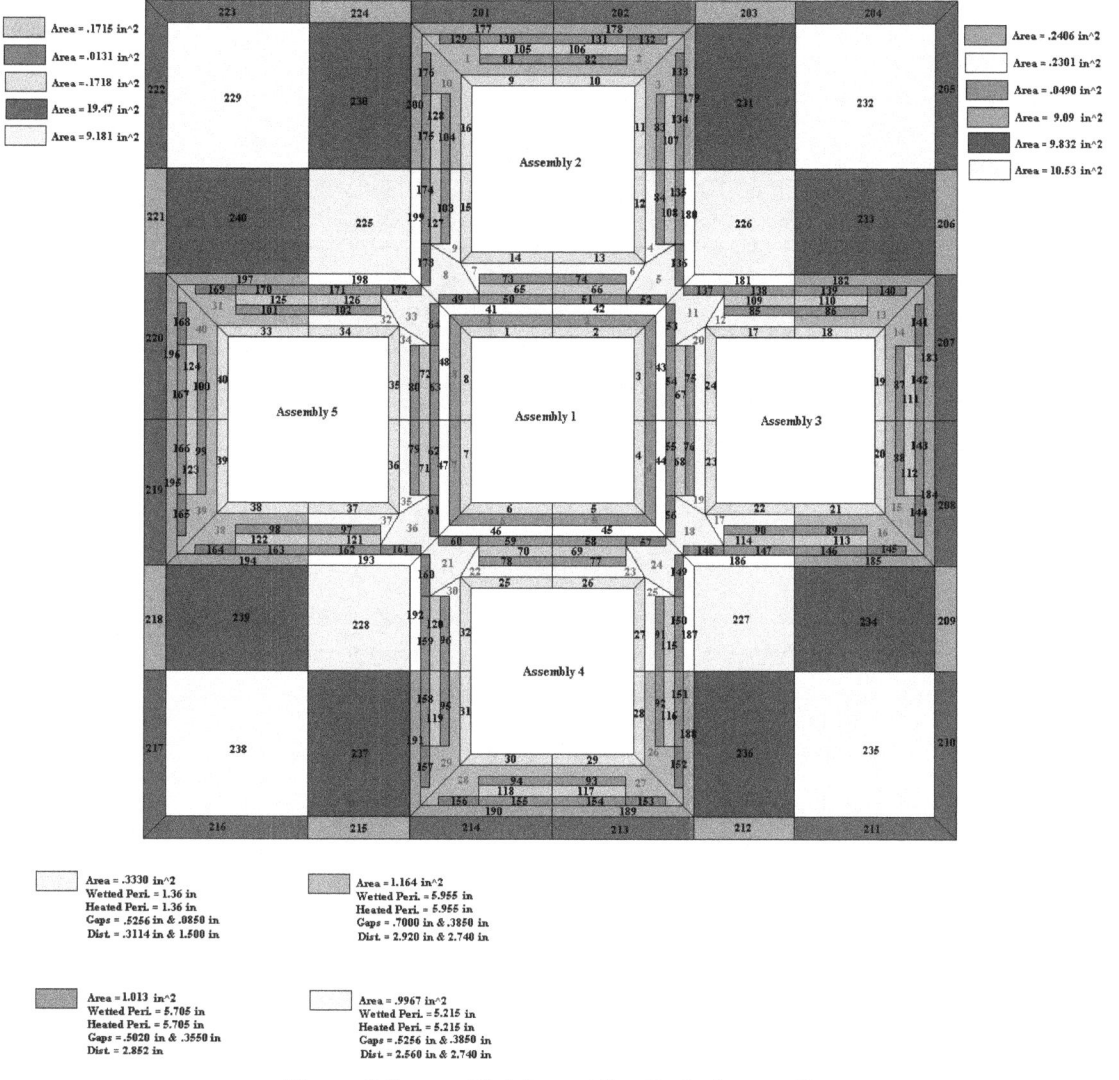

Figure 6.8   Slab layout for the 1×4 assembly

Figure 6.9 gives the detailed nodalization of the individual rod arrays. The yellow rods indicate the location of the partial rods in a full-scale assembly. Since COBRA only allows for a regular rod array, the water rods were modeled as the interstitial space bounded by the rods highlighted with red.

113

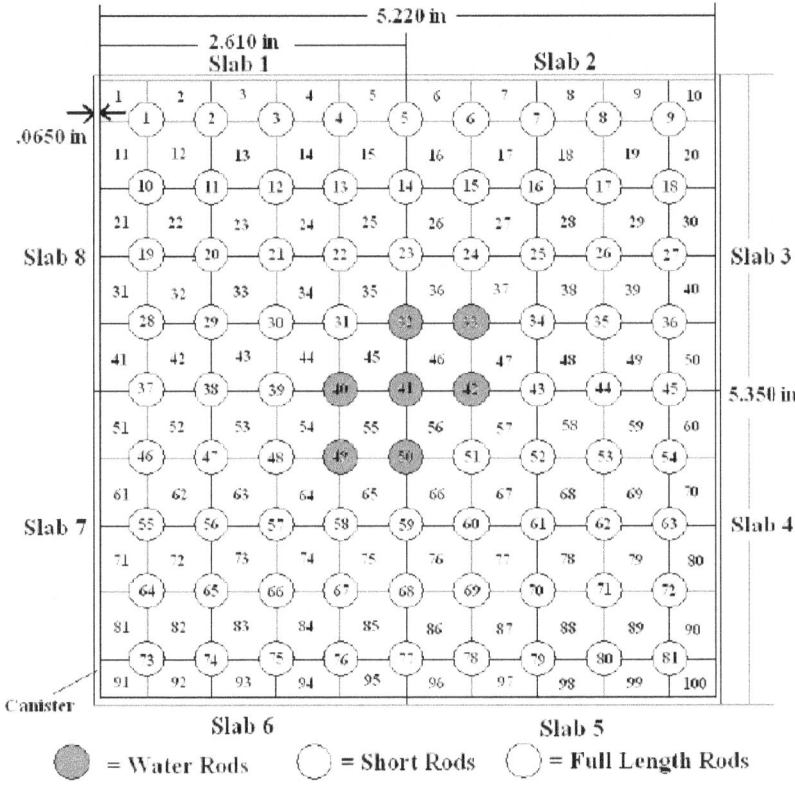

Figure 6.9   Individual rod array illustration

6.4   Results

Table 6.1 is a summary of the tests conducted with the 1×4 Incoloy assembly. Testing focused on the high contrast heating, i.e., heating of only the center bundle. Power was terminated at maximum assembly temperatures of ~873 K to prevent melting of the Boral panels, which are primarily aluminum. The apparatus was then monitored as the center assembly cooled by radiating to the peripheral assemblies. All numerical results represent average bundle temperatures.

Table 6.1   Summary of all 1×4 thermal radiation testing on the Incoloy *short stack* assembly

Date	Description	Time at Power (hrs)	Test Duration (hrs)
2/27/2006	1.5 kW - Center only	12	12
3/15/2006	3 kW - Center only	2.2	10.4
3/21/2006	3 kW - Center only	4.1	6.8
3/23/2006	5 kW - Center only	2	8.8

Figure 6.10 shows symmetrically placed peripheral bundle temperatures as a function of time for the powered test conducted on March 21, 2006. Temperatures were within 10°C for all but the hottest location at the 41-in. level. Here, the maximum difference was 15°C.

114

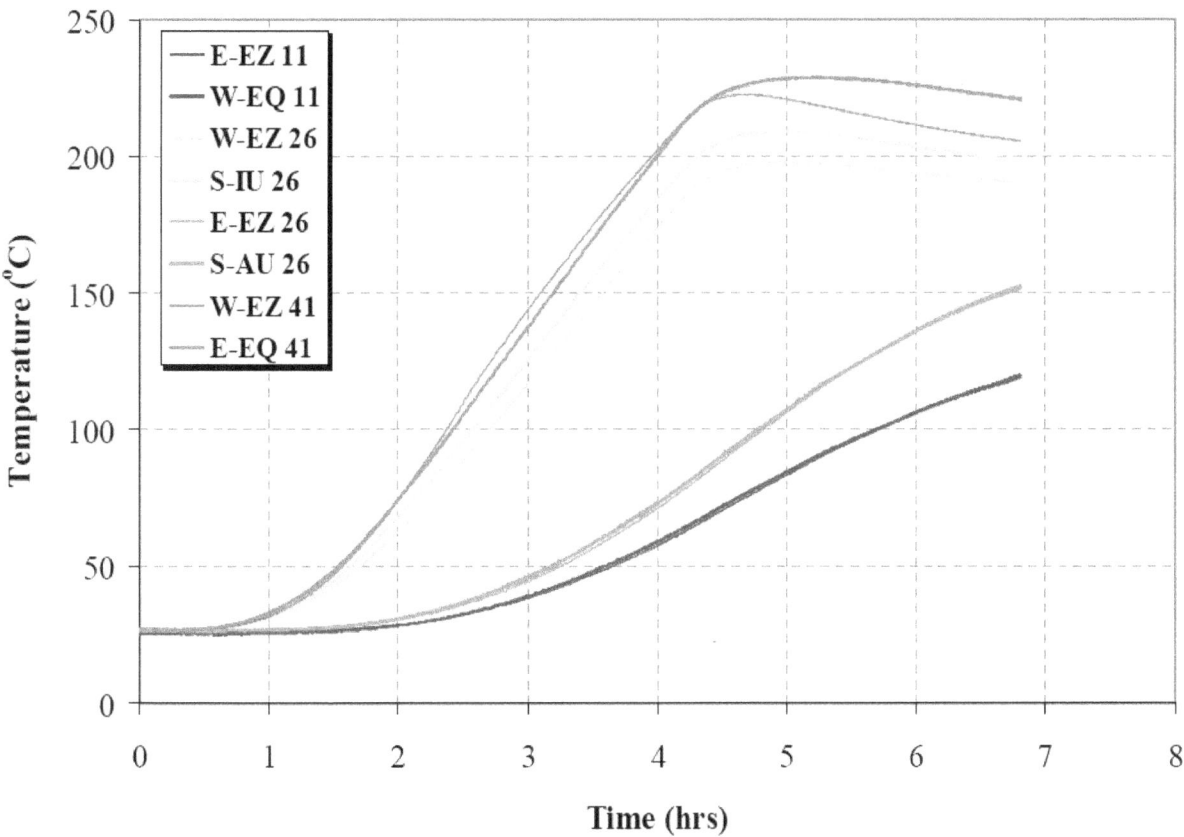

Figure 6.10   **Symmetric peripheral bundle temperatures as a function of time for the powered test conducted on 3/21/06**

Figure 6.11 through Figure 6.20 compare the experimental minimum (blue lines), maximum (red lines), and average bundle temperatures (black lines) to the predicted average bundle temperatures of COBRA (green dashed lines) and MELCOR (pink dashed lines). Discrepancies between the codes and experimental data are attributed to additional heat loss to the bottom and top surfaces, which is not treated in the numerical models.

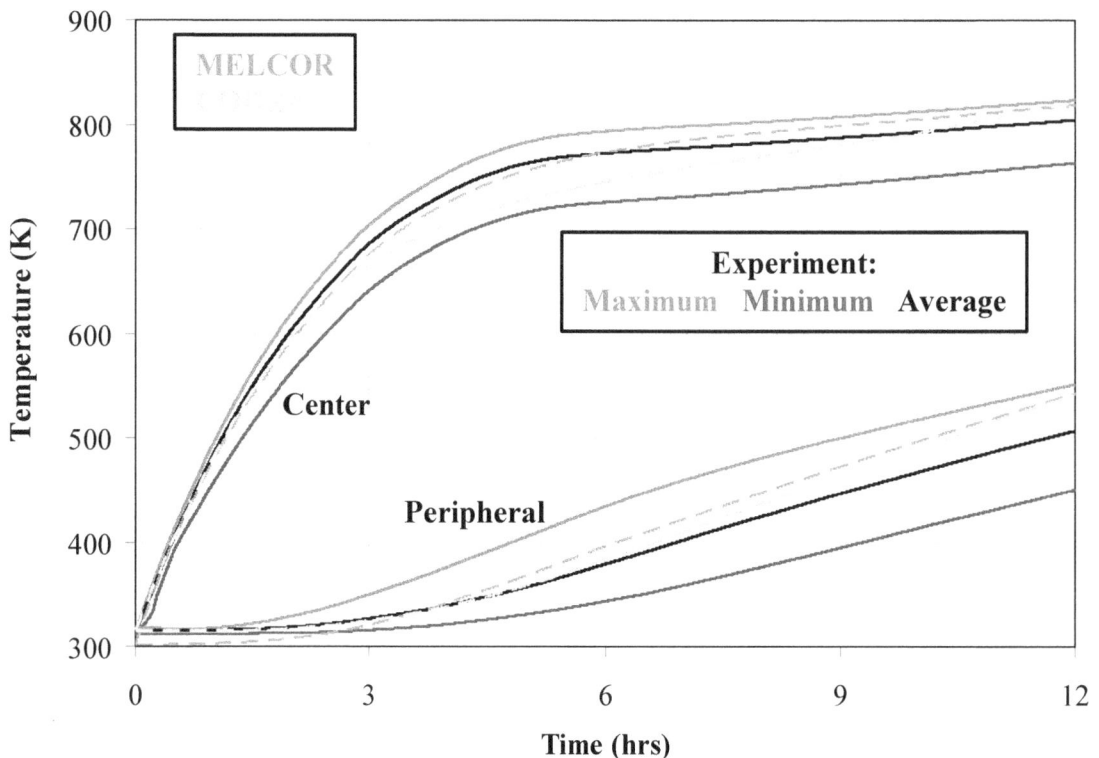

Figure 6.11   Power test 2/27/06 – 1.5 kW center assembly, 26-in. level

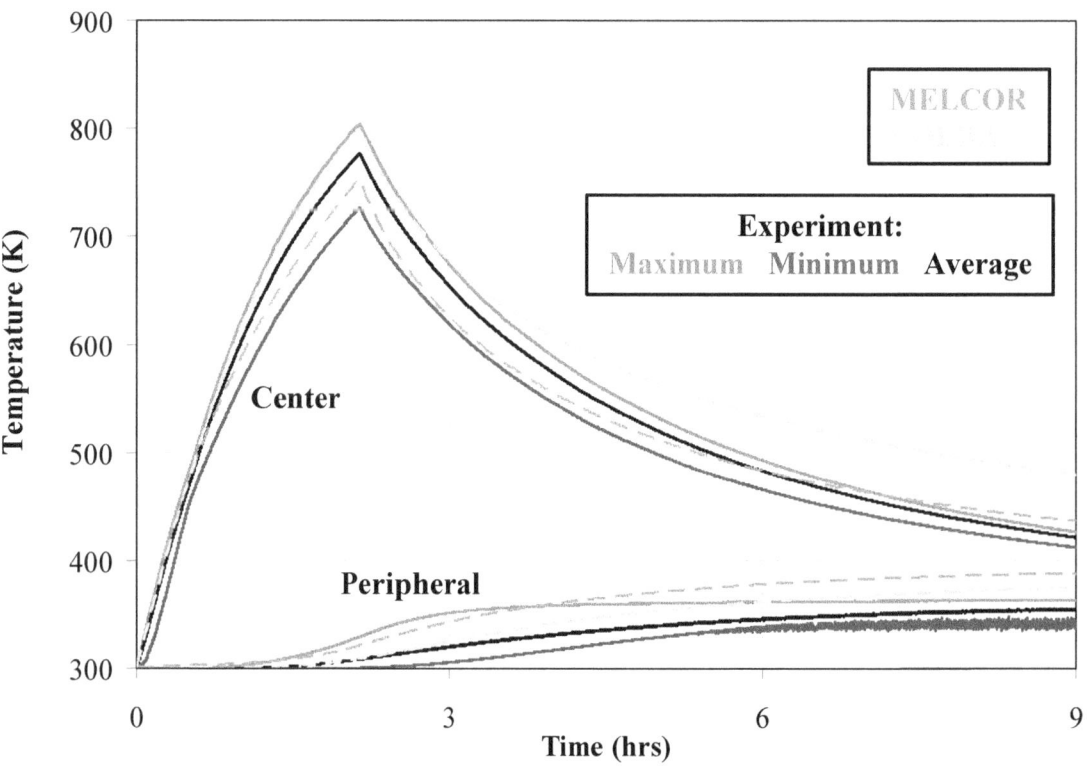

Figure 6.12   Power test 3/15/06 – 3 kW center assembly, 11-in. level

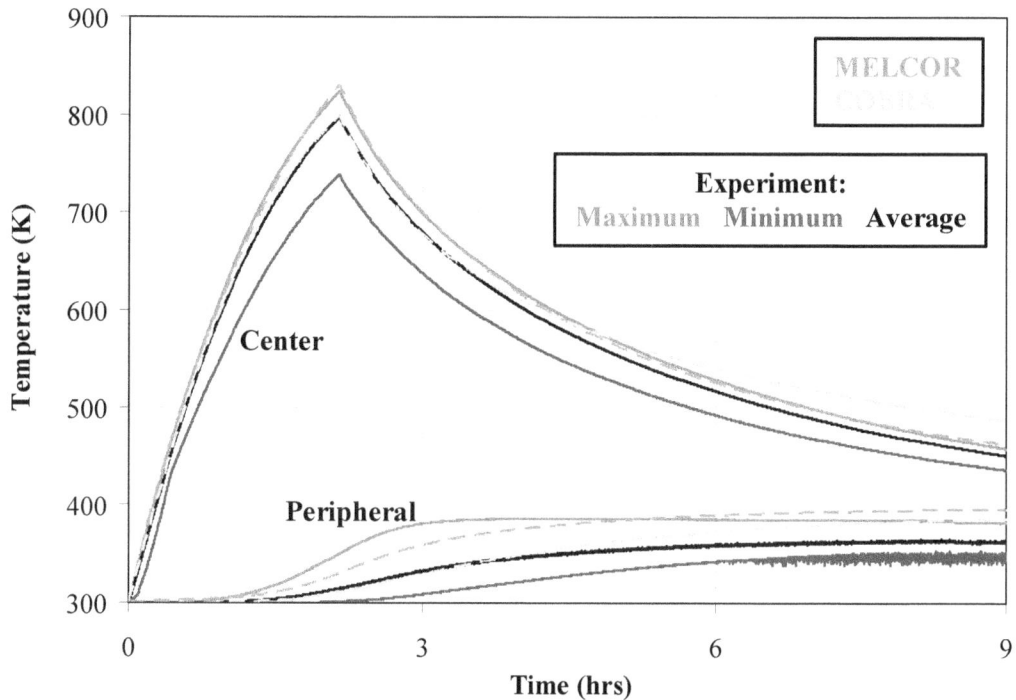

Figure 6.13   Power test 3/15/06 – 3 kW center assembly, 26-in. level

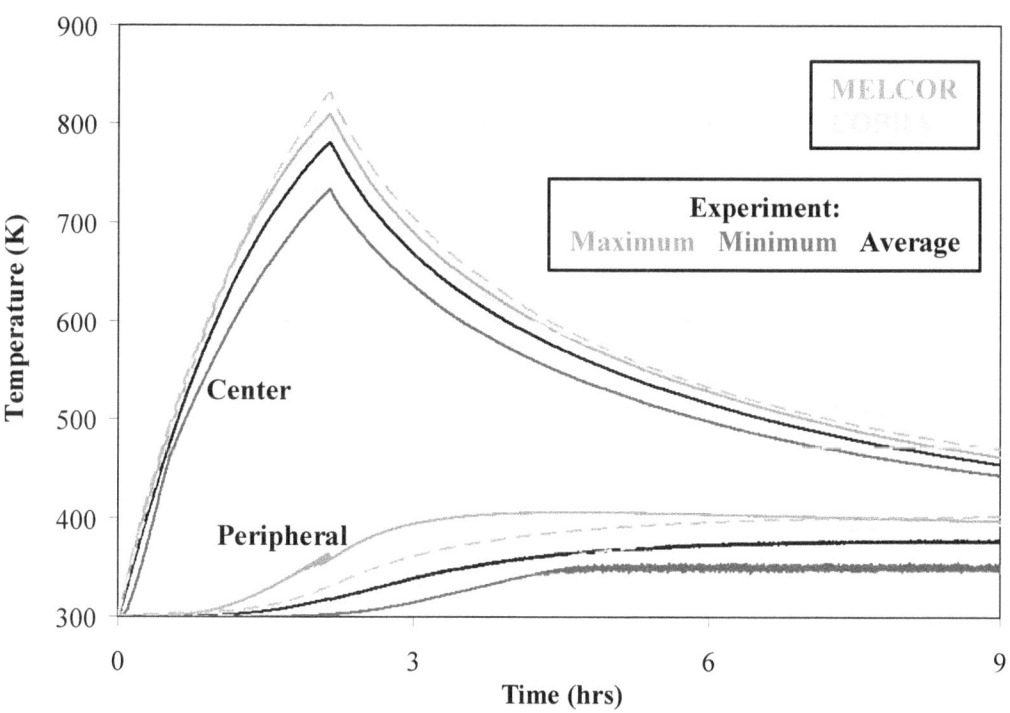

Figure 6.14   Power test 3/15/06 – 3 kW center assembly, 41-in. level

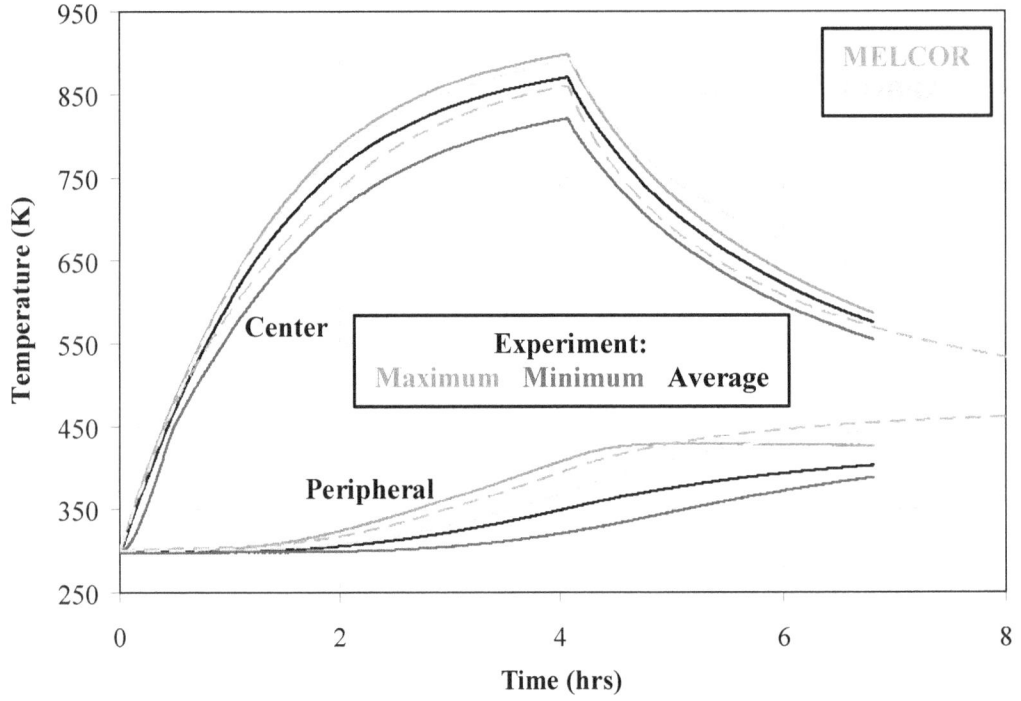

Figure 6.15   Power test 3/21/06 – 3 kW center assembly, 11-in. level

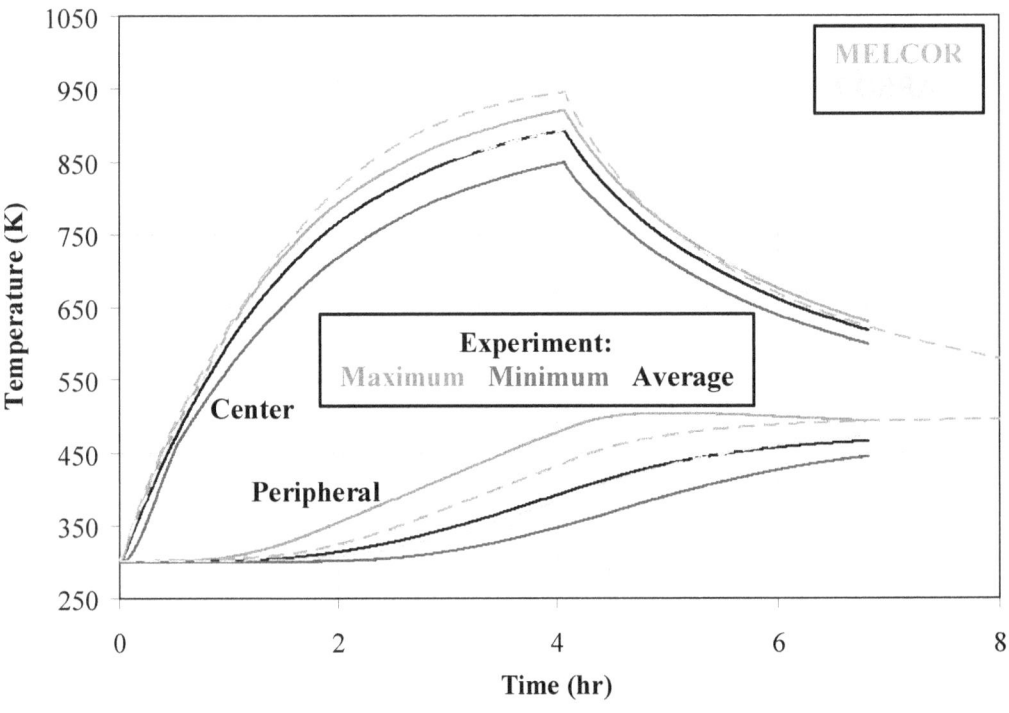

Figure 6.16   Power test 3/21/06 – 3 kW center assembly, 26-in. level

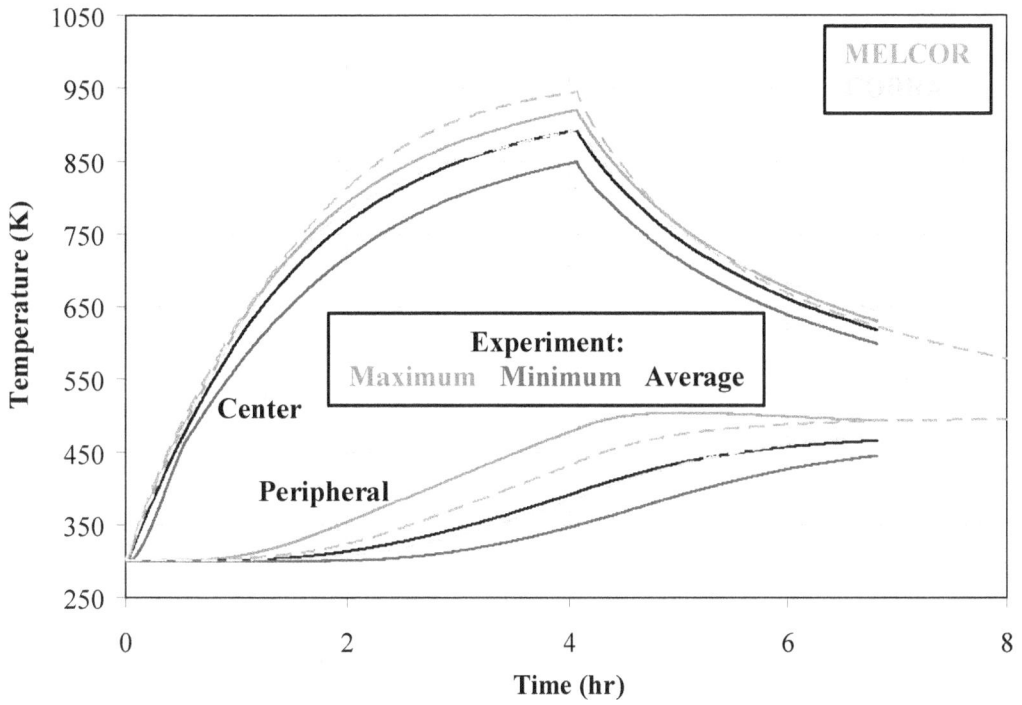

Figure 6.17   Power test 3/21/06 – 3 kW center assembly, 41-in. level

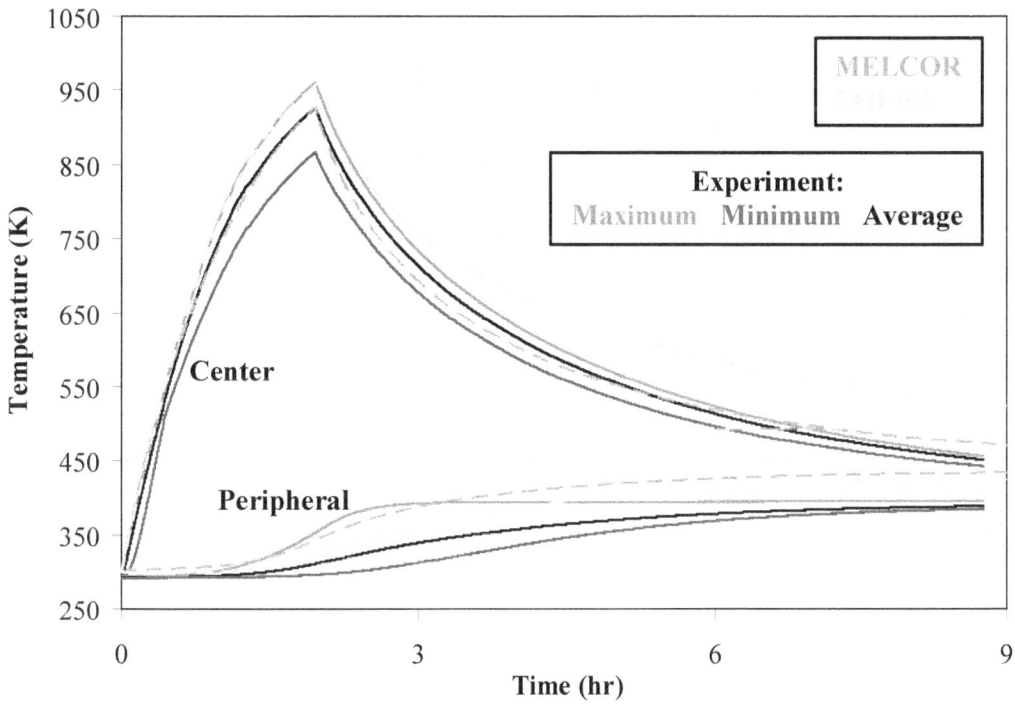

Figure 6.18   Power test 3/23/06 – 5 kW center assembly, 11-in. level

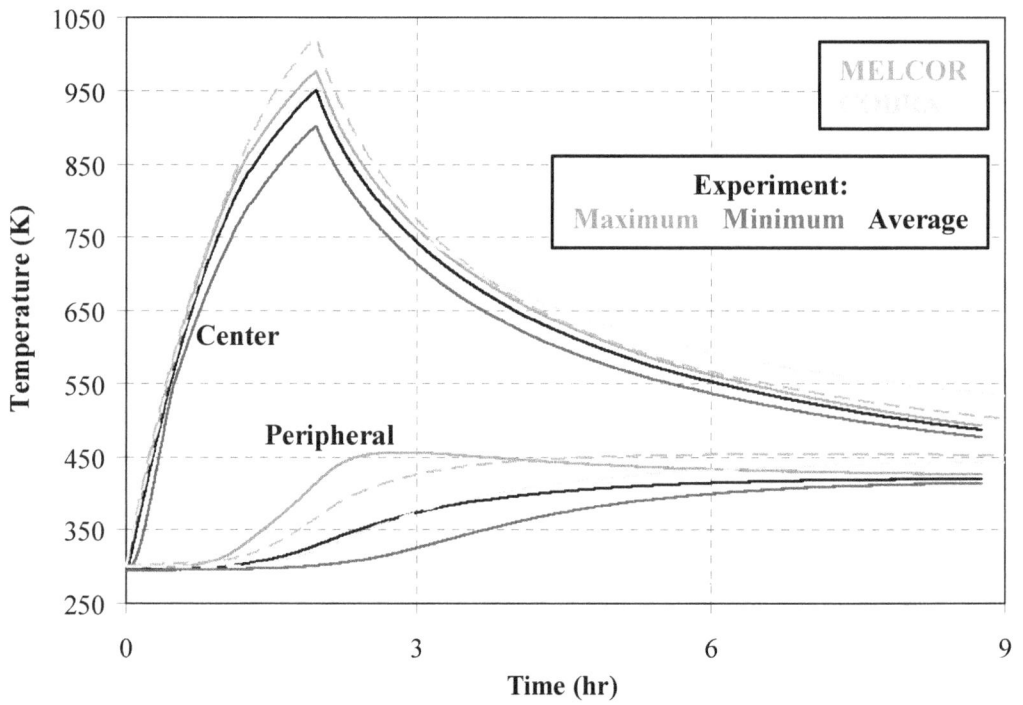

Figure 6.19   Power test 3/23/06 – 5 kW center assembly, 26-in. level

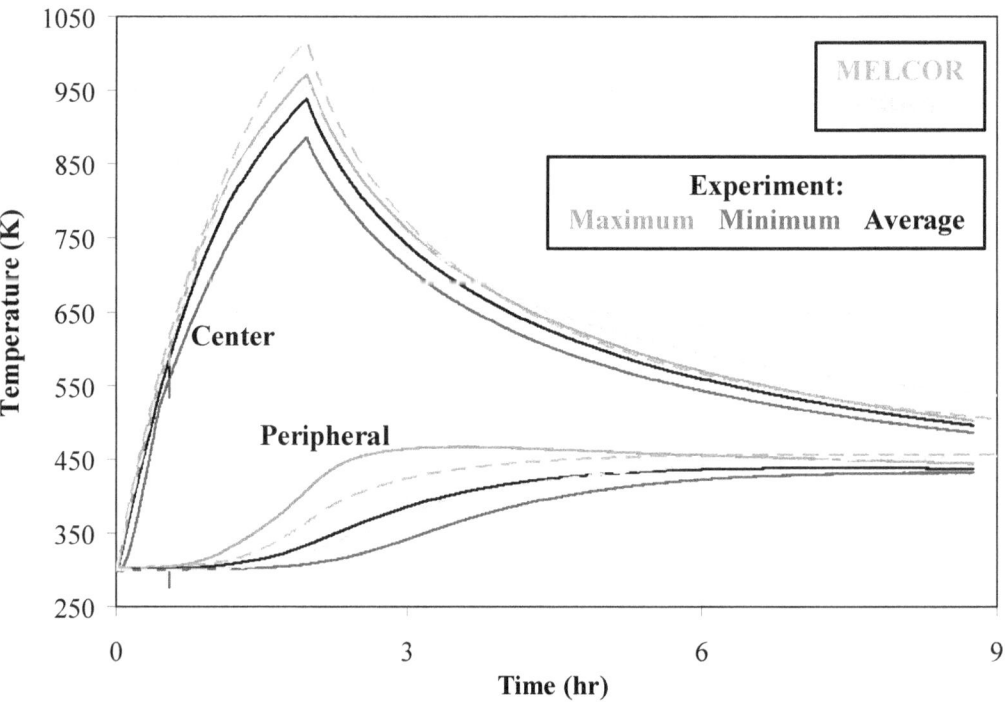

Figure 6.20   Power test 3/23/06 – 5 kW center assembly, 41-in. level

6.5   Model Sensitivity to Emissivity

Sensitivity to emissivity of the Zircaloy and stainless steel surfaces inside the 1×4 thermal radiation models is presented in the following pages. All reported values correspond to average bundle temperatures. Comparisons were made to the data collected during the powered test on March 21, 2006, 3 kW input power to the center assembly. The cases were chosen based on the following criteria:

- MELCOR Baseline: $\varepsilon_{Zirc} = 0.325$, $\varepsilon_{SS} = 0.256 + 3.47 \times 10^{-4}$ (T-616.5) (where T is in units of °F) (Figure 6.21 through Figure 6.23).

- COBRA Baseline: $\varepsilon_{Zirc} = 0.43$, $\varepsilon_{SS} = 0.43$ (Figure 6.24 through Figure 6.26).

- Maintain ε_{eff}: $\varepsilon_{Zirc} = 0.62$, $\varepsilon_{SS} = 0.33$ (Figure 6.27 through Figure 6.30).

- Set $\varepsilon_{Zirc} = 0.80$, hold $\varepsilon_{SS} = 0.43$ (Figure 6.31 through Figure 6.33).

Effective emissivity is defined for a two surface gray enclosure by the following formula:

$$\varepsilon_{eff} = \left(\frac{1-\varepsilon_{Zirc}}{\varepsilon_{Zirc}} + \frac{1}{F_{12}} + \frac{1-\varepsilon_{SS}}{\varepsilon_{SS}} \right)^{-1}$$

Here, F_{12} is the view factor between the Zircaloy channel box and stainless steel pool cell. Assuming $F_{12} \approx 1$, simplifies the expression to the following:

$$\varepsilon_{eff} = \left(\frac{\varepsilon_{Zirc} \varepsilon_{SS}}{\varepsilon_{Zirc} - \varepsilon_{Zirc} \varepsilon_{SS} + \varepsilon_{SS}} \right)$$

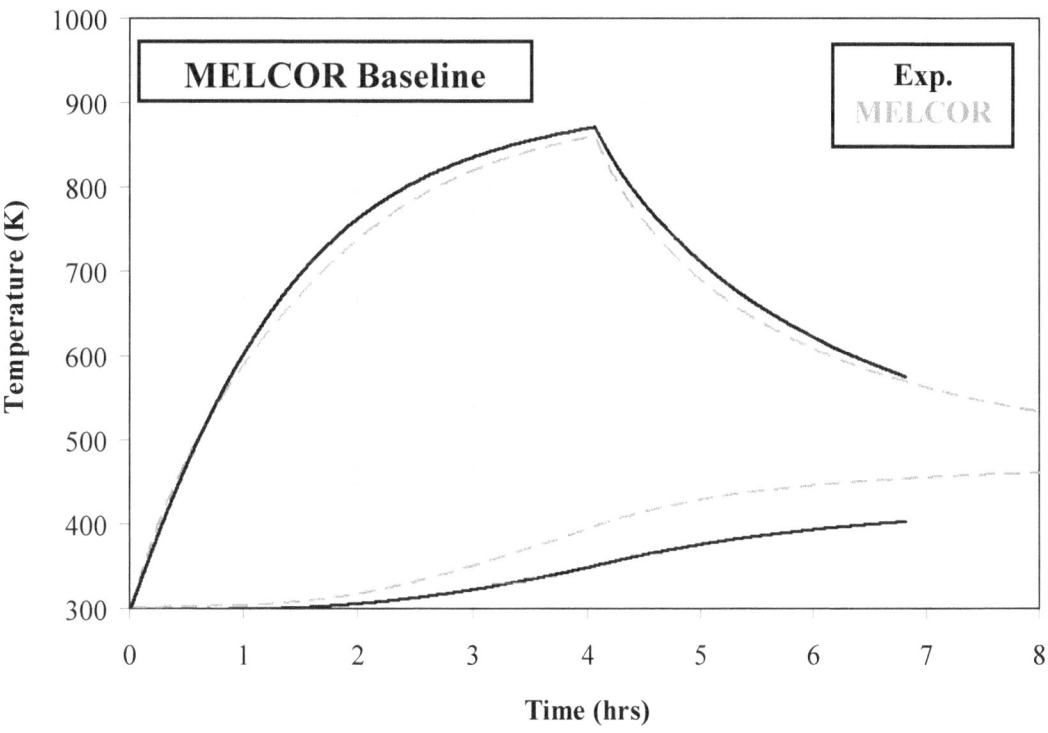

Figure 6.21   **MELCOR baseline emissivity for power test 3/21/06 – 3 kW center assembly, 11-in. level**

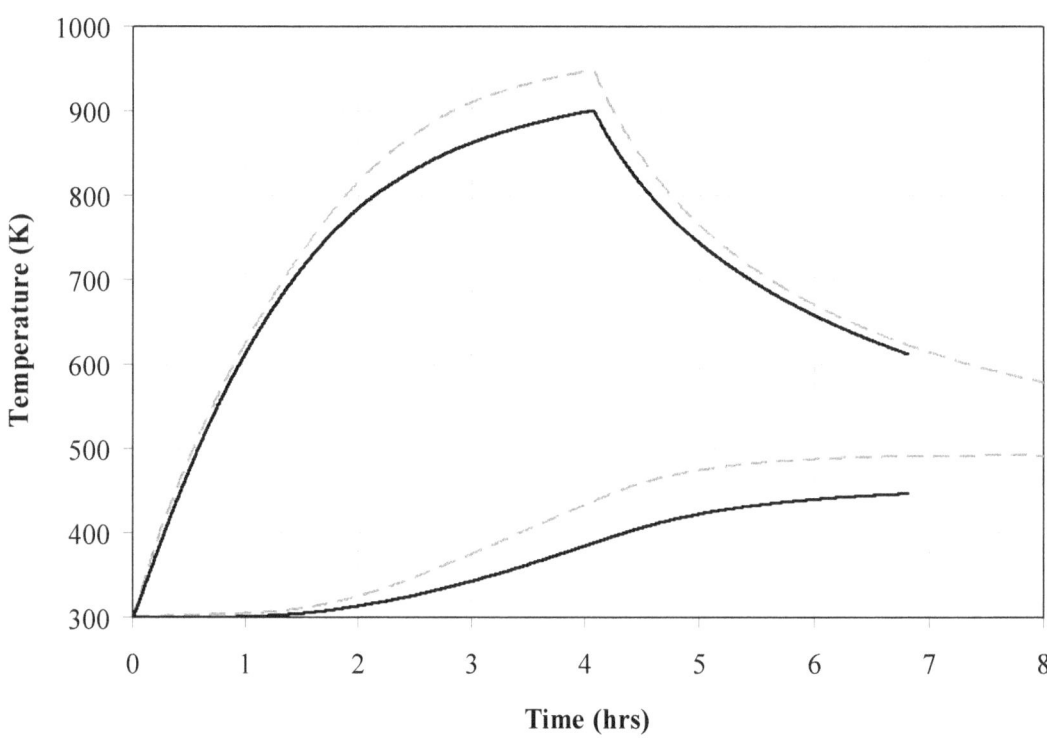

Figure 6.22   **MELCOR baseline emissivity for power test 3/21/06 – 3 kW center assembly, 26-in. level**

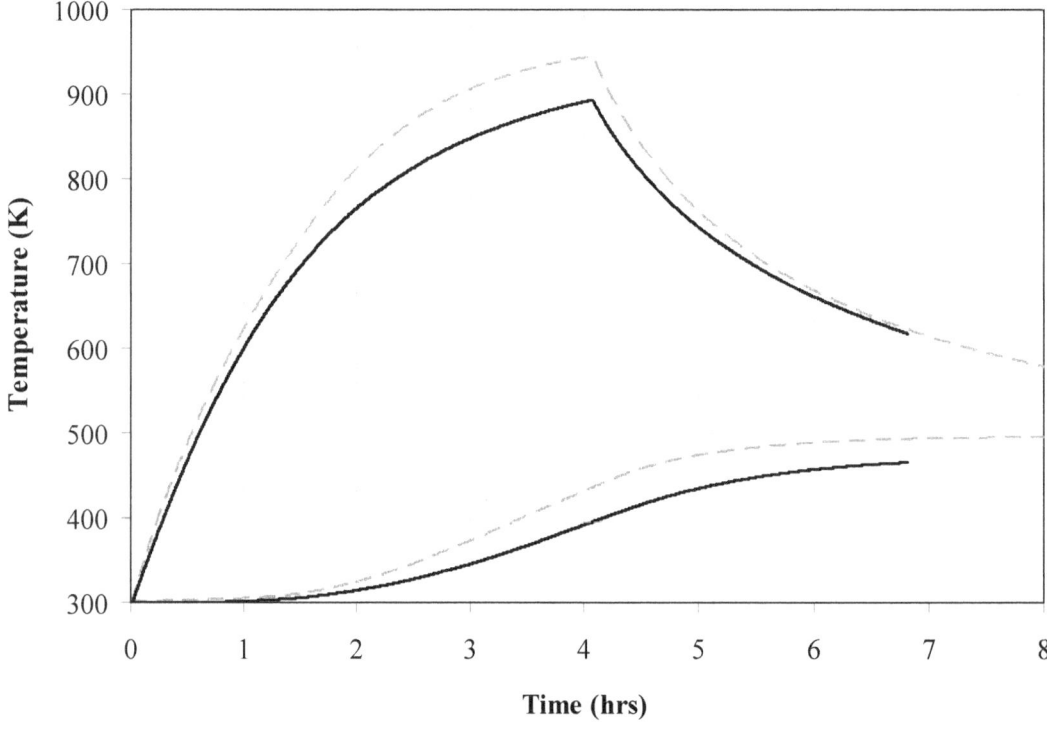

Figure 6.23   **MELCOR baseline emissivity for power test 3/21/06 – 3 kW center assembly, 41-in. level**

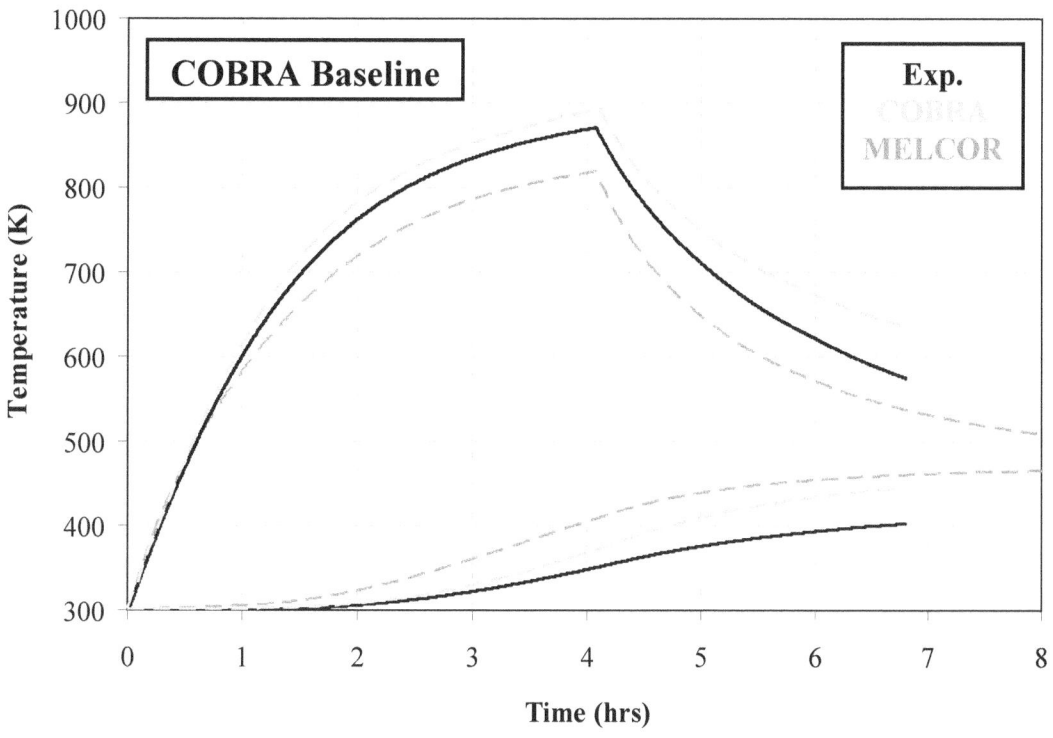

Figure 6.24　　　COBRA baseline emissivity for power test 3/21/06 – 3 kW center assembly, 11-in. level

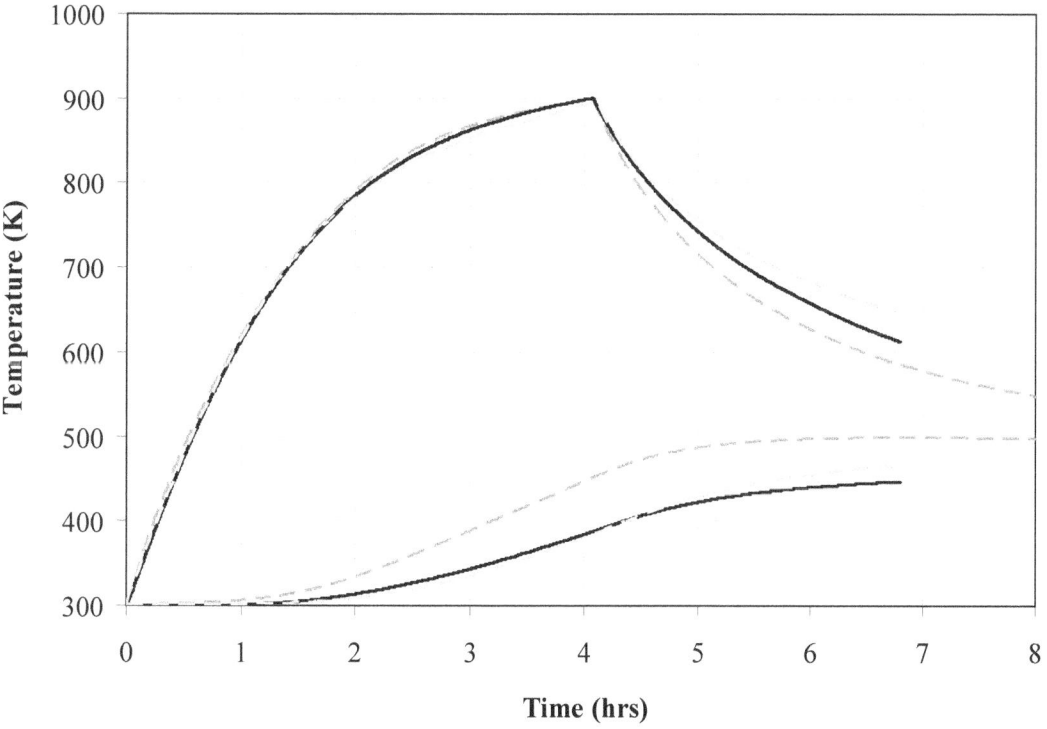

Figure 6.25　　　COBRA baseline emissivity for power test 3/21/06 – 3 kW center assembly, 26-in. level

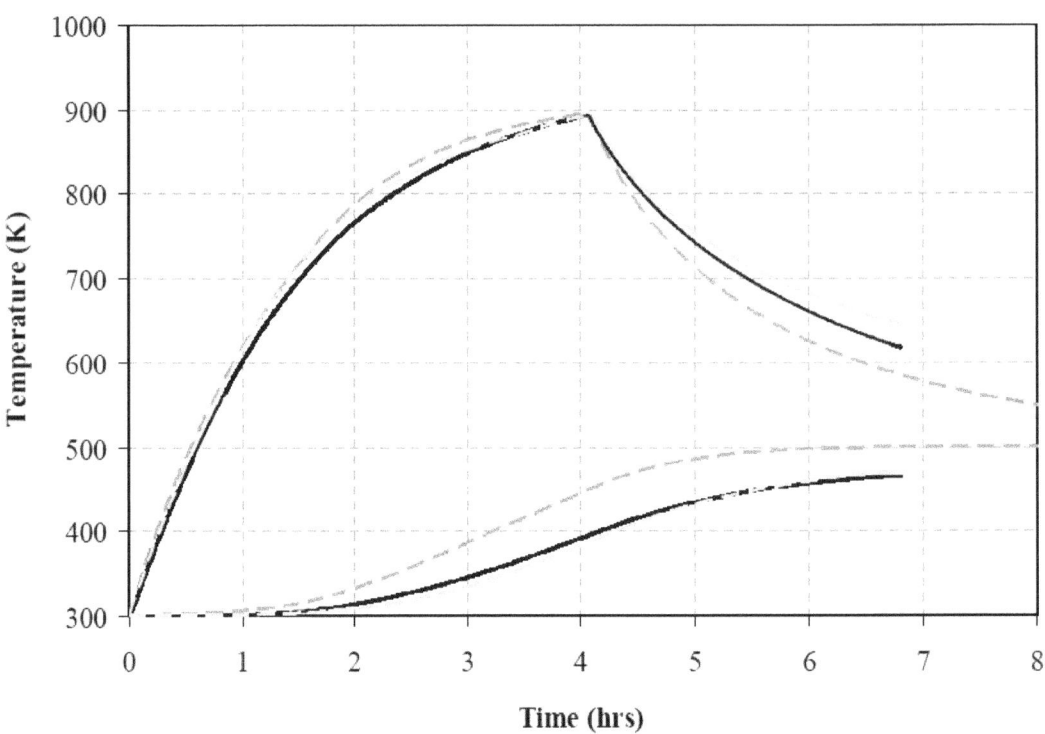

Figure 6.26   **COBRA baseline emissivity for power test 3/21/06 – 3 kW center assembly, 41-in. level**

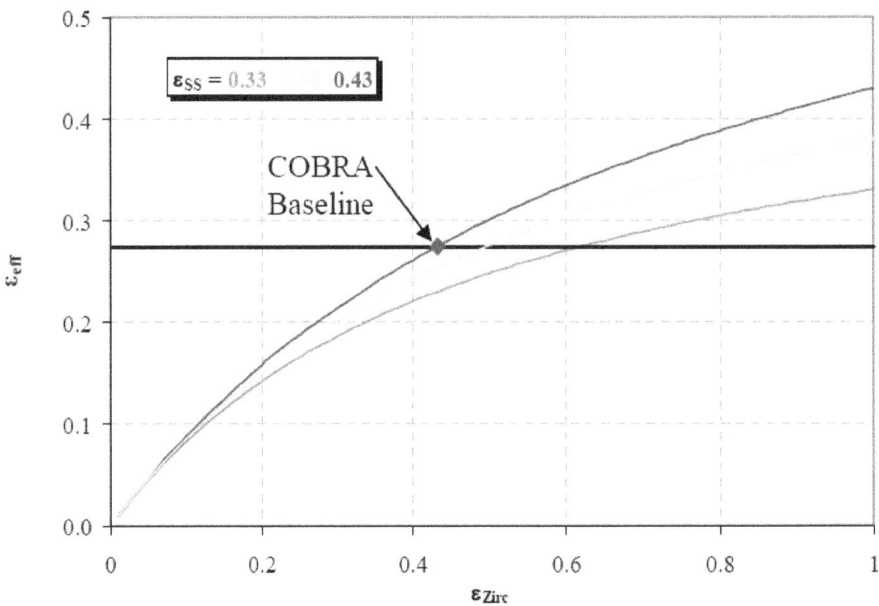

Figure 6.27   **Effective two-surface emissivity as a function of the Zircaloy emissivity for stainless steel emissivities of 0.33, 0.38, and 0.43**

Note: The COBRA baseline was set at ε_{SS} and $\varepsilon_{Zirc} = 0.43$.

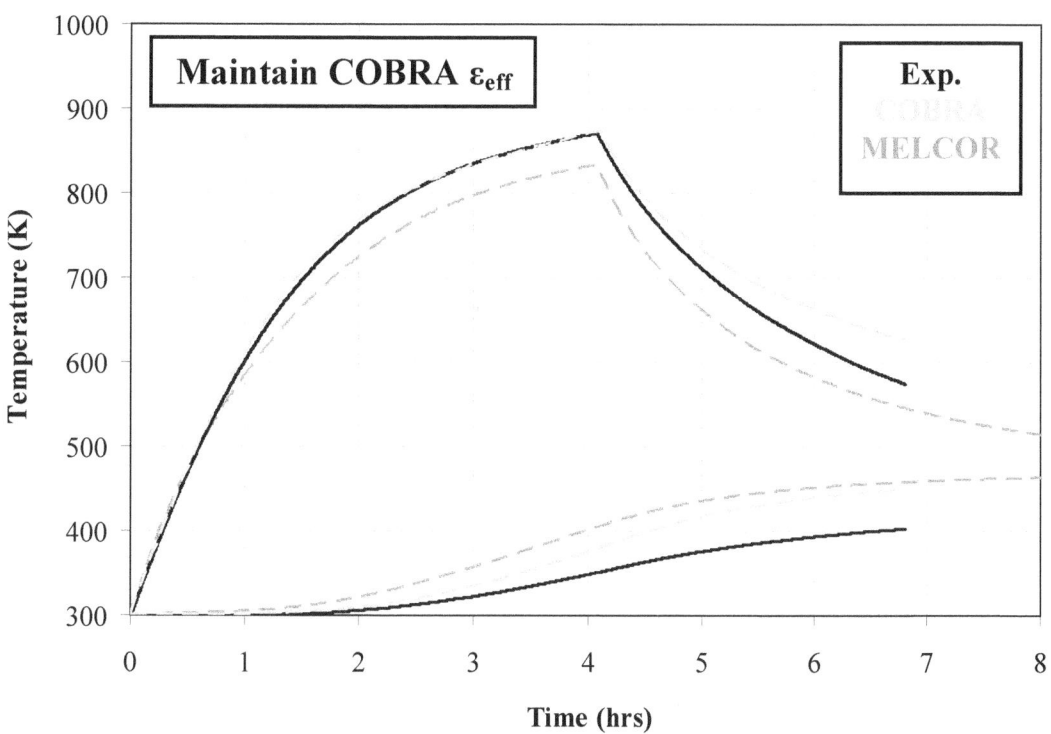

Figure 6.28   Maintain effective emissivity for power test 3/21/06 – 3 kW center assembly, 11-in. level

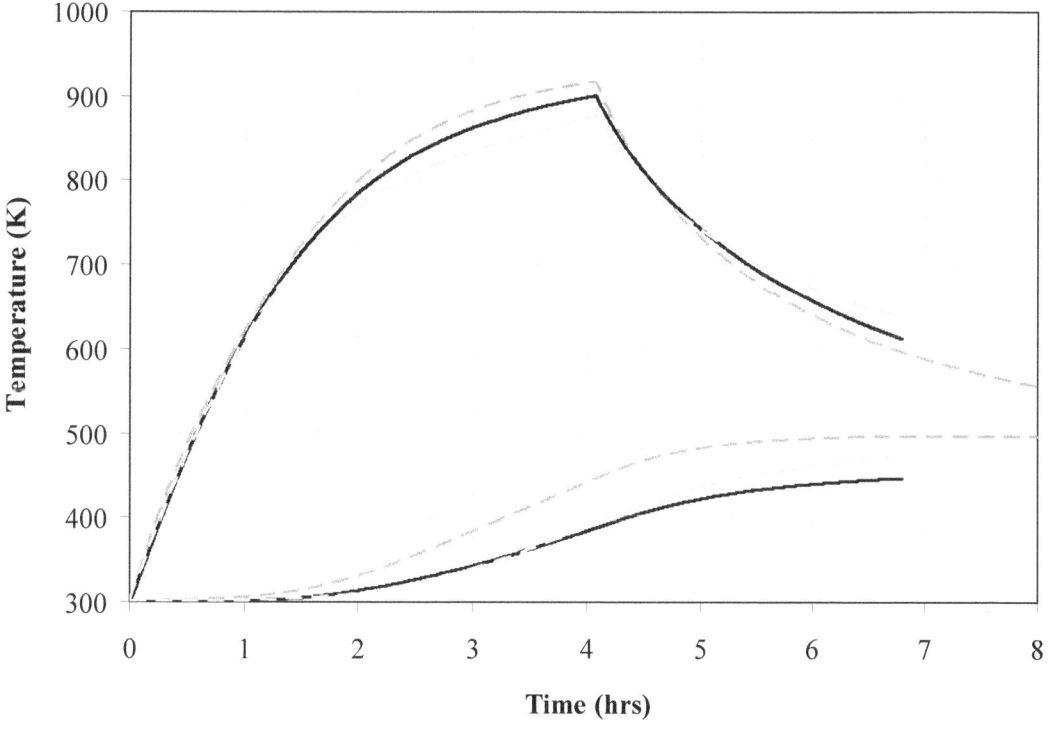

Figure 6.29   Maintain effective emissivity for power test 3/21/06 – 3 kW center assembly, 26-in. level

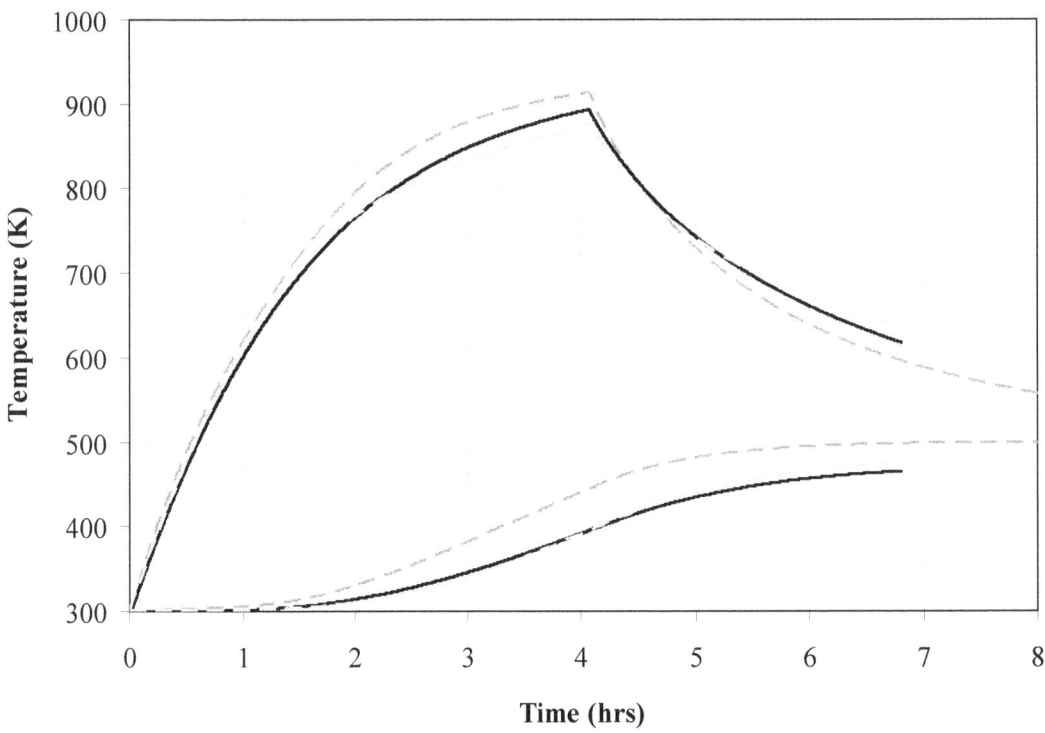

Figure 6.30   Maintain effective emissivity for power test 3/21/06 – 3 kW center assembly, 41-in. level

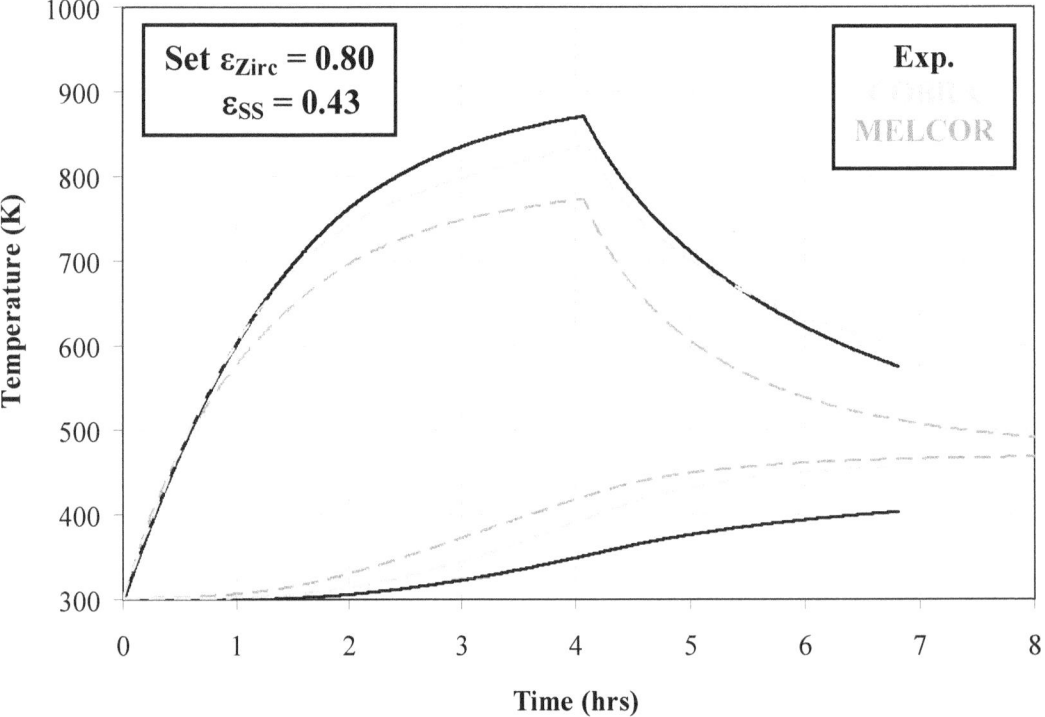

Figure 6.31   Increase Zircaloy emissivity to 0.80 and set stainless steel emissivity to 0.43 for power test 3/21/06 – 3 kW center assembly, 11-in. level

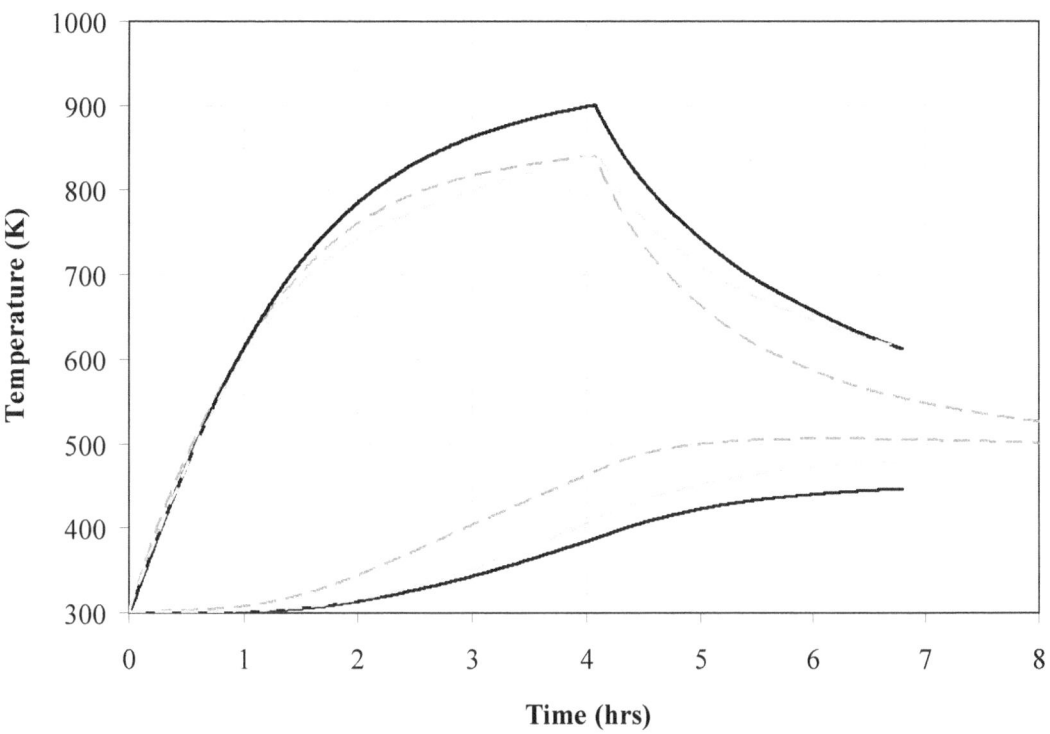

Figure 6.32   **Increase Zircaloy emissivity to 0.80 and set stainless steel emissivity to 0.43 for power test 3/21/06 – 3 kW center assembly, 26-in. level**

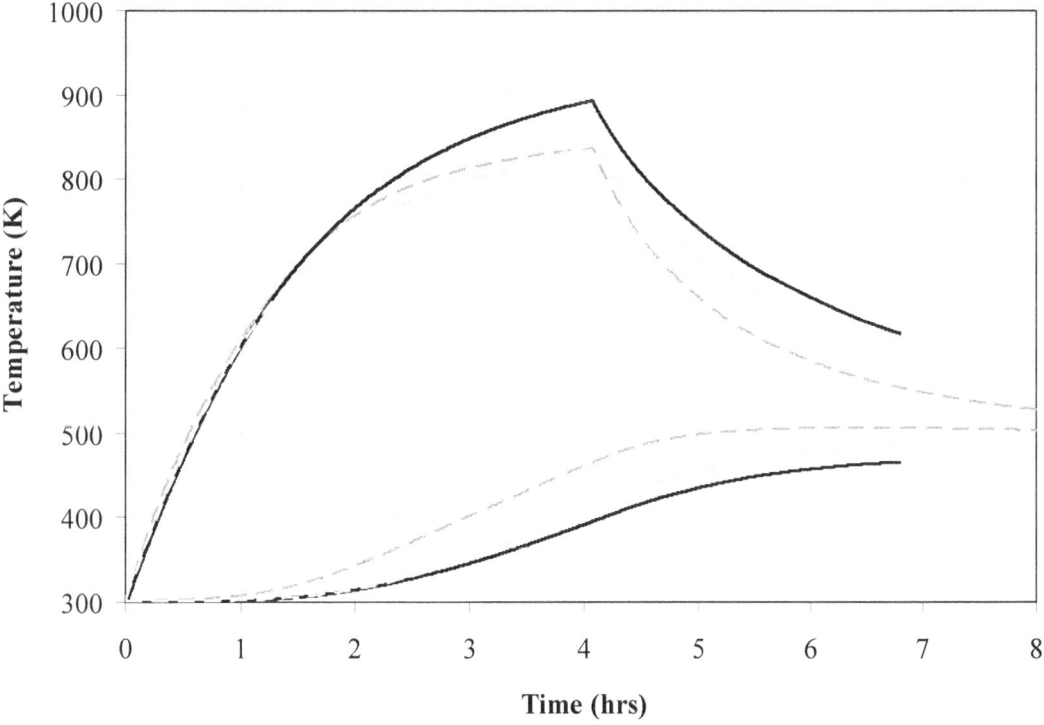

Figure 6.33   **Increase Zircaloy emissivity to 0.80 and set stainless steel emissivity to 0.43 for power test 3/21/06 – 3 kW center assembly, 41-in. level**

6.5.1 Additional COBRA Model Modifications

Additional efforts were made to reconcile the differences in the COBRA model to the experimental data. In particular, the temperature profile in the peripheral assembly was examined. Figure 6.34 shows the evolution of the COBRA model along with two temperature profiles from the powered test from March 21, 2006. The temperature profiles were taken along the axis of highest temperature gradient.

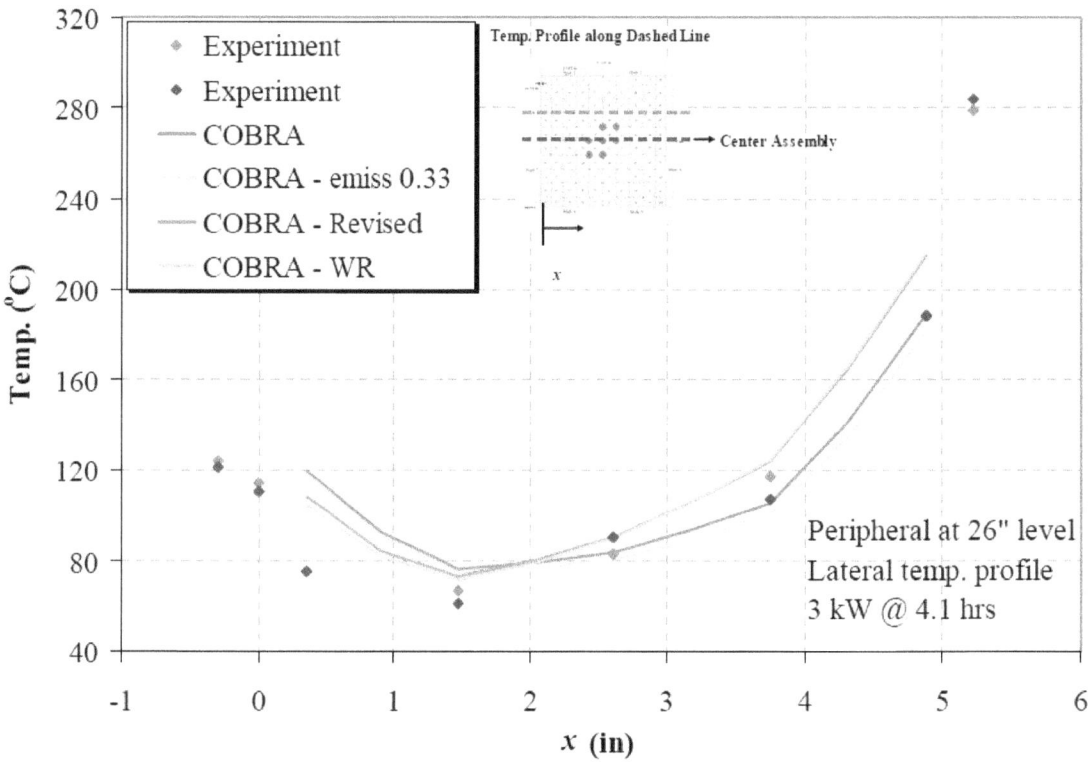

Figure 6.34   **Peripheral temperature profiles at 4.1 hours elapsed test time for the powered test conducted on 3/21/06**

The red line shows the original COBRA baseline model. This model agrees well toward the hotter side of the assembly but over predicts the temperature at the opposite side. The next model iteration raised the Zircaloy emissivity and lowered the stainless steel emissivity to maintain the effective two-surface emissivity (green line) as discussed in the previous section. This change in emissivity had very little effect on the thermal response of the system. Next, the hydraulic loss of the two-rod spacers was added (pink line). This alteration hindered the convective cell which forms inside the peripheral assemblies. The result was elevated temperatures toward the side facing the center assembly and lower temperatures on the opposite side of the bundle. Finally, the model elements constituting the interior of the water rods were configured to prevent any axial flow (light blue line). This modification had minimal impact on the thermal response of the peripheral bundles.

7 FULL-SCALE ZIRCALOY PRE-IGNITION AND IGNITION RESULTS OF THE SFP EXPERIMENT

This report summarizes the findings of the SFP full-scale, pre-ignition and ignition experiments conducted between May 23 and June 8, 2006. The stated purpose of these investigations was to determine the thermal and hydraulic response for the validation of the MELCOR severe accident analysis code. The full-scale Zircaloy assembly was configured for the open bypass and closed drain holes flow arrangement. Pre-ignition testing was conducted to verify the thermal and hydraulic response of the assembly against previous heated experiments. These tests were conducted at power inputs of 1800 to 3000 W. The ignition test (5000 W) was conducted after feedback to MELCOR provided agreement with pre-ignition tests. The MELCOR pre-test predictions of the ignition experiment were accurate to within 40 K and 5 minutes of ignition.

7.1 Apparatus

The basic configuration of this apparatus was very similar to the full-length Incoloy assembly. Figure 7.1 shows the experimental apparatus before testing. A prominent feature of this setup was the side-mounted sensors, which were not included in the full-length Incoloy assembly. The largest of these instruments were the *in situ* oxygen sensors mounted at 2-ft. intervals along the extent of the assembly. Details of these sensors are provided later. The smallest sensors are TCs and light pipes that penetrated through the pool cell wall and monitored the Zircaloy canister wall. The type B and S TCs were placed roughly every 6 in. starting at $z = 54$ in., as measured from the top of the bottom tie plate. Silica light pipes were mounted between these TCs, again in 6-in. increments. Finally, an infrared (IR) spectrometer was positioned at the $z = 139.5$-in. level to observe the Zircaloy channel box through a calcium fluoride window.

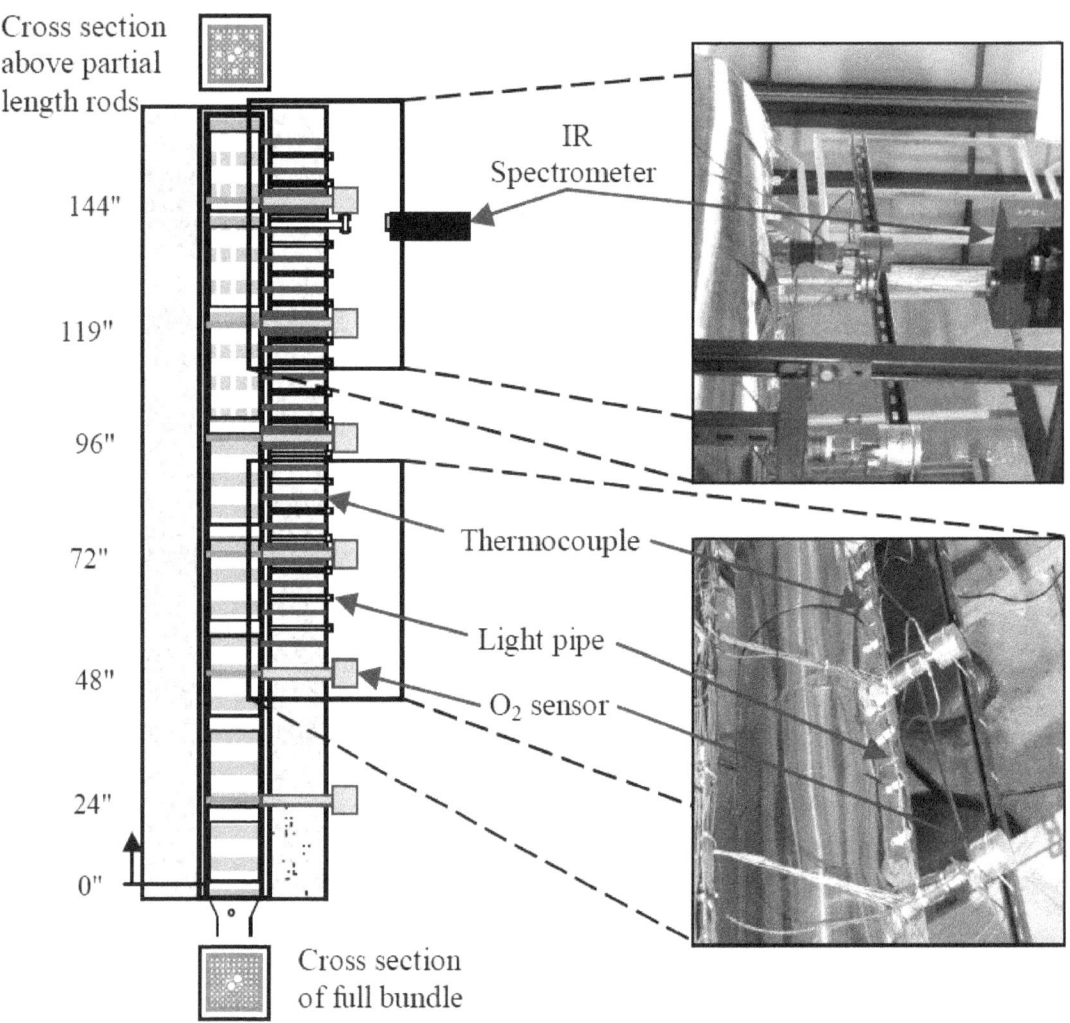

Figure 7.1   Schematic of externally mounted sensors

The Oxyfire high temperature *in situ* oxygen sensors were made by Marathon Sensors Inc., Cincinnati, Ohio. Figure 7.2 schematically shows the sensor flow paths and ceramic felt gasket seal location. The sensing element is at the tip of a hemispherically closed 3/8-in. diameter zirconia tube. The sensor exploits the phenomena that at high temperatures (>600°C), oxygen can readily diffuse through zirconia. Platinum coatings on the inside and outside surfaces of the zirconia tube tip allow the potential of the oxygen diffusion to be measured. The resulting potential follows the Nertz equation,

$$V = 0.0215 \ T \ \ln(C_{ref}/C)$$

where

V is the potential produced in mV,

T is the sensor tip temperature in K,

C_{ref} is the oxygen concentration in the reference gas, and

C is the measured oxygen concentration.

130

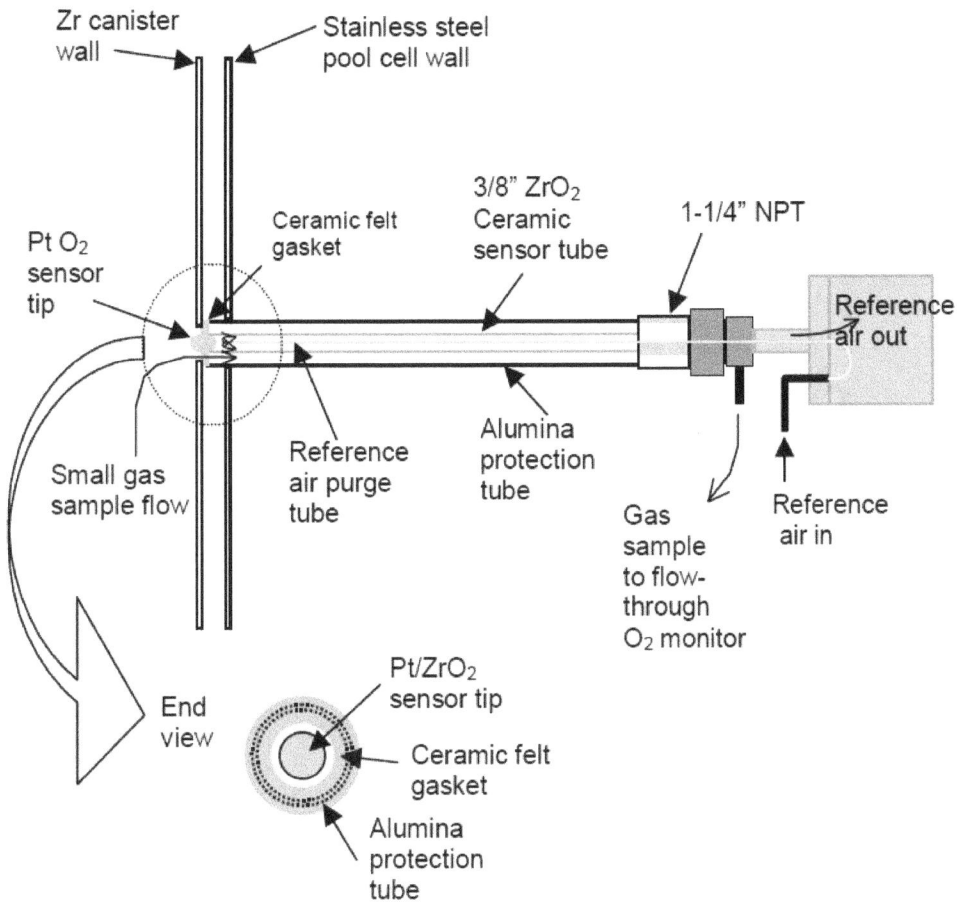

Figure 7.2   *In situ* oxygen sensor seal and flow schematic

An integral component of the sensor is a type B TC located on the outside surface of the zirconia tube adjacent to the edge of the platinum coating. This TC provides a measure of the sensor tip temperature.

As shown in Figure 7.2, the zirconia sensor tube is concentrically located inside an alumina protection tube. The Oxyfire sensors used were custom fabricated so that the protection tube ended 3/16 in. before the end of the sensor tip. This allowed for the sensor tip to protrude approximately 1/16 in. inside the Zircaloy channel box through a 3/4-in. diameter hole. The hole in the channel box was oversized to ensure that the sensor tip did not contact any metallic parts, which would result in shorting of the signal. A ceramic felt gasket was used to seal the gap between the end of the protection tube and the outer Zircaloy channel box wall.

An additional port on the Oxyfire sensor allowed access to the annular gas space between the alumina protection tube and the zirconia sensor tube. This port allowed a small gas sample flow to be drawn from inside the Zircaloy canister and into an external flow-through type oxygen monitor providing a backup oxygen measurement. The oxygen monitors used were Model 65 made by Advanced Micro Instruments Inc., Huntington Beach, California. Note that the oxygen sensor at the 24-in. level was damaged during installation and was abandoned in place. No oxygen measurements or samples were taken at this level.

Figure 7.3 shows a plan view schematic of the Oxyfire sensor installation on the experimental assembly. The main support for the sensor is provided by a 1-1/2-in. stainless steel half-coupling welded to the pool cell. Large Swagelok fittings and 1-1/2-in. stainless steel tubing were used to extend the access port out past the insulation. The final Swagelok ferrule was compressed while exerting an axial force on the sensor. This helped compress the ceramic felt gasket between the end of the alumina protection tube and the Zircaloy canister wall.

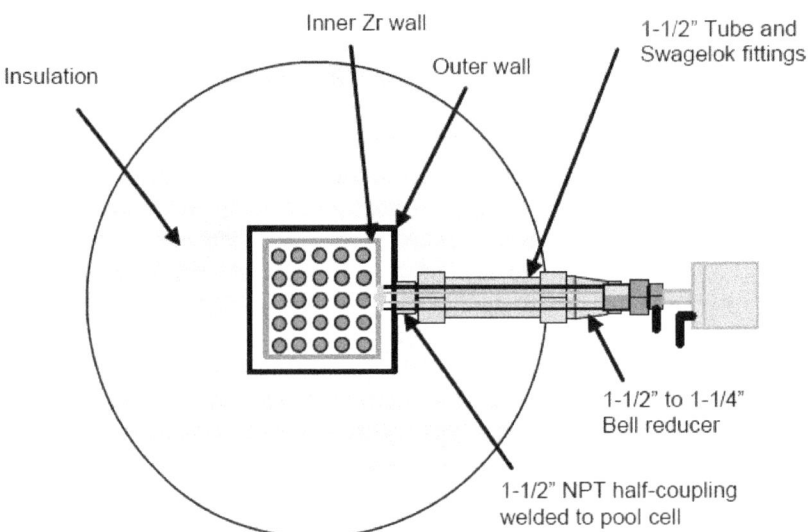

Figure 7.3   Plan view schematic of oxygen sensor installation

Figure 7.4 shows the general layout of all internal K-type TCs. The naming convention follows the dual-alpha character grid as labeled in the figure. The axial location is given after the rod location of the TC. An additional character W is added as a precursor to the TC name for those TCs attached to water rods. The TCs along an axial array also have an extra character A at the end of the name for distinction.

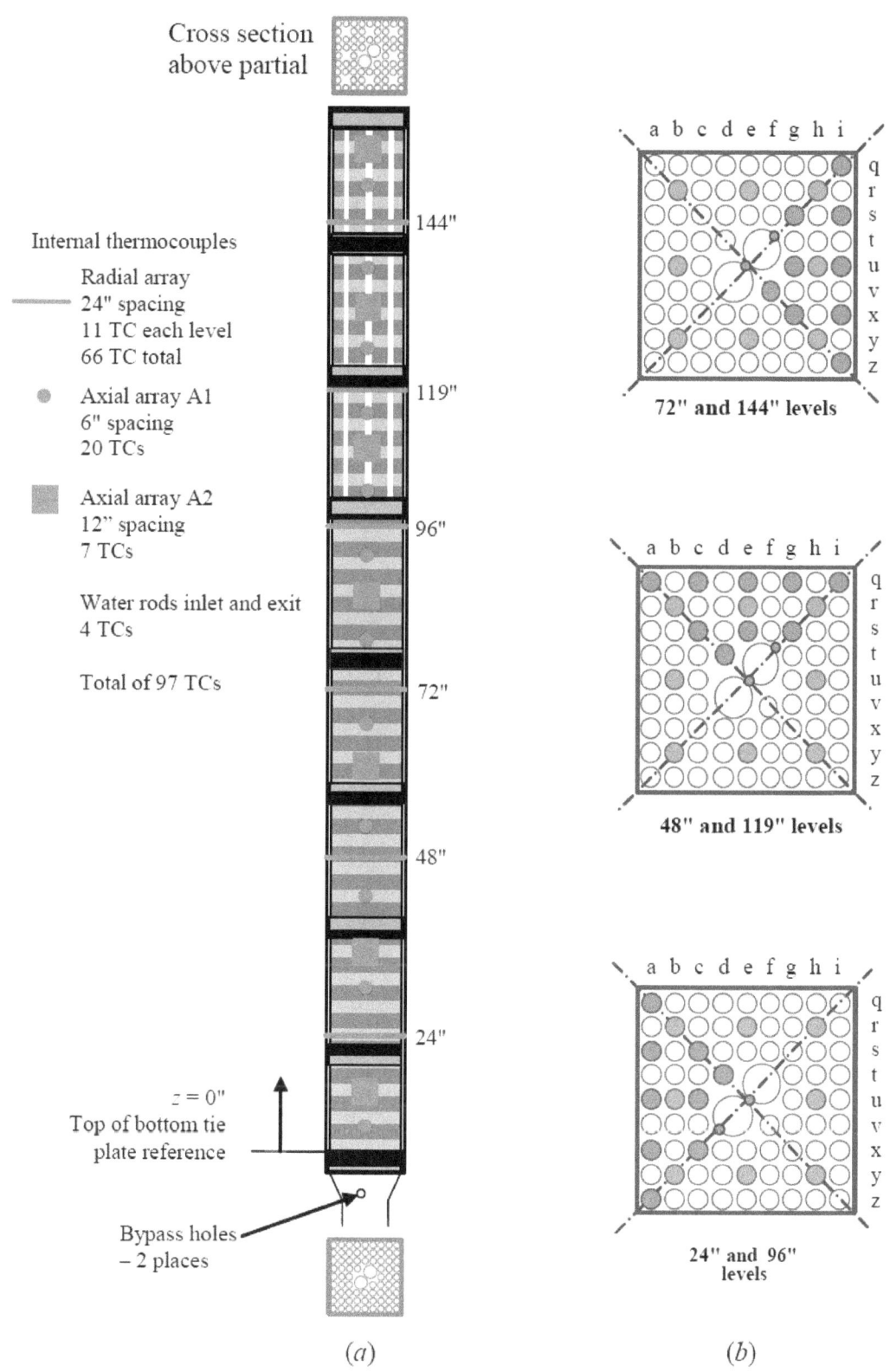

Figure 7.4   Experimental mock fuel assembly showing internal, as-built (a) axial and (b) lateral TC locations

Figure 7.5 gives the locations of the two hot wire anemometers placed inside the inlet of the assembly. These hot wires measure the induced flow rate of air into the apparatus. The unheated flow calibrations of these hot wires are shown in Figure 7.6. Here, the input volumetric flow rate into the assembly is plotted as a function of the measured velocities of the hot wires.

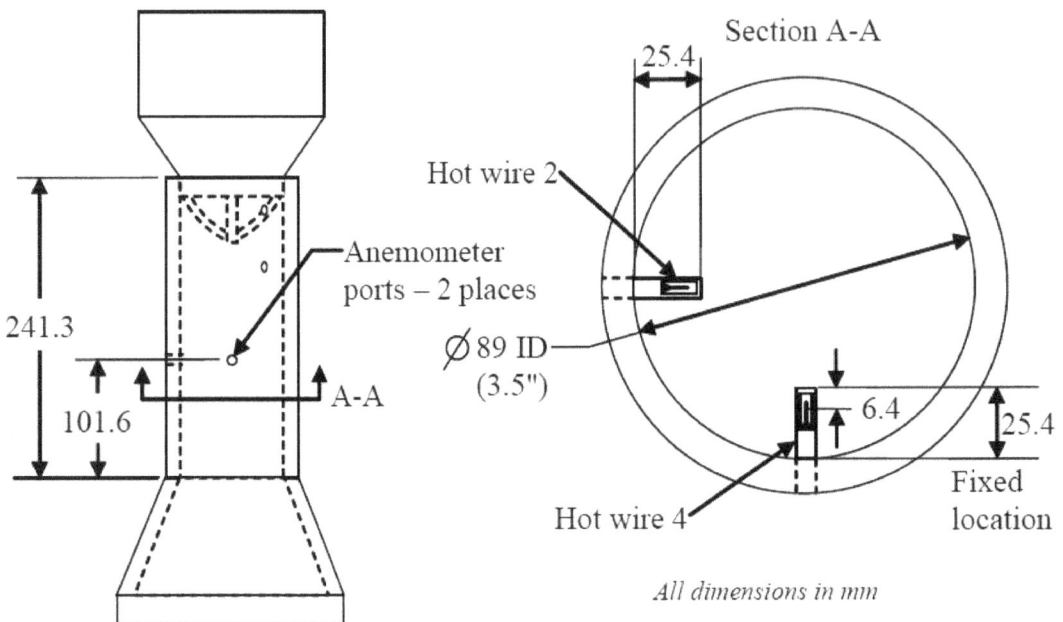

Figure 7.5   Schematic showing the layout of the assembly inlet hot wires 2 and 4

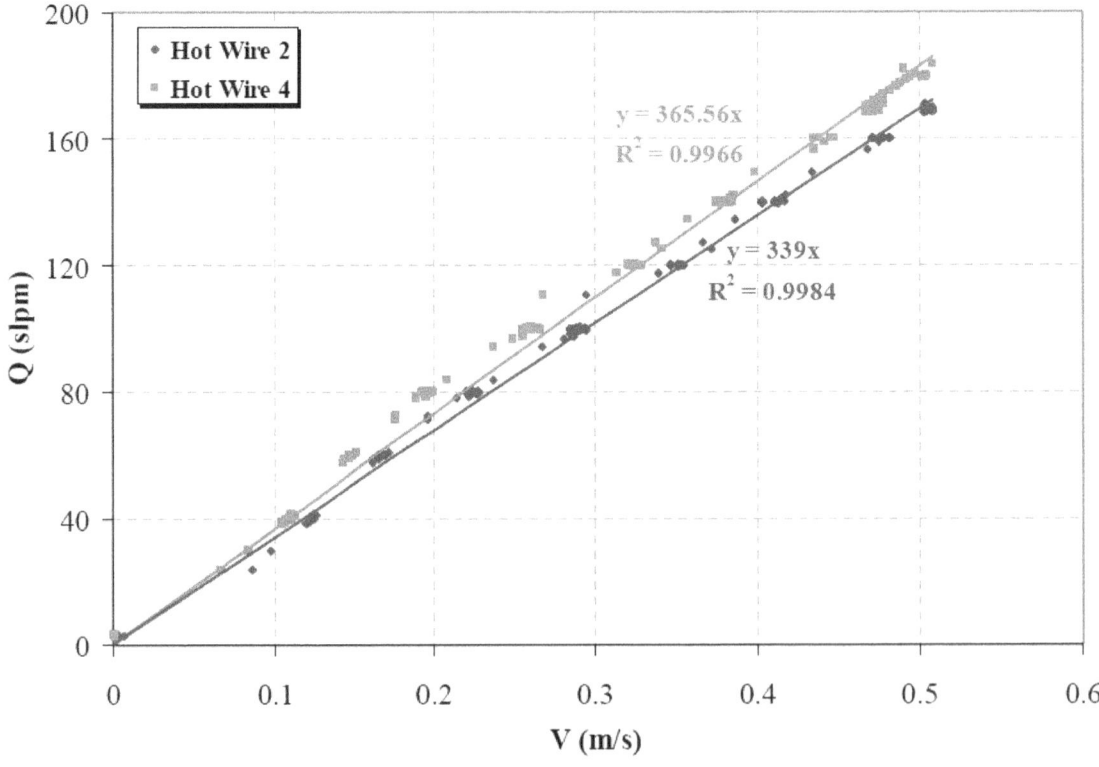

Figure 7.6   Inlet hot wire calibrations for hot wire 2 (diamonds) and hot wire 4 (squares)

7.2 Pre-Ignition Testing and Analysis

7.2.1 Comparison of Thermal Results

Figure 7.7 gives the thermal response of the test assembly as a function of time for the pre-ignition test with a power input of 2250 W. As with the previous Incoloy full-length testing, the assembly reached steady state near the bottom first and continued to heat at the top. The location of the hottest portion of the assembly also moved upward during heat-up.

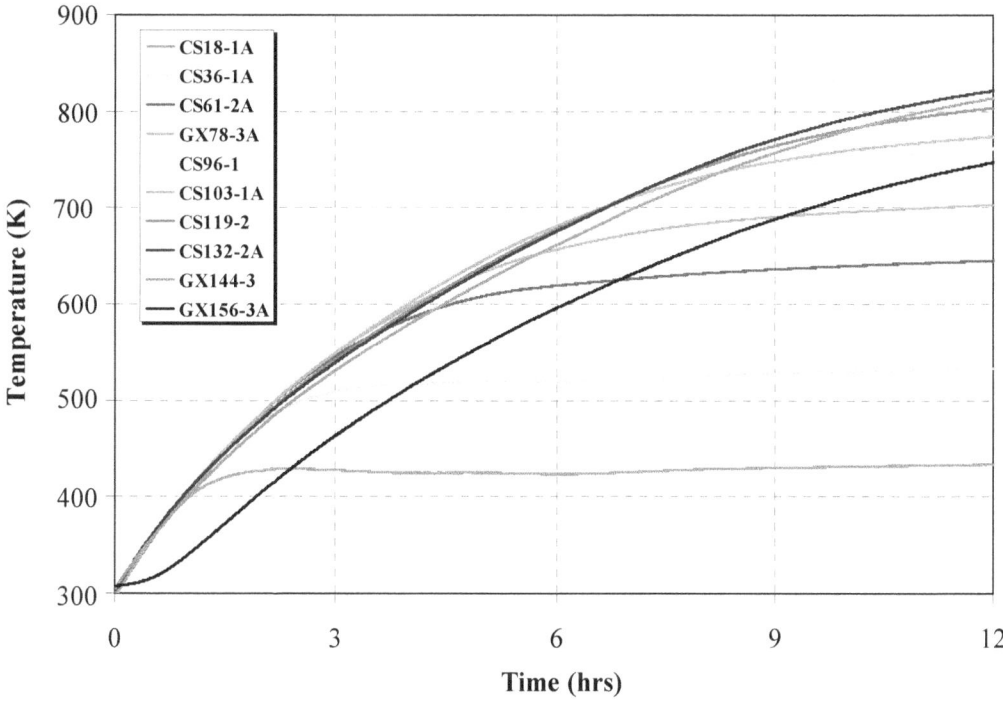

Figure 7.7   **Temperature as a function of time for the pre-ignition test with a power input of 2250 W**

Figure 7.8 shows the initial MELCOR model results for the same test as depicted in Figure 7.7. The MELCOR model was within 28 K at all times during the test. However, the predicted temperatures were consistently higher than those in the experiment were. This disparity was most likely due to the initial MELCOR model under predicting the induced flow rate, which is discussed next.

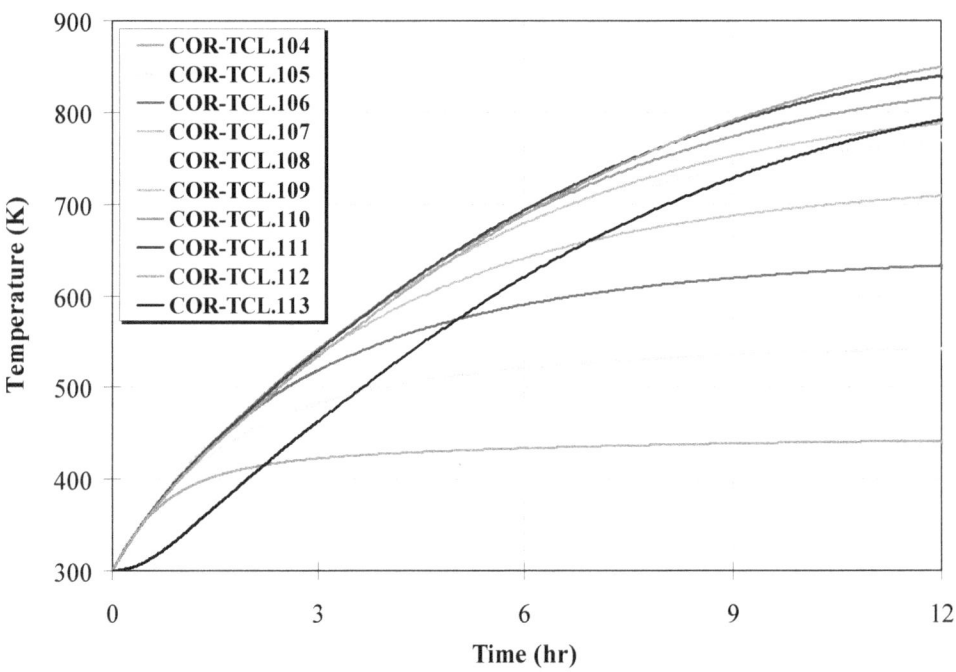

Figure 7.8   **MELCOR predictions for temperature as a function of time for the pre-ignition test with a power input of 2250 W**

7.2.2   Comparison of Air Flow Results

The induced flow rate into the assembly and bypass pressure drop as a function of time are shown in Figure 7.9 for the powered test of 2250 W. Here, the MELCOR model under predicts the flow rate by up to 13 slpm. The flow resistances in the model were based on the previous Incoloy experiments. However, the penetrations of the oxygen sensors into the bundle of the apparatus (Figure 7.2 and Figure 7.3) appeared to provide additional flow paths from the bundle to the annulus, despite being sealed with alumina felt. Attempts were made to resolve this issue on the apparatus and within MELCOR and are described later in this section.

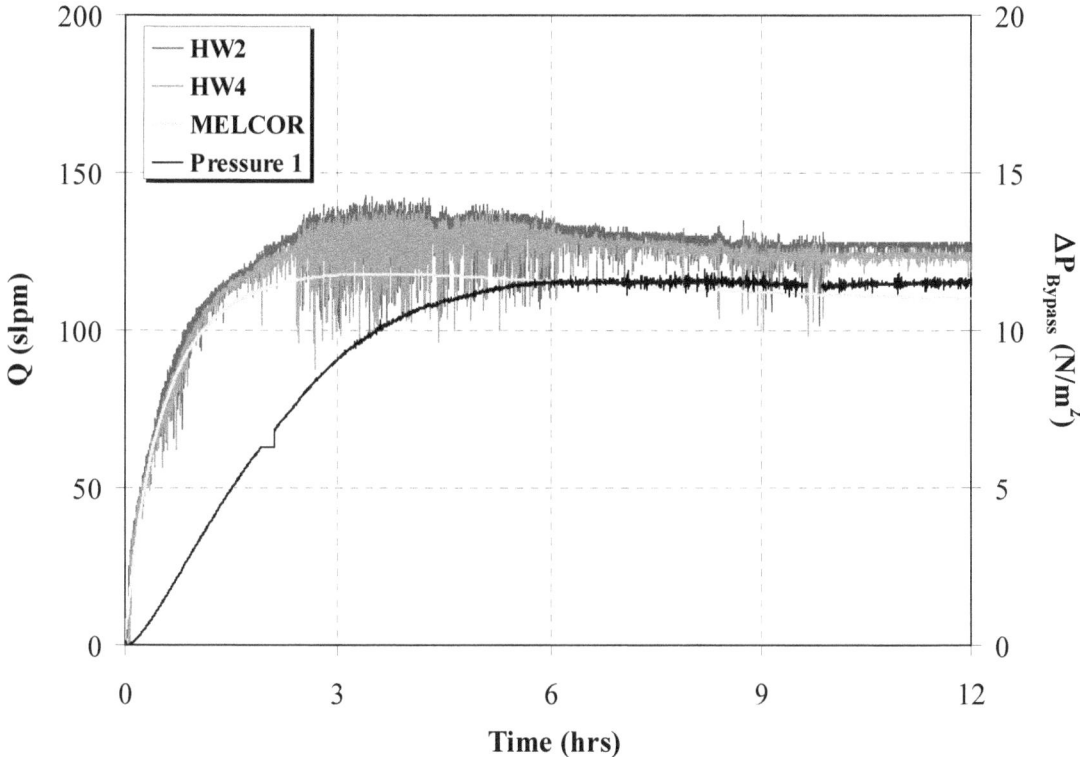

Figure 7.9 Flow rate and pressure drop across the bypass as a function of time for hot wire 2 (HW2) (blue), hot wire 4 (HW4) (red), MELCOR (green), and pressure gage (black)

Initial attempts to model the hydraulic behavior of the Zircaloy apparatus under predicted the inlet flow rates. Prior to the final ignition test, the trends from the pre-ignition testing showed a decreasing assembly flow rate with increasing assembly power. This was also shown in the Incoloy assembly but perhaps exaggerated in the MELCOR calculations. As shown in Figure 7.10, the calculated trends from MELCOR using the best-estimate flow resistance trended with the original Incoloy correlation but were well below the new Zircaloy assembly data.

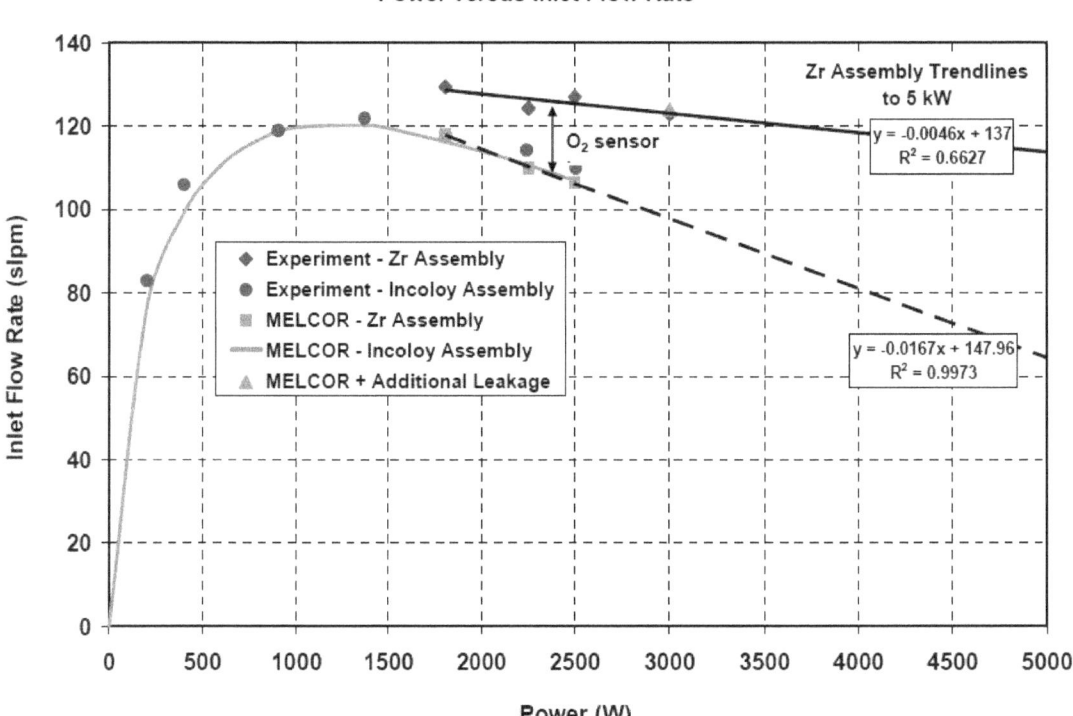

Figure 7.10 **Comparison of pre-ignition MELCOR predictions and experimentally measured assembly inlet flow rates as a function of bundle power**

The higher flow rate was attributed to leakage around the oxygen sensors. During pre-test repairs of the oxygen sensors, pressure drop measurements indicated leakages around the sensor penetrations through the canister. Recall that the oxygen sensors were placed at 2 ft., 4 ft., 6 ft., 8 ft., 10 ft., and 12 ft. Following the 3-kW test, the MELCOR model was adjusted to match the higher measured flow in the 3-kW test (see red triangle at 3000 W in Figure 7.10). Rather than introduce small leaks at each sensor location, the bypass leakage through the nosepiece was increased until the overall leakage matched the 3-kW test. This enhanced (bypass) leakage (to the annulus) model was used for the 5-kW pre-test calculation.

7.3   Ignition Test and Analysis

7.3.1   Thermal Data

The thermal response of the apparatus at different axial locations as a function of time is shown in Figure 7.11. The assembly first reached ignition between the 96- and 103-in. levels at ~7.3 hours. The burn front then propagated downward through the assembly. Noise in the temperature readings above ~1000°C indicates failure of the type K TCs.

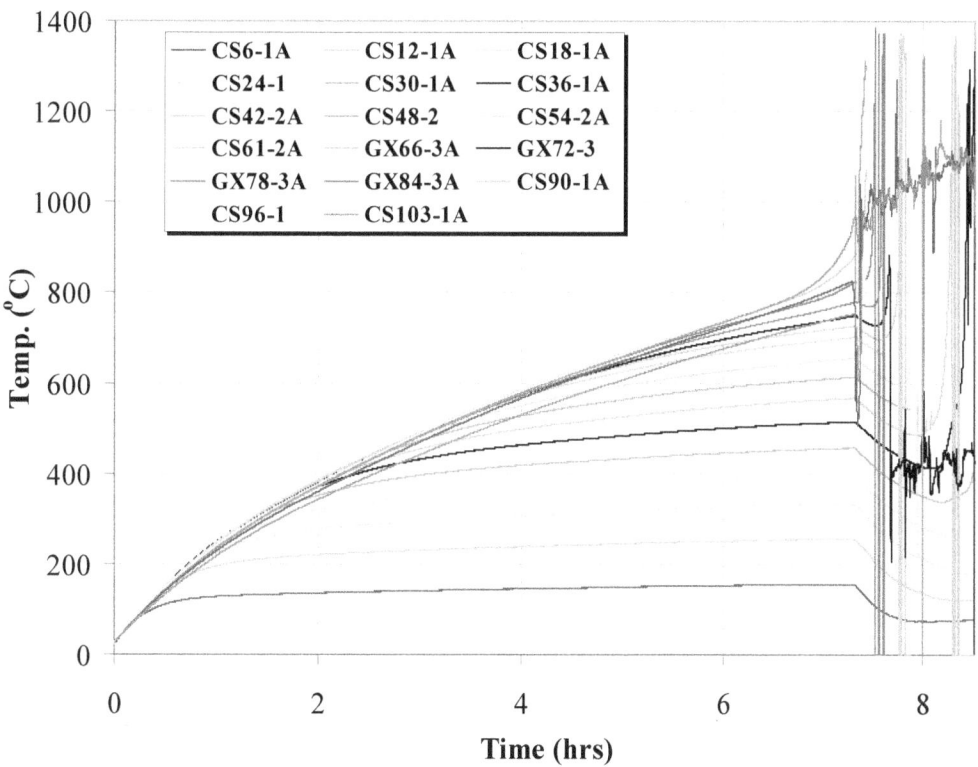

Figure 7.11   **Temperature as a function of time at different axial levels during the ignition test**

7.3.2   Oxygen Data

Two independent oxygen measurements were made at each of the 2-ft. intervals from the 4-ft. elevation to the 12-ft. elevation. These measurements were made by an *in situ* sensor located just inside the bundle region and by a gas sample drawn into a flow-through oxygen monitor. Details are provided previously in the Apparatus section. Two additional gas samples were drawn from the top of the assembly with the oxygen content measured by two flow-through type monitors. One gas sample was taken from the bundle region and the other was taken from the annular region.

Figure 7.12 shows the sensor signal and temperature response for the sensor located at the 144-in. level. The sensor response is not valid for determining the oxygen concentration until the sensor temperature is greater than 600°C. Figure 7.13 compares the oxygen concentration at the 144-in. level determined from the *in situ* sensor signal and temperature measurement, via the Nertz equation, with the corresponding response from the flow-through oxygen monitor. Both measurements closely agree on the timing of the precipitous drop in oxygen at 7.42 hours that is indicative of Zirconium combustion at or below this level. The close agreement on the ignition timing was also evident at each of the lower levels analyzed. After oxygen concentration dropped at the 144-in. level, the *in-situ* sensor consistently read near zero oxygen concentration while the flow-through monitor registered about 2% oxygen. This is attributed to small leaks of air into the gas sample stream drawn into the flow-through analyzer. The 210-mV signal generated by the *in situ* sensor can only be a result of oxygen molecules diffusing through the ceramic matrix of the sensing element. Therefore, false positive measurements are improbable leading the sensor manufacturer to state in the manual, "…basic sensor theory negates the possibility that the sensor will give an incorrect, low reading."

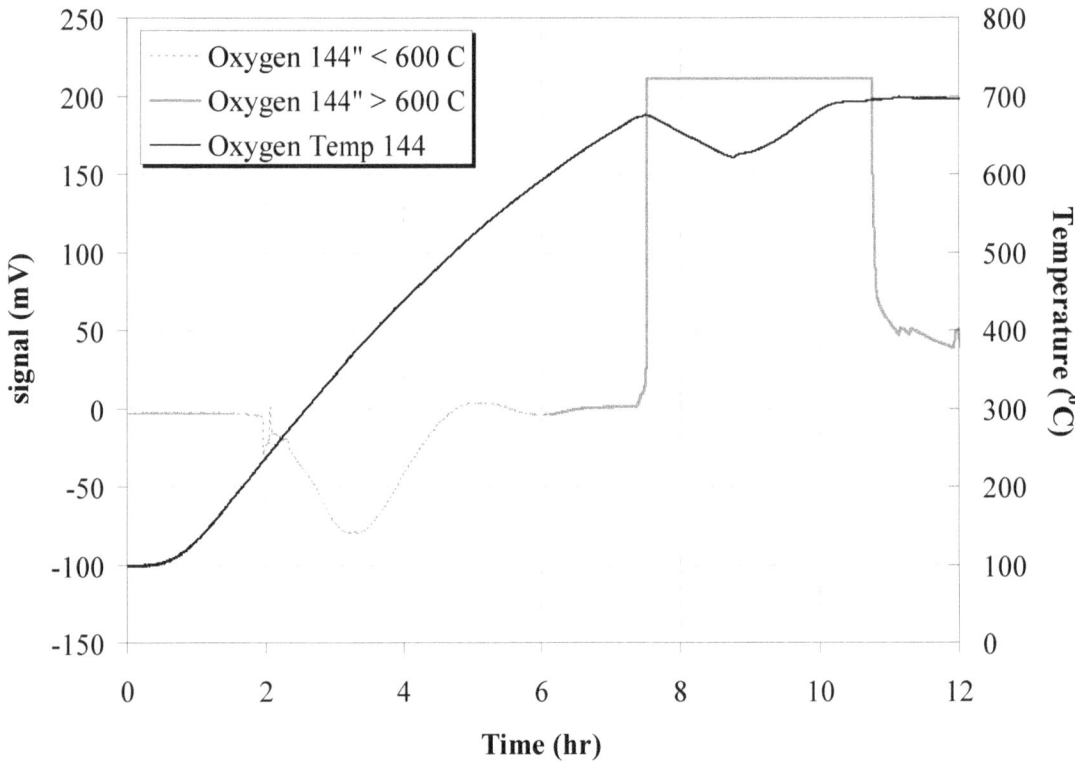

Figure 7.12   **Signal and temperature response for the oxygen sensor located at** z **= 144 in.**

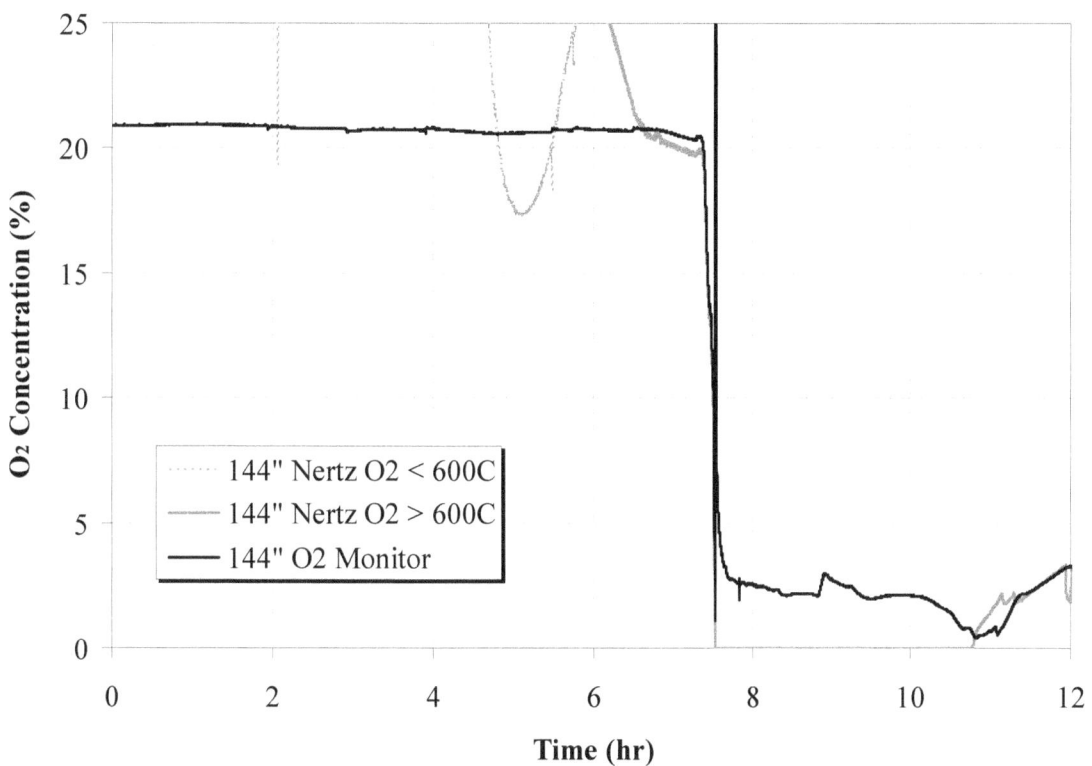

Figure 7.13 **Comparison of the measured oxygen concentrations measured by the *in situ* oxygen sensor and the flow-through oxygen monitor**

Figure 7.14 shows the response for all of the flow-through oxygen monitors. Table 7.1 gives the time at which an oxygen concentration of 15% was measured by each of the oxygen sensors and monitors. The oxygen measurement at the top bundle indicates that ignition occurred at 7.28 hours.

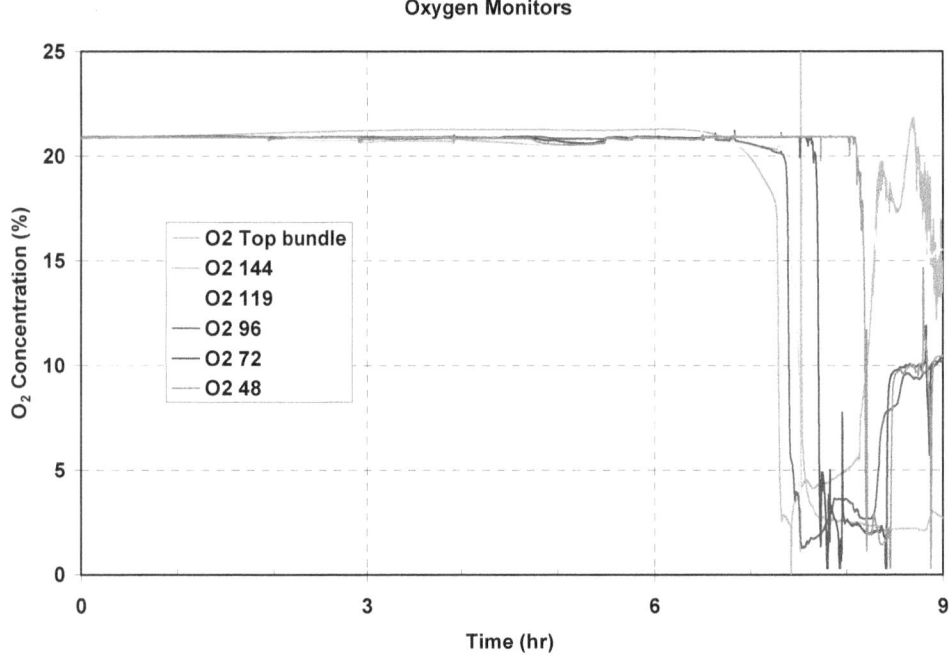

Figure 7.14   Measured oxygen concentration from the monitor devices

Table 7.1   Elapsed experimental time to reach 15% oxygen concentration for the oxygen sensors and monitors

Location	O_2 Sensor time to 15% O_2 (hr)	O_2 Monitor time to 15% O_2 (hr)	Average time to 15% O_2 (hr)	Apparent burn rate (in/hr)
Top bundle	-	7.277	7.277	-
Top annulus	-	7.419	7.419	-
144 in	7.418	7.425	7.422	-
119	7.436	7.427	7.432	-
96	7.397	7.387	7.392	80.11
72	7.684	7.699	7.692	51.15
48	8.143	8.178	8.161	-
-	-	-	average	65.63

7.3.3   Comparison of Air Flow Results

The results of pre-test calculations with and without the added oxygen sensor leakage are shown in Figure 7.15. The green line shows the best-estimate total induced flow rate based on the previous full-length Incoloy experiments and SNL standards laboratory flow/pressure drop characterization of the nosepiece bypass holes. The black line shows the pre-test calculation at 5 kW with the sensor leakage added to the bypass holes, which did not extrapolate well from 3 kW to 5 kW. Attempts to repair the oxygen sensor seals were made between the pre-ignition testing and the ignition test and may have increased the leakage.

Both MELCOR models show a sharp downturn in flow rate at the time of ignition. This decrease is due to the rapid heating and expansion of the air in the assembly, producing a choking effect. The test data indicate an increase in the inlet flow rate immediately following ignition. This behavior suggests that the leakage into the annulus increased at this time. As the stainless steel pool cell melted away at the point of ignition, the pool rack and insulation slumped downward. This event pulled the oxygen sensors down as well, possibly corrupting the seal at the sensor tip. Attempts to refine the MELCOR model to more accurately portray the sensor leakages are discussed next.

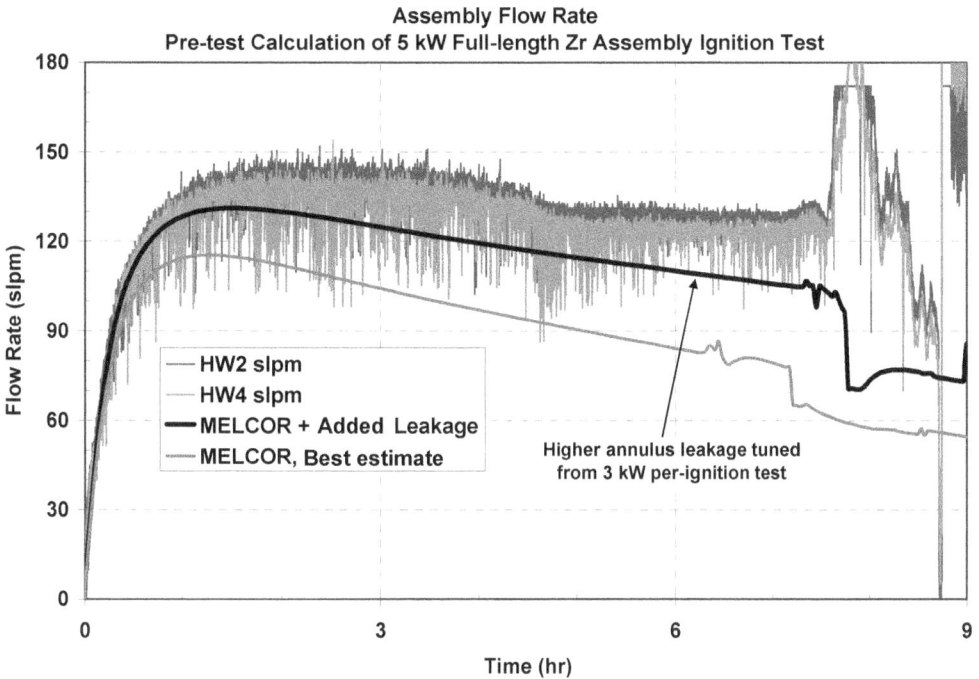

Figure 7.15   **Comparison of MELCOR pre-test prediction with and without sensor leakage versus the measured assembly inlet flow rate**

Individual leakage paths were introduced for the post-test calculation for each oxygen sensor at the 2-ft., 4-ft., 6-ft., 8-ft., 10-ft., and 12-ft. elevations. A leakage of 1.5×10^{-4} m^2/sensor was needed to match the measured flow data (Figure 7.16). This represents a ~0.070-in. annular gap around each sensor.

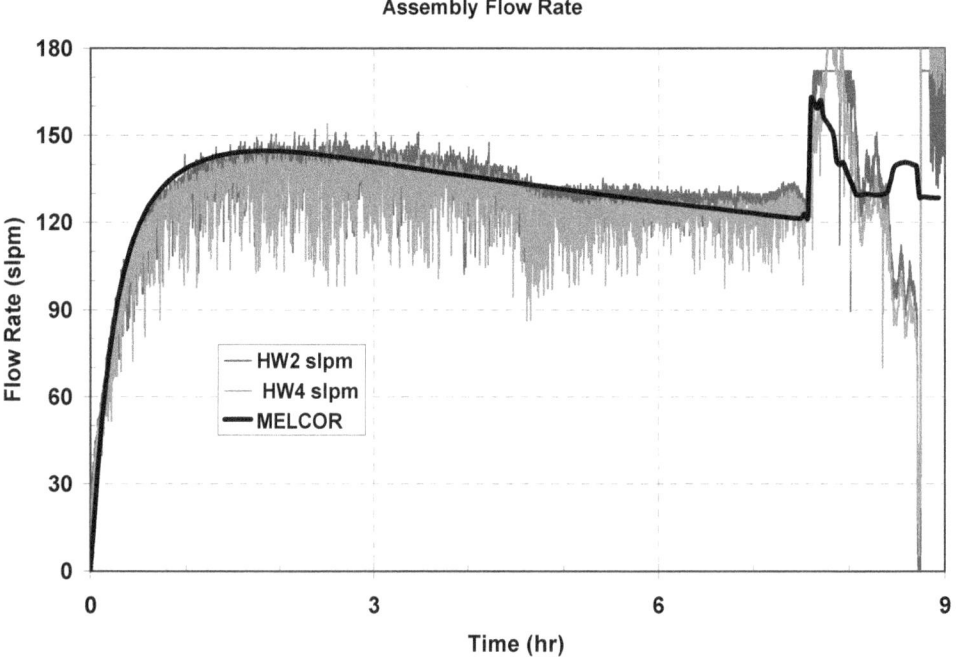

Figure 7.16   **Comparison of MELCOR post-test prediction with additional sensor leakage versus the measured assembly inlet flow rate**

The measured assembly inlet flow rate achieved a steady value at ~4.7 hours. The peak measured temperature at 4.7 hours was 900 K. The temperature steadily rose thereafter until 6.9 hours when the temperature rise increased slightly near the ignition point and then dramatically at ~7.3 hours as the breakaway phenomena occurred. Between 4.7 hours to ignition, oxidation became increasingly more important. However, the oxygen sensors and the MELCOR calculations did not measure a significant reduction in the oxygen concentration until close to ignition. Hence, the steady flow rate does not seem to be related to any thermal or chemical phenomena. This response remains unresolved, but the increased leakage MELCOR result tracks the overall magnitude of the flow reasonably during this period but not the flattening trend (Figure 7.16).

A sudden increase in flow was measured at 7.59 hours. The most probable cause was the catastrophic failure of the oxygen sensor seal to the canister at the time of ignition. To address this issue, the MELCOR leakage flow area was increased by a factor of 10 at all sensor locations at 7.56 hours (results shown in Figure 7.17) to represent this assertion. Consequently, instead of decrease in flow rate at the start of ignition (see pre-test result in Figure 7.15), the flow increased to almost the value measured in the experiment. Examination of the calculated flows showed the additional leakage flow came primarily from the 2-ft. elevation (Figure 7.17). There were smaller outflows from the assembly to the annulus at the 4-ft., 6-ft., and 12-ft. elevations. Near the ignition front (8 ft. and 10 ft.), the leakage was inward, which fed the oxidation reaction. Hence, a larger opening(s) of the oxygen sensor(s) could have caused the sharp increase in flow rate. However, the MELCOR model suggests that the most dramatic impact is near the bottom of the assembly (especially, the 2-ft elevation). At 7.59 hours, the measured ignition front was between the 96-in. and 72-in. elevations, which are significantly above the bottom of the assembly. Consequently, a few residual questions remain:

144

- Through numerous sensitivity calculations, replication of the magnitude or duration of the high flow period was not duplicated. The flow rate increased to >180 slpm from 7.7 to 8 hours. This increase could be the result of many factors, but apparently, there was little to no blockage.

- Based on the MELCOR calculations, the most significant impact to the inlet flow rate comes from the lowest sensor. The failed oxygen sensor seal was most likely near the ignition front (72 in. to 96 in.). However, the postulated loss of oxygen sensor seals seems reasonable given the observed slumping of the pool cell and insulation as discussed earlier in this section.

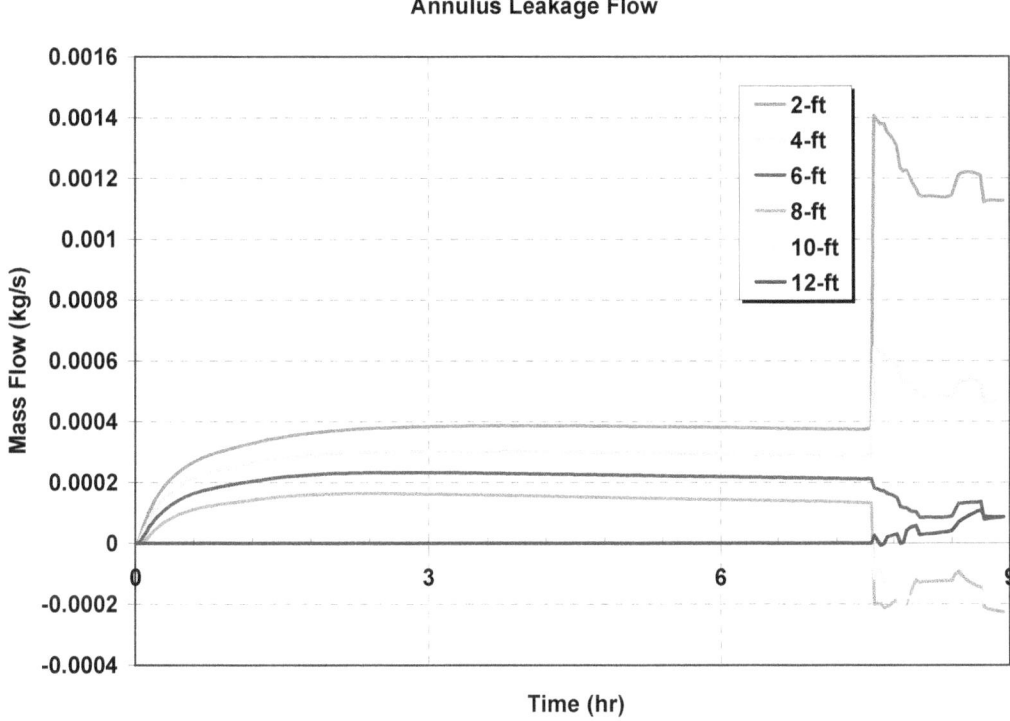

Figure 7.17   Calculated leakage flow from the post-test enhanced leakage model

7.3.4   Comparison of Thermal Results

The MELCOR model predicted the same burn rate as in the experiment. The measured ignition front started at ~96 in. at 7.3 hours and moved to 42 in. by 8.3 hours, or 54-in/hr. The MELCOR ignition front started calculation in the 117-in. to 131-in. cells at 7.5 hours and moved to the 77-in. to 96-in. cells by 8.5 hours, or 54-in/hr. Here the entire length of the nodes is considered to be burning; thus, the MELCOR model burned between the 131-in. and 77-in. levels in the stated timeframe.

- Here much smaller nodes in MELCOR would have better tracked the ignition front. Furthermore, more energy is required to ignite a 19-in. node versus a continuously downward moving front.

The measured ignition front started at ~96 in. at 7.3 hours versus 117 in. to 131 in. at 7.5 hours in the MELCOR calculation. Possible reasons for this difference are presented next.

- The initial breakaway oxidation lifetime value cannot be input in MELCOR. When the assembly flow rate was increased, the ignition timing increased from 7.3 to 7.5 hours. There was a small oxidation layer (i.e., estimated maximum of 15 μm at the top of the assembly). Consequently, some pre-oxidation effects could have effected the location of ignition.

- As shown in Figure 7.18 and Figure 7.19, the calculated and measured pre-ignition temperature responses were relatively close across a broad axial height. Hence, subtle local factors may have influenced the exact ignition location. Again, the MELCOR breakaway correlation seems to accurately predict the physics governing the experiment. In the final comparison calculations, an offset was added to the lifetime function to move ignition from 7.5 hours to 7.3 hours.

- Sensitivity calculations were performed. When the initial oxide layer above the partial rods was increased to 24 μm, the difference between the temperatures above and below the top of the partial rods was lessened (i.e., <20 K at the time of ignition). However, the 117-in. to 131-in. nodes still transitioned to breakaway first.

- In the MELCOR model, the magnesium oxide and cladding in the unheated plenum at the top of the partial rods is accurately modeled. However, it is not possible to only power the full-length rods and not the partial rods in this region. For a sensitivity calculation, the power distribution was changed to also heat the plenums above the partial rods. The results did not change the location of ignition.

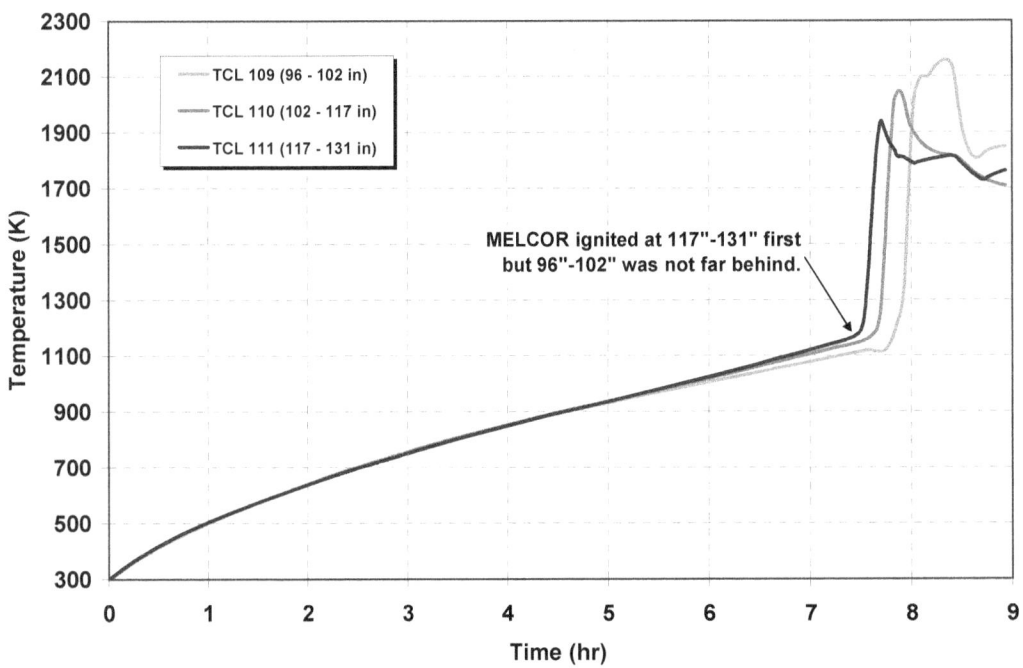

Figure 7.18   Calculated temperature response from the posttest enhanced leakage model

Median Axial Temperature Data from Full-Length Zr Assembly Test

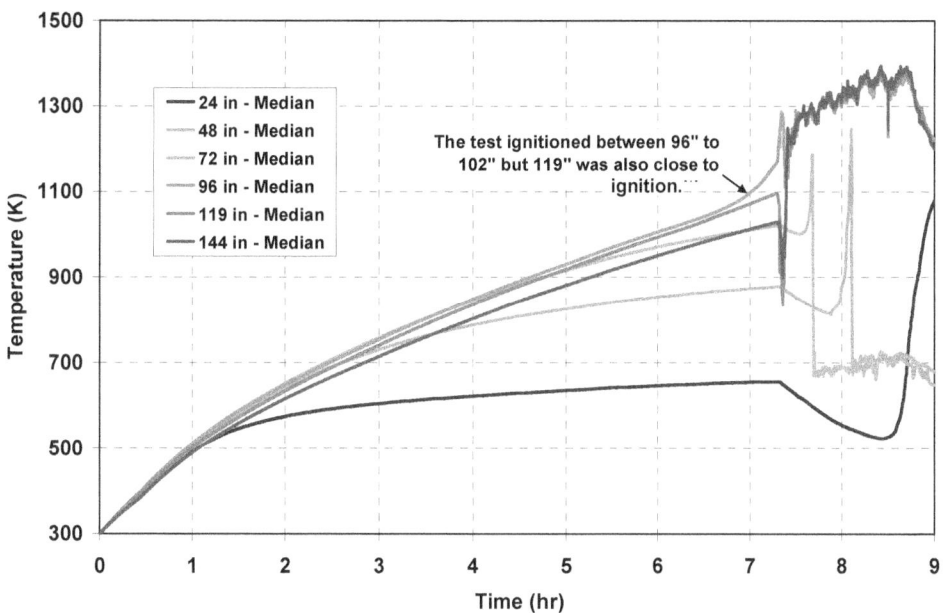

Figure 7.19   Measured median temperature response

Due to the complex nature of breakaway and perhaps stochastic or other factors, the location of ignition is not accurately predicted by MELCOR. The role of nitrogen appears critical to the onset of breakaway oxidation in which nitrogen actually enhances the magnitude of the oxidation rate versus pure oxygen. SEM of frozen cladding samples just prior to the onset of breakaway shows a very erratic and complex grain matrix that has been disrupted by the nitrogen.

Finally, the assembly heat balance in MELCOR for this test/configuration leads to higher temperatures than in the test near the top portion of the apparatus. Hence, ignition in the model at 117 to 131 in. was inherent of the range of sensitivities explored. However, in real SFP assemblies there will be an axial power profile that will skew the power toward the center of the assembly. The model cannot capture the nuances of the test apparatus such as the heat balance near the top of the partial rods, scratches in rods near the grid spacer, or some localized power variances that led to ignition between 96 in. and 102 in.

7.4   MELCOR Post-test Comparison Summary

Changes from the pre-test prediction are summarized by the following statements:

- The leakage rate was increased between the canister and the annulus to the equivalent of 1.5×10^{-4} m^2/sensor.

- A factor of $10\times$ further increase in leakage rates was assumed at all sensor locations at 7.59 hours.

- A small adjustment to the breakaway lifetime correlation was applied to move ignition from 7.5 hours to 7.3 hours. The inlet flow rate as a function of time is plotted in Figure 7.20 for both the experiment and the final post-test MELCOR model, which reflects the changes outlined earlier.

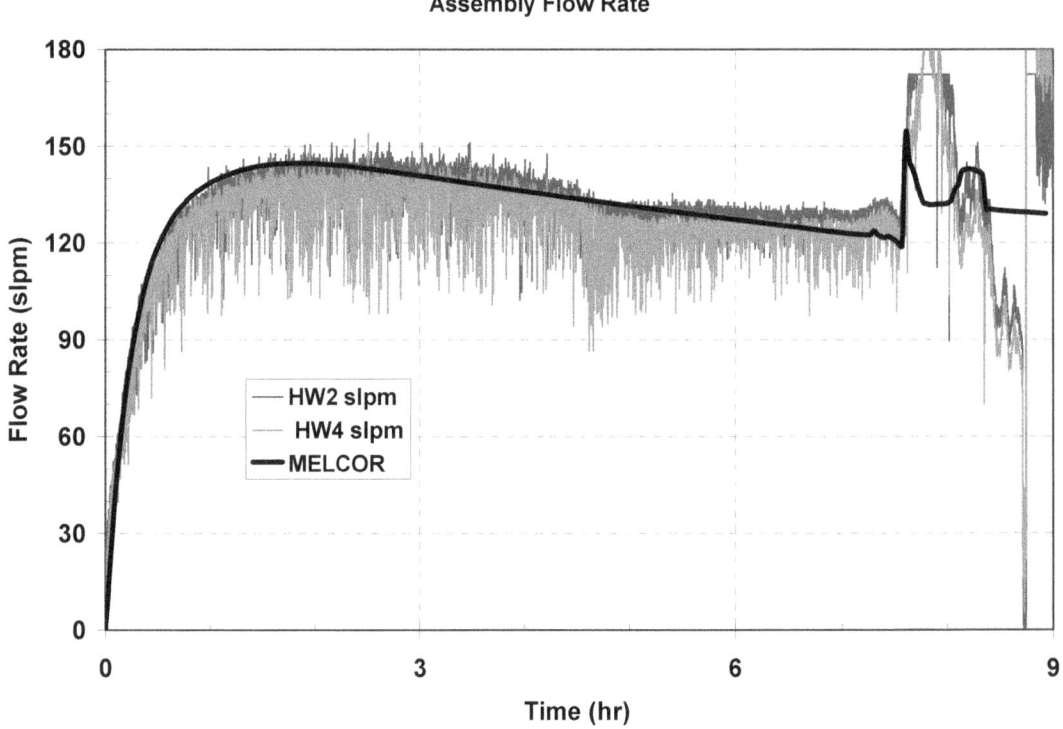

Figure 7.20 **Flow rate as a function of time for the experiment hot wires and the final post-test MELCOR model**

The following are some various additional observations:

- Calculated power failure occurred at PCT = 1445 K and 7.31 hours (almost identical to measured failure at 1444 K and 7.31 hours). The maximum temperature along the assembly is shown in Figure 7.21 for the experiment and the final post-test MELCOR model.

- Interestingly, the heaters failed at 7.31 hours at a PCT of 1444 K (almost the identical point as Heater Design Test 2 and the best-estimate value for the full-length Zircaloy ignition test).

- Calculations were terminated at 9 hours. In the experiment, the argon purge was started at 8.7 hours. The assembly inlet flow damper was closed a few minutes before.

- MELCOR was primarily burning downward at cell 108, but cell 111 (117 in. to 131 in.) was also slowly burning at the end of the calculation (Figure 7.22).

- As discussed earlier, the ignition front started lower in the experiment. Hence, no ignition occurred in the calculation at the lower levels (see Figure 7.23 and Figure 7.24). The 58-in. to 77-in. cells (107) were rapidly approaching ignition at the end of the calculation (Figure 7.25).

- The pre-ignition heat-up rates at the 48-in. to 119-in. elevations were in good agreement with the data (see Figure 7.24 through Figure 7.27). The 24-in. data is also well bounded by the two MELCOR cells around this location.

- The pre-ignition temperatures at 144 in. were below the calculation (Figure 7.28). The heat loss appears to have been under estimated in the calculation.

Maximum Cladding Temperature

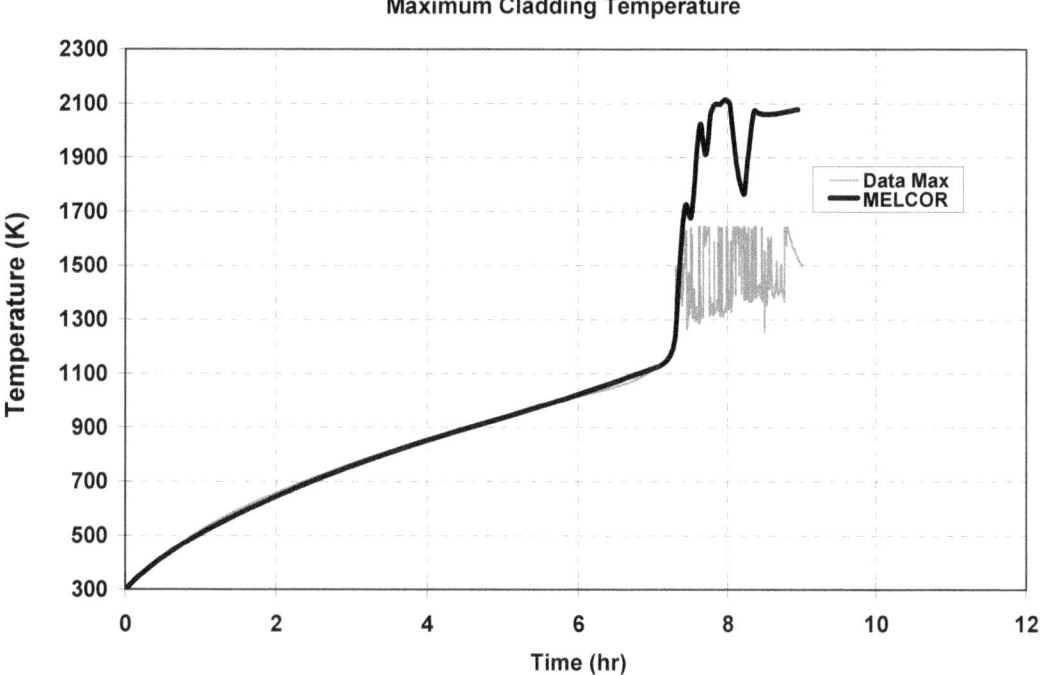

Figure 7.21   PCT for both the ignition experiment and the final post-test MELCOR model

Breakaway Lifetime Function

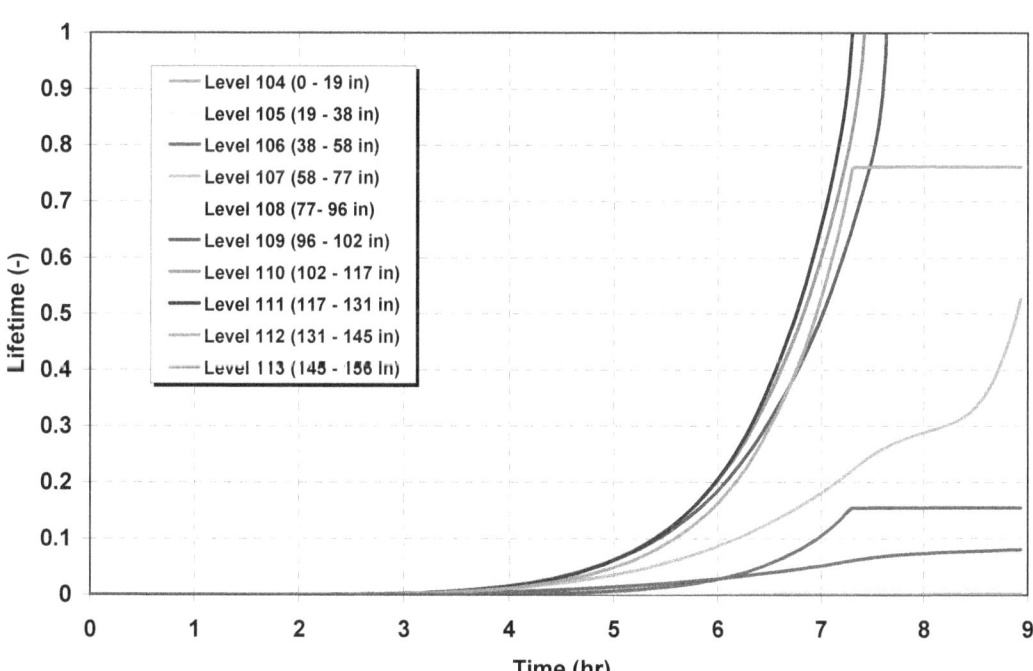

Figure 7.22   Breakaway lifetime function as a variable of time in the MELCOR ignition model

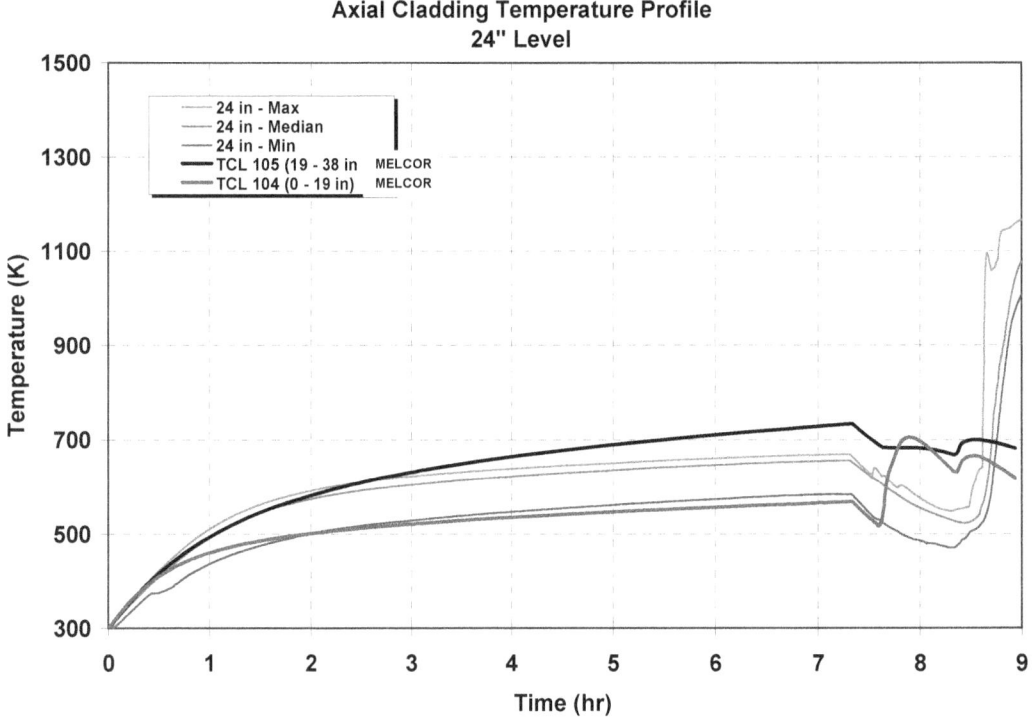

Figure 7.23   Cladding temperatures at z = 24 in. for both experiment and MELCOR

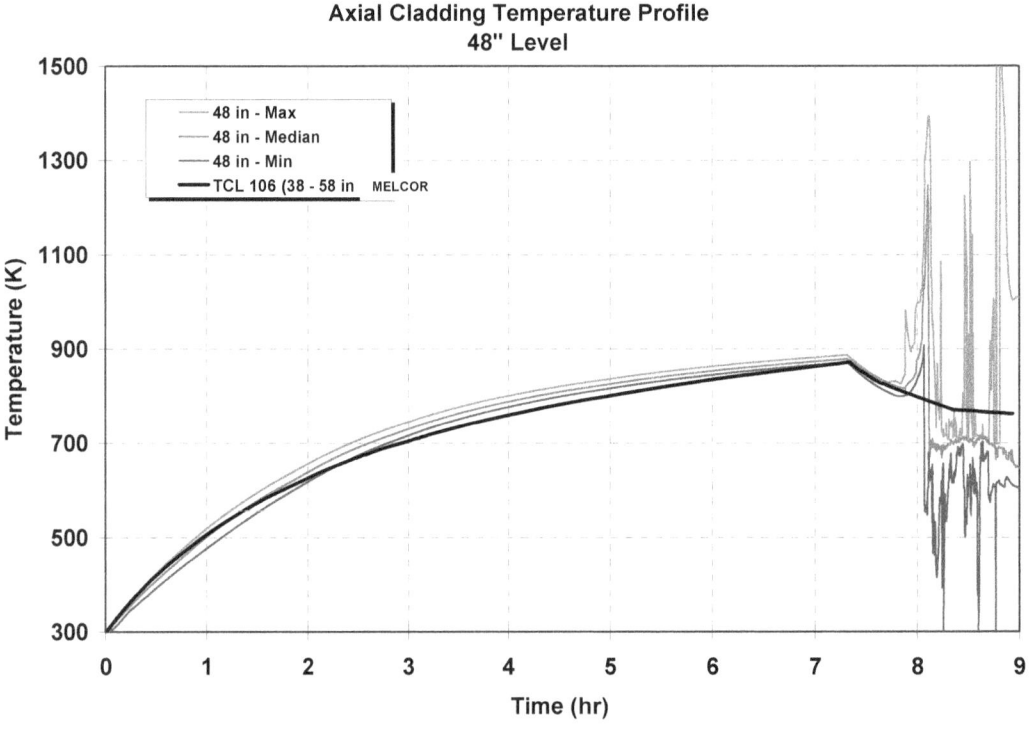

Figure 7.24   Cladding temperatures at z = 48 in. for both experiment and MELCOR

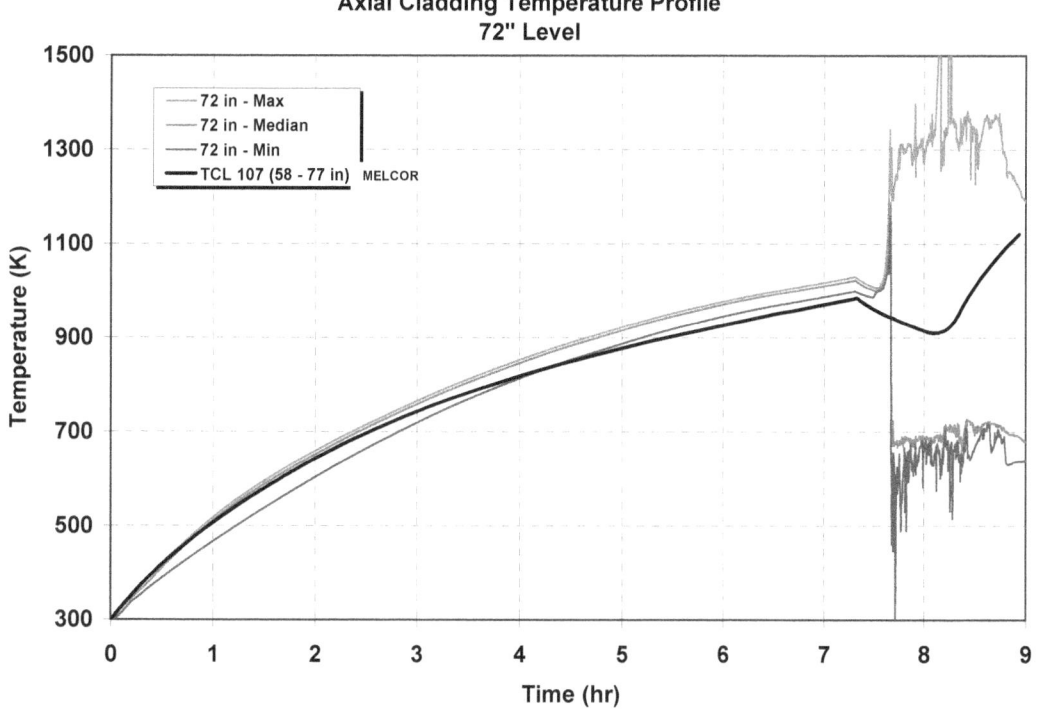

Figure 7.25   Cladding temperatures at $z = 72$ in. for both experiment and MELCOR

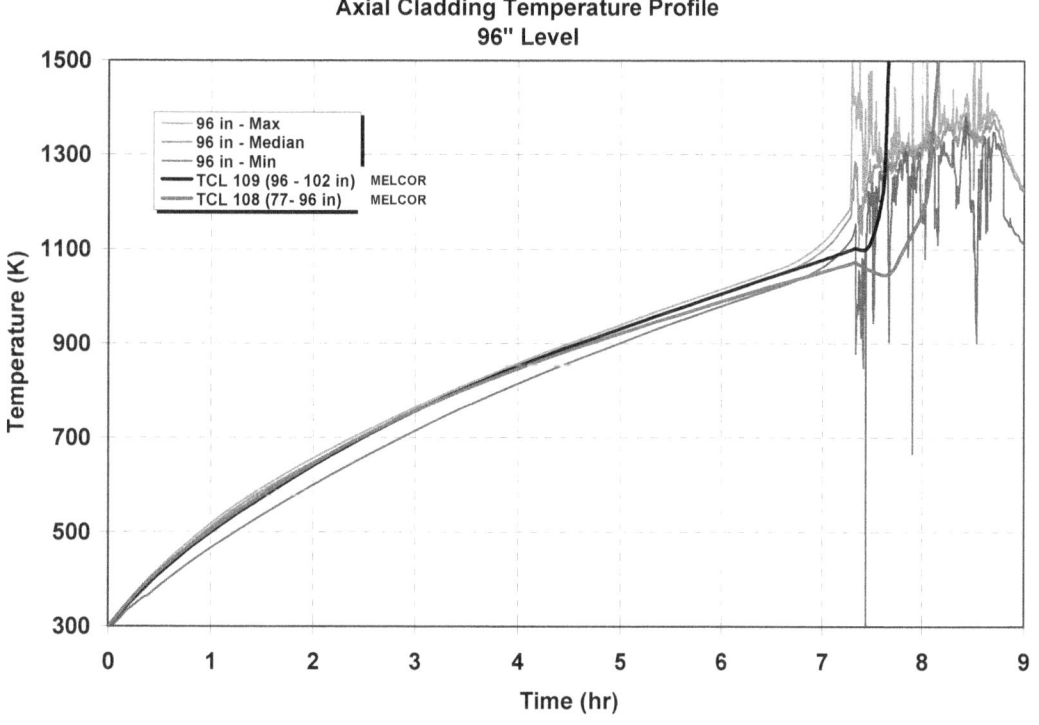

Figure 7.26   Cladding temperatures at $z = 96$ in. for both experiment and MELCOR

151

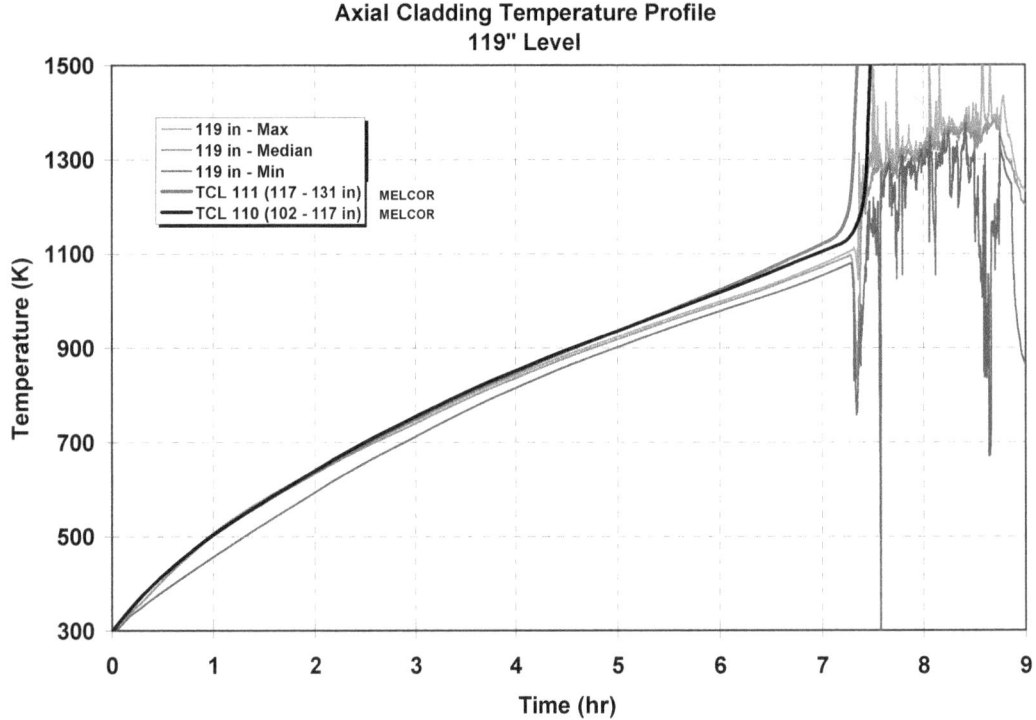

Figure 7.27   Cladding temperatures at z = 119 in. for both experiment and MELCOR

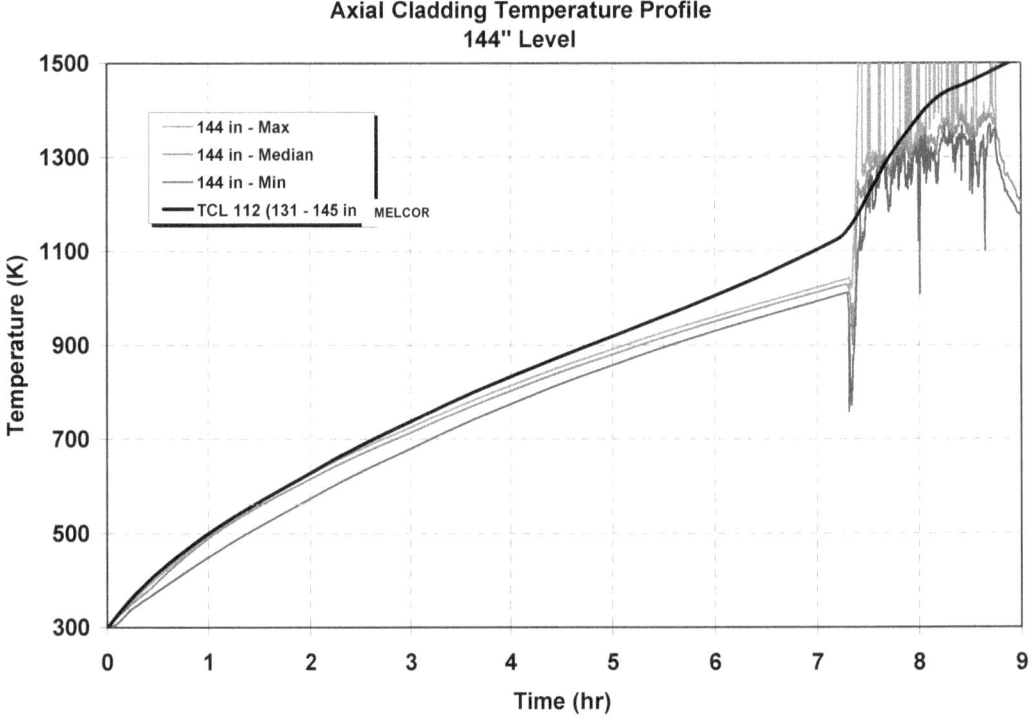

Figure 7.28   Cladding temperatures at z = 144 in. for both experiment and MELCOR

7.5   Discussion

The initial pre-test MELCOR model predicted the PCT of the assembly to within 40 K at all times. The model also predicted the time of ignition to within 5 minutes. However, the model under predicted the assembly inlet flow rate by up to 20 slpm. This difference was likely due to leakage around the ceramic felt used to seal the oxygen sensors to the channel can.

The post-test MELCOR model accounted for the leakages as flow areas of 1.5×10^{-4} m^2/sensor (or ~0.070-in. annular gap around the seal). These leakage areas were increased to 1.5×10^{-3} m^2/sensor at ignition in order to duplicate the probable catastrophic failure of the sensor seals. In addition, a small adjustment was made to the breakaway lifetime correlation to move the ignition time to the experimentally observed 7.3 hours.

Finally, Figure 7.29 shows photos of the assembly after ignition. The top left shows the top of the assembly. The pool cell (dark gray) has slumped downward from its initial height, which matched that of the channel can (whitish gray). The bottom left image is of the top 4 ft. of the assembly as viewed from below. The damage to the pool cell is clearly evident. The two photos on the right show the internal damage to the rod bundle in the vicinity of the initial ignition.

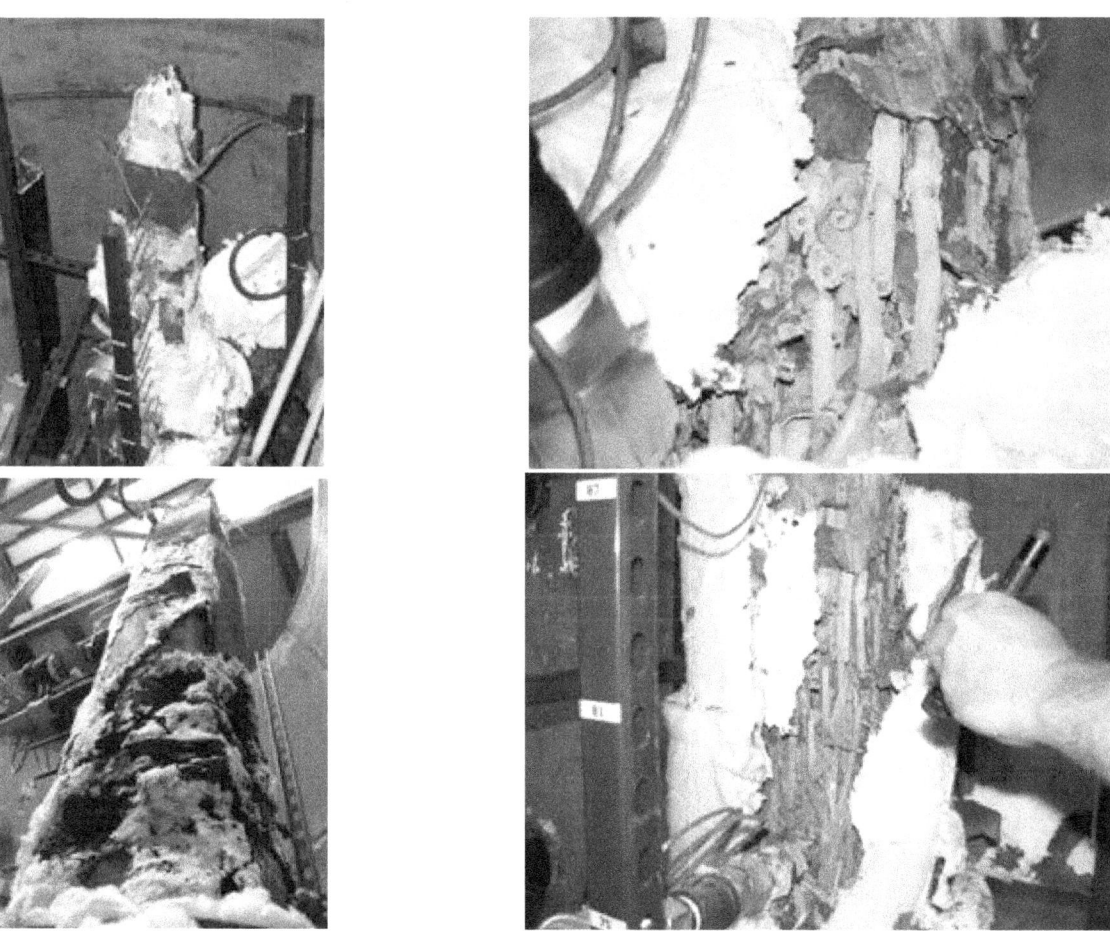

Figure 7.29   Postmortem of the assembly after ignition

8 1×4 ZIRCALOY PRE-IGNITION AND IGNITION RESULTS OF THE SFP EXPERIMENT

This report summarizes the findings of the SFP heated radial experiments conducted between August 1 and August 31, 2006. The stated purpose of these investigations was to validate the MELCOR severe accident analysis code. The apparatus consisted of five shortened assemblies, recreating 1.22-m (4-ft.) sections of a SFP in a 1×4 arrangement. For this testing, only the center assembly was electrically heated up to 4.87 kW. Air flow rates and inlet temperatures were controlled at the inlets of the center and peripheral assemblies separately.

8.1 Assembly Design

The 1×4 Zircaloy assembly was designed to study the potential propagation of a cladding fire from a centrally located hot bundle to its cooler neighbors. The assembly segments were located in an analogously shortened but otherwise highly prototypic 3×3 pool rack fabricated by Holtec Inc. This piece of hardware included prototypic Boral neutron absorption plates between cells. The Boral plates consisted of boron impregnated aluminum. Air flow and temperature were prescribed using metered flow rates into several air heaters. The entire assembly height was approximately 52 in. with a heated length of 48 in. The apparatus simulated the top 48 in. of the fully-populated bundle in a full-length assembly . The simulation of this bundle region was chosen based on the fact the full length Zircaloy experiment ignited at the top of the fully populated bundle region and burned downward.

8.1.1 General Construction

Figure 8.1 shows different views of the experimental apparatus. An exploded schematic is pictured on the left with details of the center assembly construction highlighted. The actual center assembly is shown under construction in the upper right photo. A completed peripheral assembly stands behind the center assembly. The lower photo depicts the flow arrangement underneath the apparatus. Copper tubing supplied the air ovens (MHI Inc., Model LTA750), which then delivered heated air to the apparatus.

Details of the flow arrangements are provided in Figure 8.2. Two flow paths are formed in each assembly, the bundle and the annular region formed by the pool rack and the channel box. Due to the higher flow rate in the center assembly bundle, two air ovens were employed to provide the heated air. Flow rates were controlled by a series of mass flow controllers (MKS Instruments Inc. Model 1559A). The air to the four peripheral annuluses was fed by two air ovens. After these ovens, the flow of each air heater was split into half with each stream entering a single peripheral annulus.

Photographs of the completed apparatus are shown in Figure 8.3. The apparatus was surrounded by 6 in. of Fiberfrax Durablanket® type S insulation with stainless steel shim at the 3- and 6-in. insulation levels serving as radiation barriers (Figure 8.3a). The corner pool cells of the rack were also filled with insulation. Finally, the air lines going from the air ovens to the apparatus were also insulated (Figure 8.3b).

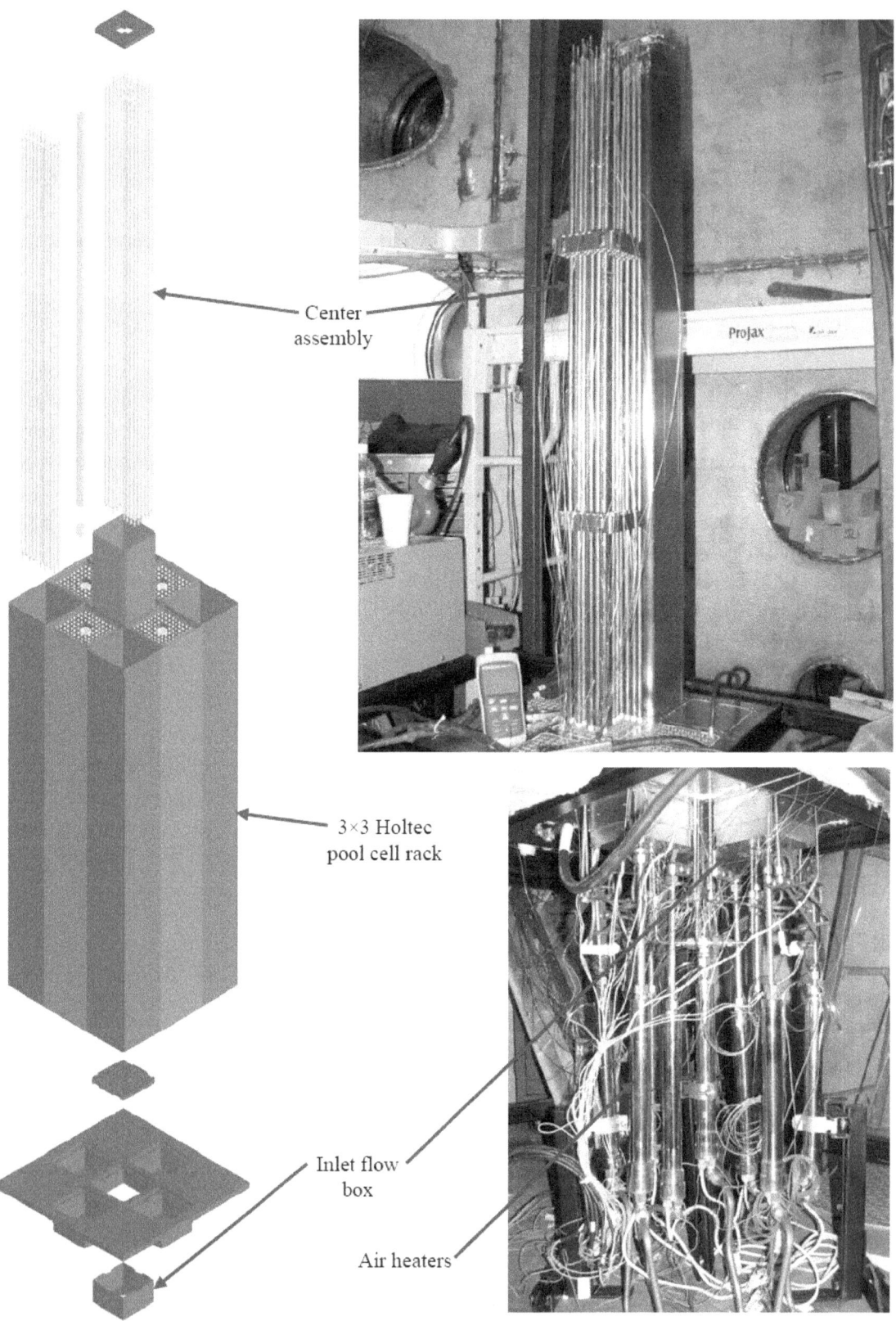

Figure 8.1 Left: Expanded schematic of the apparatus construction. Upper right: Photo of the center assembly under construction. Lower right: Photo of the undercarriage of the apparatus

156

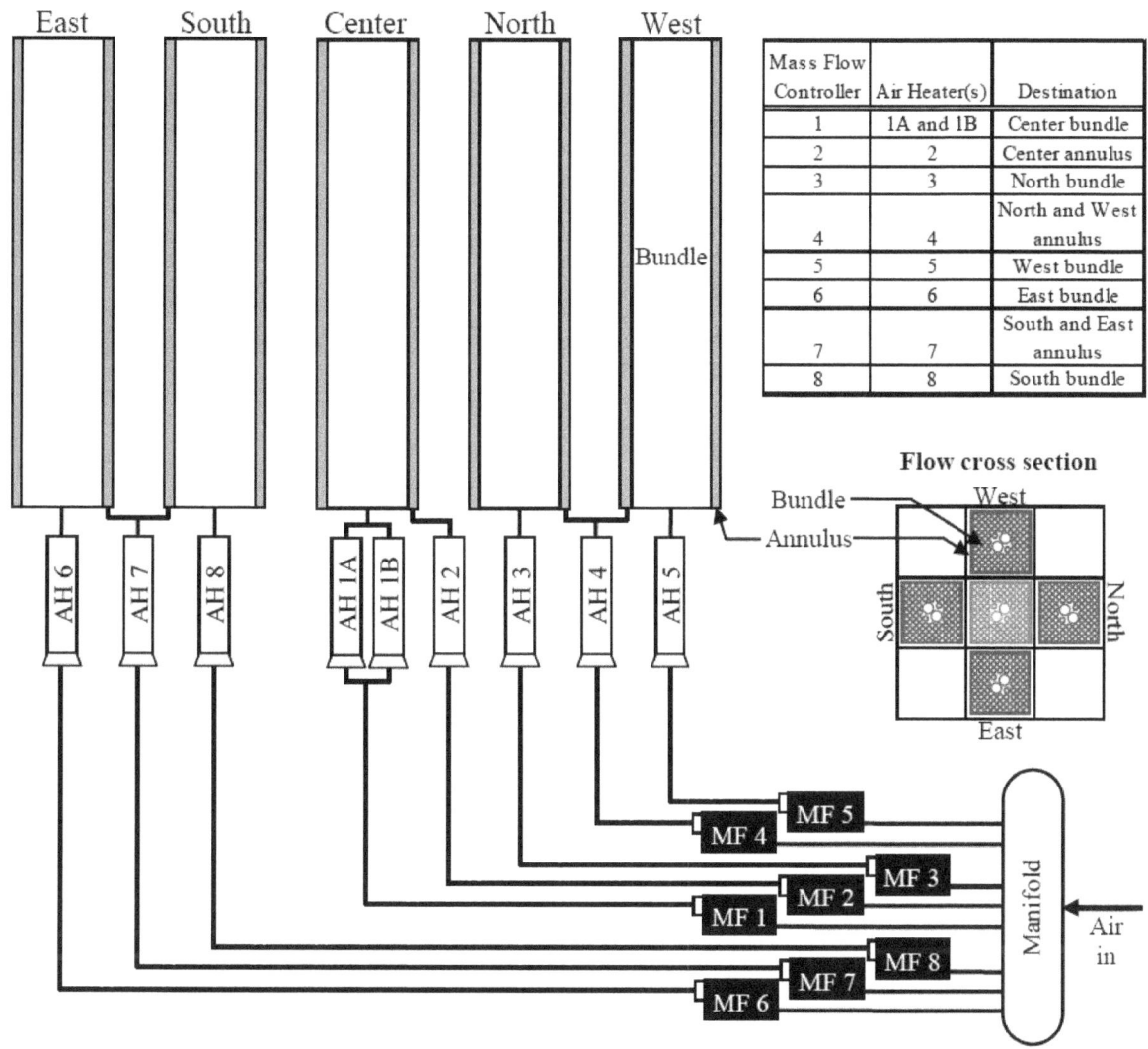

Mass Flow Controller	Air Heater(s)	Destination
1	1A and 1B	Center bundle
2	2	Center annulus
3	3	North bundle
4	4	North and West annulus
5	5	West bundle
6	6	East bundle
7	7	South and East annulus
8	8	South bundle

Figure 8.2   Flow diagram of mass flow controllers, air heaters, and assemblies

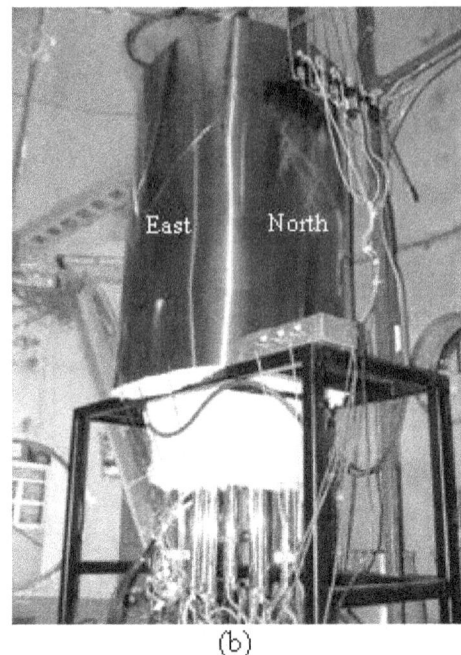

(a)   (b)

Figure 8.3   **Photographs of the final assembly with insulation applied around (a) the perimeter and (b) the lower portion**

Note: The corner cells were filled with insulation.

8.1.2   TC Layout

The type K TC (73 K to 1523 K temperature range) configuration for the ignition test was based on the 1×4 Incoloy testing experiences (Section 6.1.1). Figure 4.14 depicts the general TC layout of the 1×4 Zircaloy assembly. Three main TC levels were placed at heights of 10.6, 25.6, and 40.6 in. as measured from the top of the bottom tie plate. The solid circles indicate heater rods with a TC attached.

Figure 8.5 provides the location and naming convention of the TCs in the center assembly. A list of all installed TCs is presented below the figure. The naming convention first identifies the bundle, followed by the rod and the axial height. An extra identifier (W) is included for TCs attached to water rods and appears after the bundle designation.

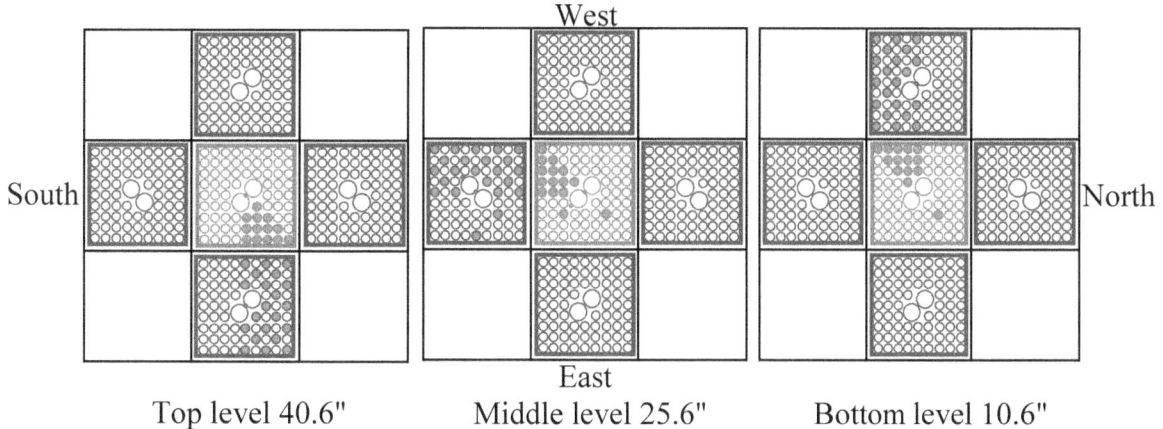

West

South   North

East

| Top level 40.6" | Middle level 25.6" | Bottom level 10.6" |

Figure 8.4   Overall type K (73K to 1523K temperature range) TC layout

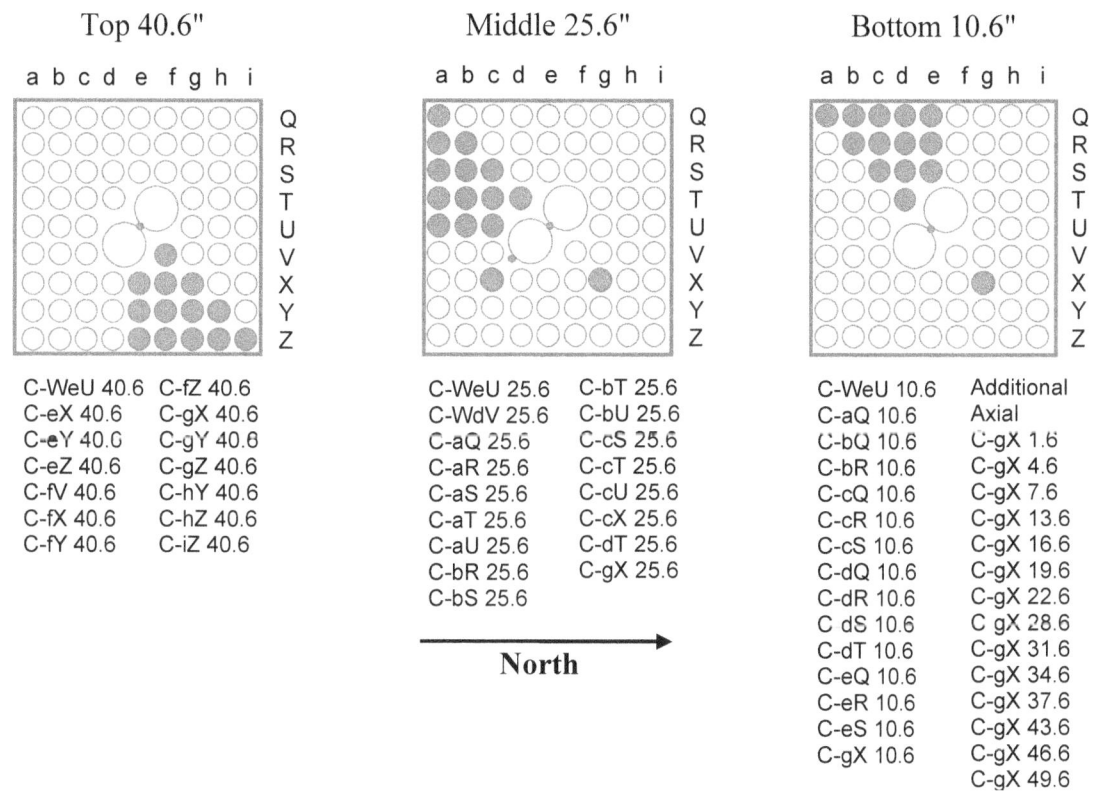

Top 40.6"   Middle 25.6"   Bottom 10.6"

a b c d e f g h i   a b c d e f g h i   a b c d e f g h i

C-WeU 40.6	C-fZ 40.6
C-eX 40.6	C-gX 40.6
C-eY 40.6	C-gY 40.6
C-eZ 40.6	C-gZ 40.6
C-fV 40.6	C-hY 40.6
C-fX 40.6	C-hZ 40.6
C-fY 40.6	C-iZ 40.6

C-WeU 25.6	C-bT 25.6
C-WdV 25.6	C-bU 25.6
C-aQ 25.6	C-cS 25.6
C-aR 25.6	C-cT 25.6
C-aS 25.6	C-cU 25.6
C-aT 25.6	C-cX 25.6
C-aU 25.6	C-dT 25.6
C-bR 25.6	C-gX 25.6
C-bS 25.6	

North →

C-WeU 10.6	Additional
C-aQ 10.6	Axial
C-bQ 10.6	C-gX 1.6
C-bR 10.6	C-gX 4.6
C-cQ 10.6	C-gX 7.6
C-cR 10.6	C-gX 13.6
C-cS 10.6	C-gX 16.6
C-dQ 10.6	C-gX 19.6
C-dR 10.6	C-gX 22.6
C-dS 10.6	C-gX 28.6
C-dT 10.6	C-gX 31.6
C-eQ 10.6	C-gX 34.6
C-eR 10.6	C-gX 37.6
C-eS 10.6	C-gX 43.6
C-gX 10.6	C-gX 46.6
	C-gX 49.6

Figure 8.5   Center assembly TC locations and naming conventions

Similar to Figure 8.5, Figure 8.6 presents the TCs located in the peripheral assemblies. Again, TCs attached to water rods are given the extra identifier (W) after the bundle identification.

In addition to the type K TCs, seven type B TCs (500K to 1923K) were used to monitor the temperature of the gas exiting the top of each assembly. The tip of each TC was positioned just

159

below the top tie plate. Each peripheral assembly had one type B TC above the bundle, the center assembly had two type B TCs above the bundle and one type B TC was located in the center assembly annulus.

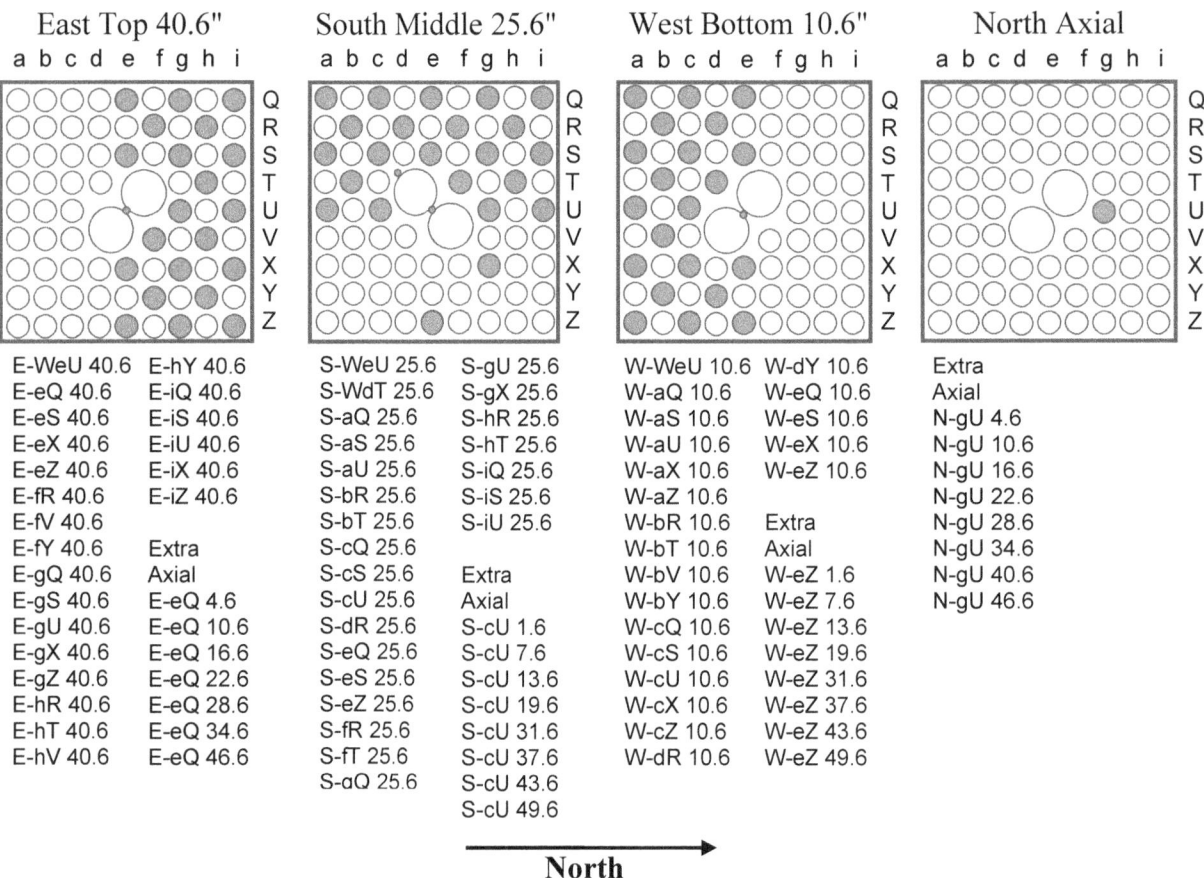

East Top 40.6"		South Middle 25.6"		West Bottom 10.6"		North Axial
E-WeU 40.6	E-hY 40.6	S-WeU 25.6	S-gU 25.6	W-WeU 10.6	W-dY 10.6	Extra
E-eQ 40.6	E-iQ 40.6	S-WdT 25.6	S-gX 25.6	W-aQ 10.6	W-eQ 10.6	Axial
E-eS 40.6	E-iS 40.6	S-aQ 25.6	S-hR 25.6	W-aS 10.6	W-eS 10.6	N-gU 4.6
E-eX 40.6	E-iU 40.6	S-aS 25.6	S-hT 25.6	W-aU 10.6	W-eX 10.6	N-gU 10.6
E-eZ 40.6	E-iX 40.6	S-aU 25.6	S-iQ 25.6	W-aX 10.6	W-eZ 10.6	N-gU 16.6
E-fR 40.6	E-iZ 40.6	S-bR 25.6	S-iS 25.6	W-aZ 10.6		N-gU 22.6
E-fV 40.6		S-bT 25.6	S-iU 25.6			N-gU 28.6
E-fY 40.6	Extra	S-cQ 25.6		W-bR 10.6	Extra	N-gU 34.6
E-gQ 40.6	Axial	S-cS 25.6	Extra	W-bT 10.6	Axial	N-gU 40.6
E-gS 40.6	E-eQ 4.6	S-cU 25.6	Axial	W-bV 10.6	W-eZ 1.6	N-gU 46.6
E-gU 40.6	E-eQ 10.6	S-dR 25.6	S-cU 1.6	W-bY 10.6	W-eZ 7.6	
E-gX 40.6	E-eQ 16.6	S-eQ 25.6	S-cU 7.6	W-cQ 10.6	W-eZ 13.6	
E-gZ 40.6	E-eQ 22.6	S-eS 25.6	S-cU 13.6	W-cS 10.6	W-eZ 19.6	
E-hR 40.6	E-eQ 28.6	S-eZ 25.6	S-cU 19.6	W-cU 10.6	W-eZ 31.6	
E-hT 40.6	E-eQ 34.6	S-fR 25.6	S-cU 31.6	W-cX 10.6	W-eZ 37.6	
E-hV 40.6	E-eQ 46.6	S-fT 25.6	S-cU 37.6	W-cZ 10.6	W-eZ 43.6	
		S-aQ 25.6	S-cU 43.6	W-dR 10.6	W-eZ 49.6	
			S-cU 49.6			

North →

Figure 8.6   **TC locations and naming conventions in the peripheral assemblies**

Finally, the locations of the TCs placed on the channel canisters and pool cell walls are identified in Figure 8.7. These TCs are labeled as either channel (Cha-) or pool cell (P-), followed by the assembly identifier (e.g. C for center, E for east), then the direction of the mounting face, next the axial height, and finally the corresponding rod location nearest the TC (if applicable). The annular region in Figure 8.7 has been exaggerated in order to show the locations of these TCs distinctly.

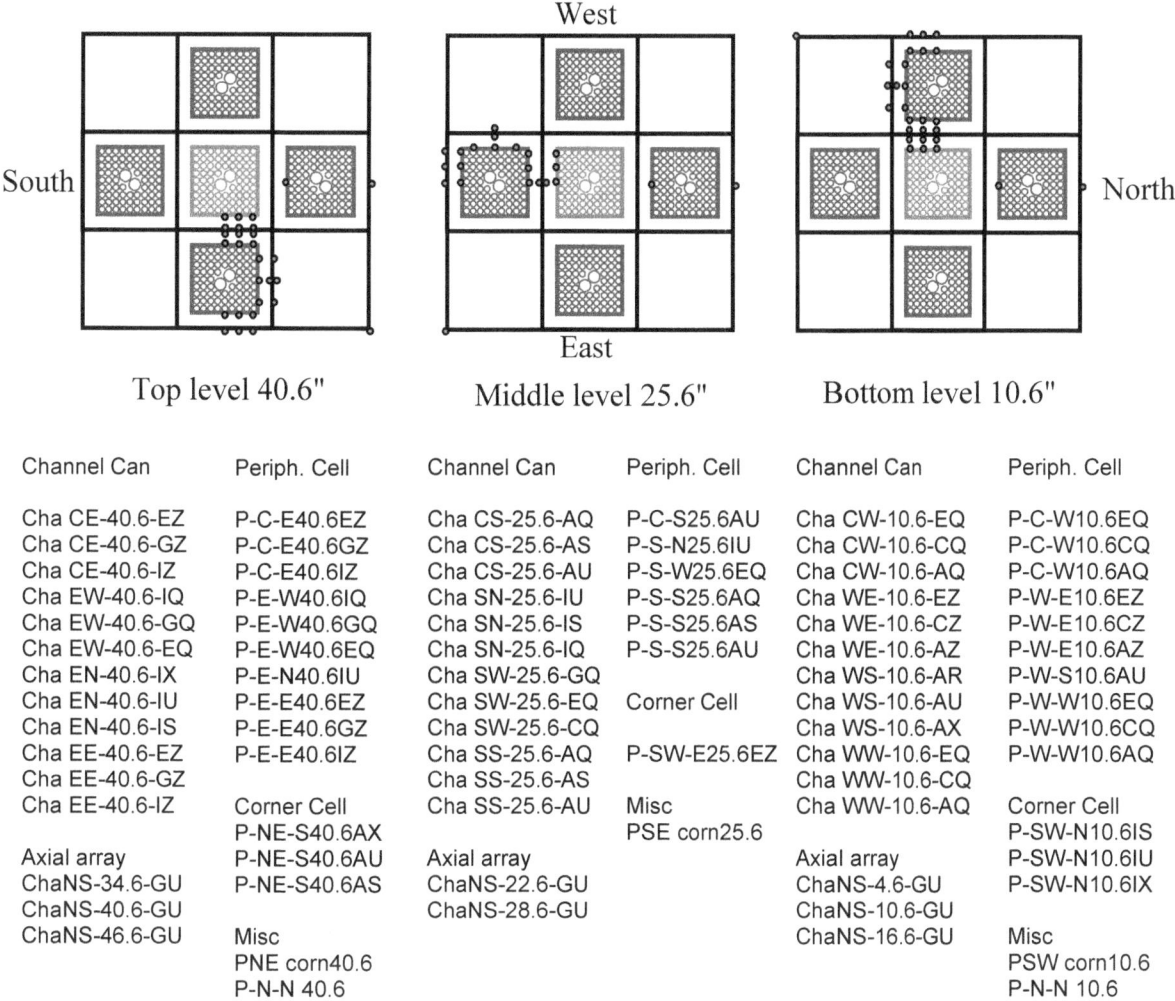

Channel Can	Periph. Cell	Channel Can	Periph. Cell	Channel Can	Periph. Cell
Cha CE-40.6-EZ	P-C-E40.6EZ	Cha CS-25.6-AQ	P-C-S25.6AU	Cha CW-10.6-EQ	P-C-W10.6EQ
Cha CE-40.6-GZ	P-C-E40.6GZ	Cha CS-25.6-AS	P-S-N25.6IU	Cha CW-10.6-CQ	P-C-W10.6CQ
Cha CE-40.6-IZ	P-C-E40.6IZ	Cha CS-25.6-AU	P-S-W25.6EQ	Cha CW-10.6-AQ	P-C-W10.6AQ
Cha EW-40.6-IQ	P-E-W40.6IQ	Cha SN-25.6-IU	P-S-S25.6AQ	Cha WE-10.6-EZ	P-W-E10.6EZ
Cha EW-40.6-GQ	P-E-W40.6GQ	Cha SN-25.6-IS	P-S-S25.6AS	Cha WE-10.6-CZ	P-W-E10.6CZ
Cha EW-40.6-EQ	P-E-W40.6EQ	Cha SN-25.6-IQ	P-S-S25.6AU	Cha WE-10.6-AZ	P-W-E10.6AZ
Cha EN-40.6-IX	P-E-N40.6IU	Cha SW-25.6-GQ		Cha WS-10.6-AR	P-W-S10.6AU
Cha EN-40.6-IU	P-E-E40.6EZ	Cha SW-25.6-EQ	Corner Cell	Cha WS-10.6-AU	P-W-W10.6EQ
Cha EN-40.6-IS	P-E-E40.6GZ	Cha SW-25.6-CQ		Cha WS-10.6-AX	P-W-W10.6CQ
Cha EE-40.6-EZ	P-E-E40.6IZ	Cha SS-25.6-AQ	P-SW-E25.6EZ	Cha WW-10.6-EQ	P-W-W10.6AQ
Cha EE-40.6-GZ		Cha SS-25.6-AS		Cha WW-10.6-CQ	
Cha EE-40.6-IZ	Corner Cell	Cha SS-25.6-AU	Misc	Cha WW-10.6-AQ	Corner Cell
	P-NE-S40.6AX		PSE corn25.6		P-SW-N10.6IS
Axial array	P-NE-S40.6AU	Axial array		Axial array	P-SW-N10.6IU
ChaNS-34.6-GU	P-NE-S40.6AS	ChaNS-22.6-GU		ChaNS-4.6-GU	P-SW-N10.6IX
ChaNS-40.6-GU		ChaNS-28.6-GU		ChaNS-10.6-GU	
ChaNS-46.6-GU	Misc			ChaNS-16.6-GU	Misc
	PNE corn40.6				PSW corn10.6
	P-N-N 40.6				P-N-N 10.6

Figure 8.7   TC locations and naming conventions in the channel canisters and pool cells

8.1.3 Oxygen Monitors

Oxygen concentration was monitored at the top of each of the five assemblies. The oxygen monitors used were Model 65 made by Advanced Micro Instruments Inc., Huntington Beach, California. Figure 8.8 shows the installation of the five monitors.

Figure 8.9 shows (a) photographic detail of the as-built sample locations and (b) a diagram with the sample location in each assembly denoted as a red circle. A small gas sample was drawn from just below the top tie plate through a ceramic tube into stainless steel tubing and through the oxygen monitor. The sampling locations were centrally located representing a bulk or average value.

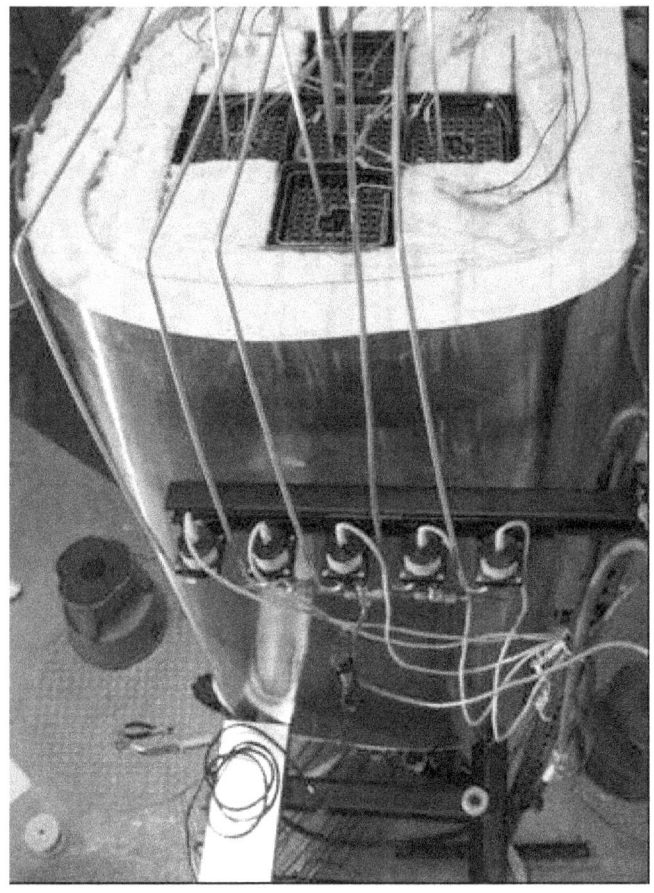

Figure 8.8   AMI Inc. Model 65 oxygen monitors

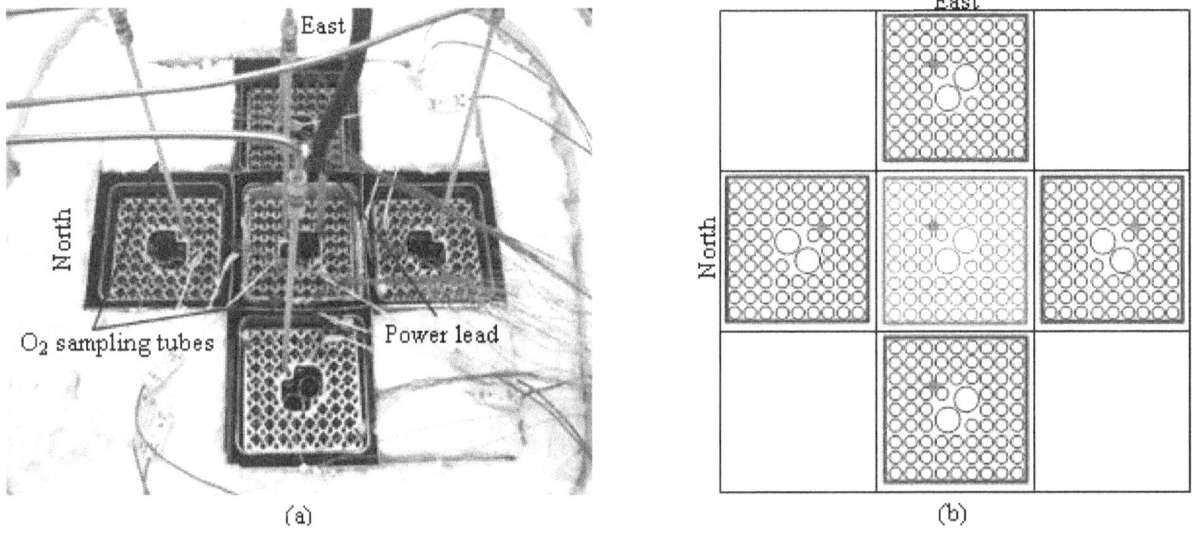

Figure 8.9   Oxygen sampling locations for each assembly as shown in (a) the as-built apparatus and (b) a schematic with sampling tube penetrations indicated in red

8.2 Ignition Test Results

8.2.1 Input Parameters

The air flow rate, temperature, and oxygen concentration input parameters for this experiment were chosen to mimic flow conditions entering the top 48 in. of the fully populated sections of five full-length assemblies arranged in a 1×4 pattern. This experimental representation does not directly simulate the lower 52 in. of fully populated sections or the upper 58 in. of partially populated sections of a full length assemblies. The center assembly represented a 17.5-day-old assembly (11.8 kW) and the four peripheral assemblies represented the average age of assemblies in the pool (0.2 kW). As shown in Figure 8.10, the center assembly was powered at 4.87 kW which represents the desired fully populated section of a full-length 11.8 kW assembly with a 1.2 peaking factor. For experimental simplicity, the rod heaters in the four peripheral assemblies were not powered. The representative power for each peripheral assembly was only 0.08 kW, which was lower than the capability of the power controller and therefore considered insignificant.

The magnitude of the other dynamically controlled input parameters were chosen based on previous experimental results (i.e., full-length Incoloy and Zircaloy assemblies) and MELCOR modeling with some additional experimental constraints. The axial oxygen concentration results from the full-length Zircaloy assembly ignition test (see Section 7.3.2) clearly showed that no oxygen depletion occurred 24 in. below the burn front meaning that atmospheric air could be used as the feed gas. As the burn front approached the bottom tie plate of the apparatus, the assumption of no oxygen depletion begins to deviate from the prototypic situation since a full length assembly would extend 52 in. below the test apparatus.

Figure 8.11 shows the flow rates for the different channels inside the apparatus imposed during the ignition test. Figure 8.12 shows the inlet temperatures for the different channels inside the apparatus during the ignition test. The flows were precisely controlled to the desired levels by programming the mass flow controllers with the LabView data acquisition program. At the time of ignition in the center assembly at about 4.9 hours, the test flow program was manually interrupted to begin a programmed ramp down in flow rate. The inlet temperatures were controlled more coarsely. The temperature controllers used only allowed for a prescribed linear ramp to a set temperature. The elevated temperatures shown in Figure 8.12 after 6 hours were due to the burn front reaching the bottom tie plate and influencing the inlet air temperature measurements.

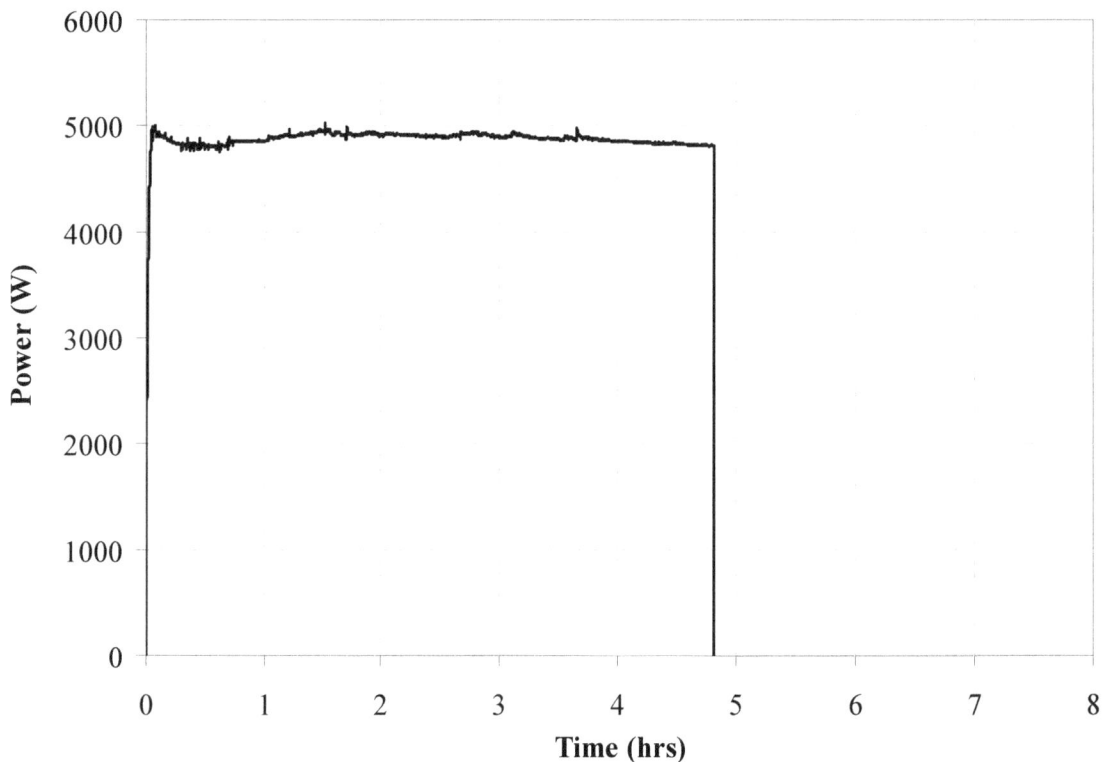

Figure 8.10   Center assembly input power for the ignition test
Note: Average power during the heated time was 4.87 kW.

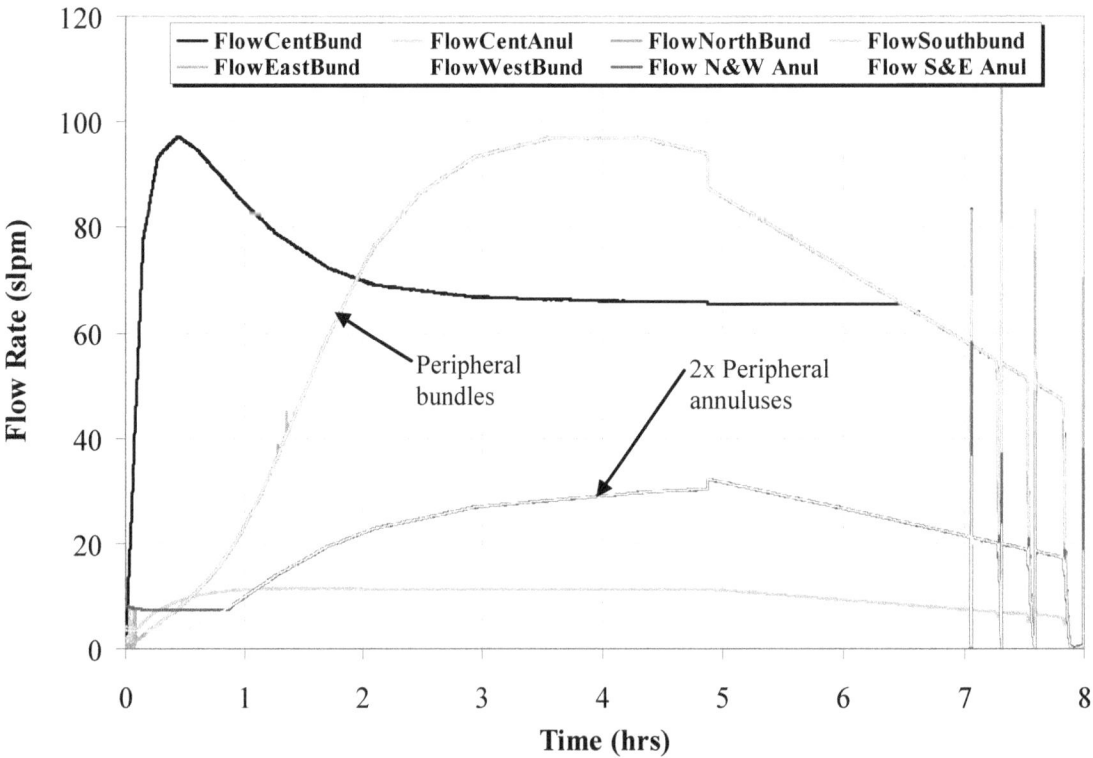

Figure 8.11   Air flow rates for the different channels inside the apparatus during the ignition test
Note: the flow into the peripheral annuluses was split into two after passing through the air oven.

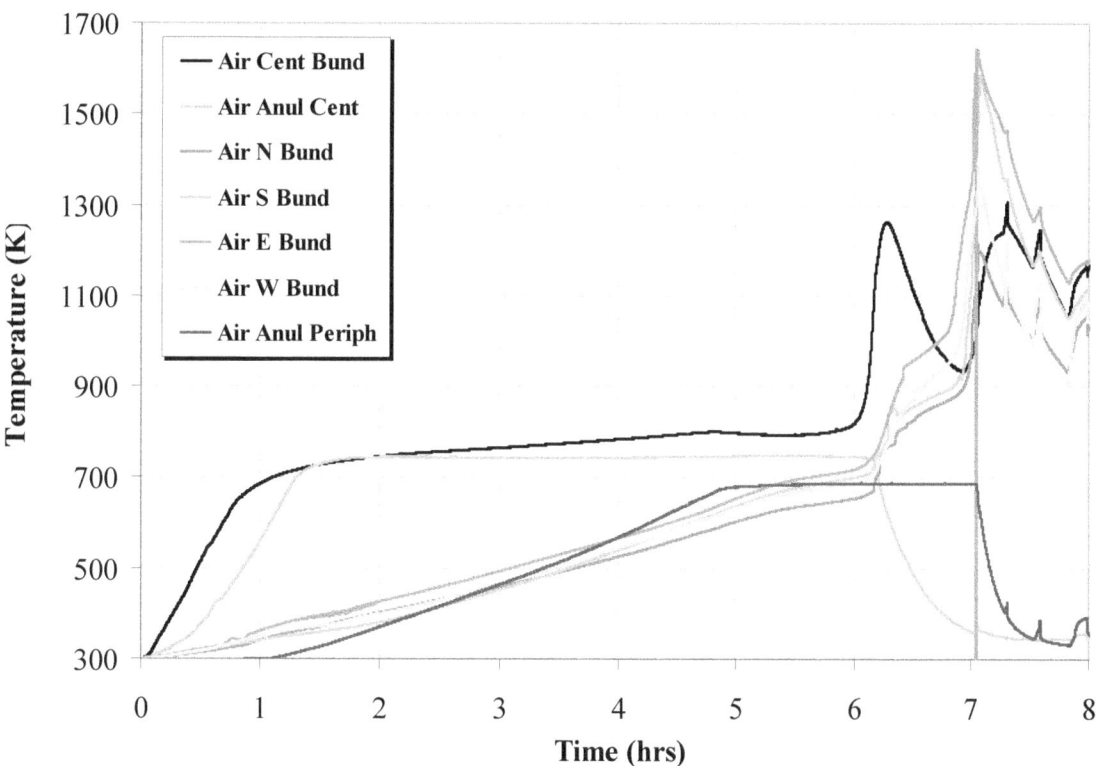

Figure 8.12   Air inlet temperatures for the different channels inside the apparatus during the ignition test

8.2.2   Oxygen Profiles

Figure 8.13 shows the response for all of the flow-through oxygen monitors except the north assembly. The oxygen monitor for the north assembly failed during installation and is not shown. Table 8.2 gives the time at which an oxygen concentration of 20%, 15%, and 5% was measured by each of the monitors considered. Also shown is the time to propagate from the center assembly to the peripheral assemblies based on the time to reach 15% oxygen. These oxygen measurements indicate that ignition occurred first in the center assembly at 4.83 hours. Figure 8.14a shows the overhead view of the apparatus at 5 hours elapsed test time. The west peripheral assembly was the first to ignite at 6.73 hours, 1.9 hours after the center assembly. The east and south peripheral assemblies were not far behind, igniting about 6 minutes later at 6.84 and 6.87 hours or about two hours after the center assembly.

Although oxygen concentration first decreased peripherally in the west assembly, the oxygen consumption rates in the west were lower than in the south and east assemblies. The oxygen concentrations dropped much steeper and to lower levels in the south and east assemblies indicating a faster radial propagation. Video evidence confirms these trends. The video shows the west brightened first followed by the south, east, and then north assemblies (see Figure 8.14b). Smoke, which indicates temperatures high enough to melt the Zircaloy, was emitted first from the east assembly followed by the south, west, and then north assemblies. Post-mortem inspection indicated that the south and east peripheral assemblies were the most heavily damaged, as discussed in more detail later.

165

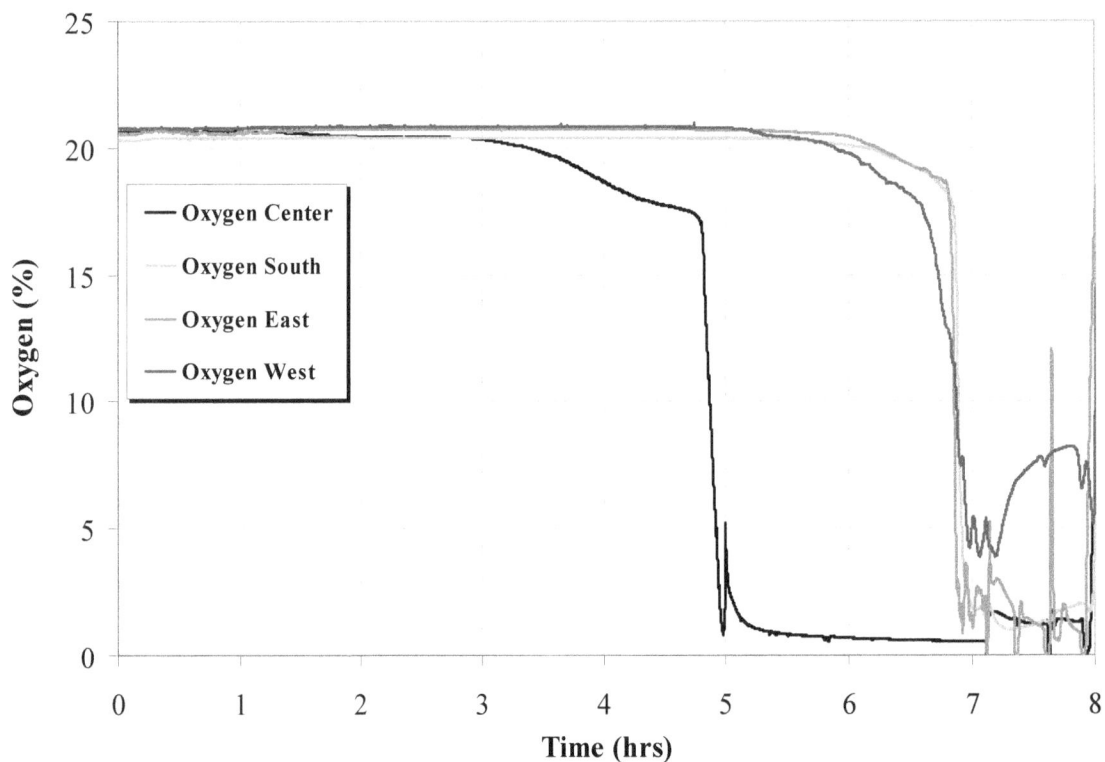

Figure 8.13   Response of all oxygen monitors as a function of time

Table 8.2   Elapsed experimental time to reach reduced oxygen concentrations for the oxygen monitors

Location	O₂ Monitor time to 20% O₂ (hr)	O₂ Monitor time to 15% O₂ (hr)	O₂ Monitor time to 5% O₂ (hr)	Radial propagation time for 15% O₂ (hr)
Center	3.39	4.83	4.92	--
West	5.88	6.73	6.96	1.90
South	6.12	6.87	6.91	2.05
East	6.22	6.84	6.86	2.01

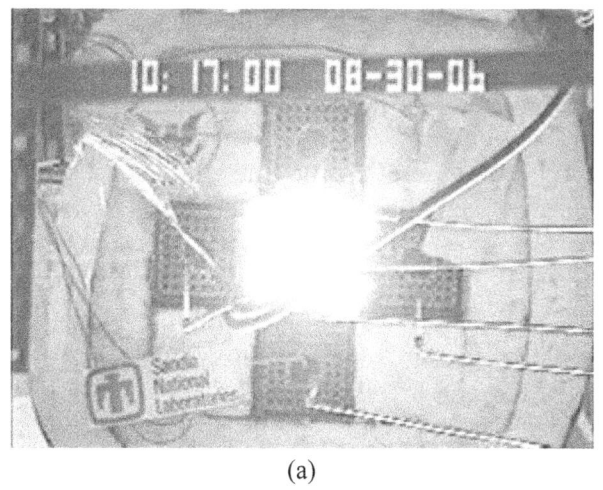 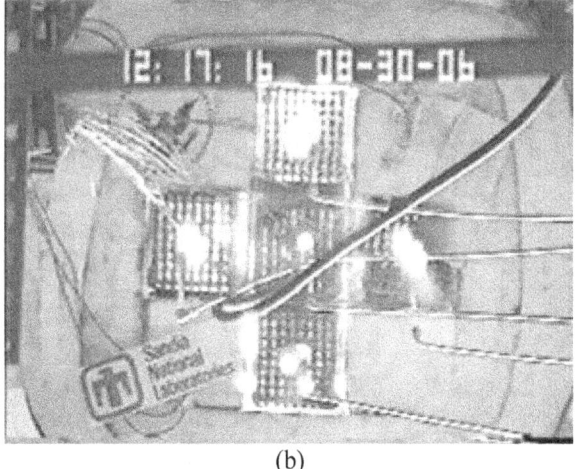

(a)   (b)

Figure 8.14 Overhead images of the test apparatus at elapsed test times of (a) 5 hours and (b) 7 hours

8.2.3 Axial Temperature Profiles

Ignition occurred in the center of the center assembly at 4.83 hours (see Figure 8.15) at levels 31.6, 34.6, and 37.6 in. The burn front moved downward in the center assembly at an overall rate of 26 in/hr. The corresponding peak temperature in the peripheral assemblies increased correspondingly, but self-sustaining ignition did not occur until after the burn front in the center assembly reached the bottom tie plate (see Figure 8.16). Based on data from the oxygen monitors, the west peripheral assembly was the first to ignite at 6.72 hours. Ignition occurred uniformly across the 10.6-in. level of the west assembly by 6.9 hours. At this time all thermocouples failed, but the burn continued according to video evidence in all assemblies for over 33 hours despite the termination of all air flows and failed attempts to quench the fire with argon gas.

Oxygen concentration is shown along with the axial temperature plots in Figure 8.15 and Figure 8.16. For the center assembly shown in Figure 8.15, the oxygen concentration begins to drop below 20% when the hottest level measured reaches 1058 K. The oxygen concentration drops sharply to 15% when the hottest level rises rapidly to 1300 K suggesting these two criteria as indicators for ignition in the center assembly. The radial gradient in the center assembly near the ignition point is fairly shallow, on the order of 100 K (see Figure 8.22) so Zircaloy temperatures in the range of 1200 K to 1300 K are needed to drop the oxygen concentration to low levels at the top of the assembly.

Figure 8.16 shows the temperature histories for the axial array in the east and west peripheral assemblies. The axial array is on the rod closest to the center assembly as shown in the inset diagram. The temperature of this rod increased sharply in response to the high temperature from the burn in the center assembly. As the burn front proceeded downward in the center assembly, the corresponding temperature on the axial array in the peripheral assembly peaked to higher and higher temperatures. However, self-sustaining ignition did not occur even though the temperatures were peaking at 1200 K to 1300 K. These temperatures decreased once the burn front in the center assembly passed by. As will be discussed in some more detail below, this is due to the steep radial temperature profile present in the peripheral assemblies. The additional oxidation energy was not sufficient to overcome the radiative heat loss when cool neighboring rods within the assembly were present. Once the radial temperature gradients were flattened significantly, ignition did occur. Ignition occurred simultaneously at the 1.6-in., 4.6-in., and 7.6-in. levels at 6.8 hours and then burned upwards. Once ignition happened at the bottom of the assembly, all the TCs failed so it is not possible to provide any burn rate information.

For the east and west peripheral assemblies shown in Figure 8.16, the situation is complicated by the steep radial temperature gradients. Figure 8.17 shows a detail of the west peripheral assembly along with the oxygen concentration near the time of the assembly ignition. When the hottest rod closest to the center assembly (W-EZ-10.6) first reaches 1300 K, the coolest rod farthest from the center assembly (W-EQ-10.6) is just over 700 K or almost 600 K cooler. Only the oxygen in the air passing near the hottest rods is consumed. The oxygen in the air flowing through the cooler regions bypasses the hottest rods and keeps the oxygen levels elevated when measured at the top. When the rod temperatures at the cooler side of a peripheral assembly approach 1200 K, oxygen concentration then begins to drop sharply.

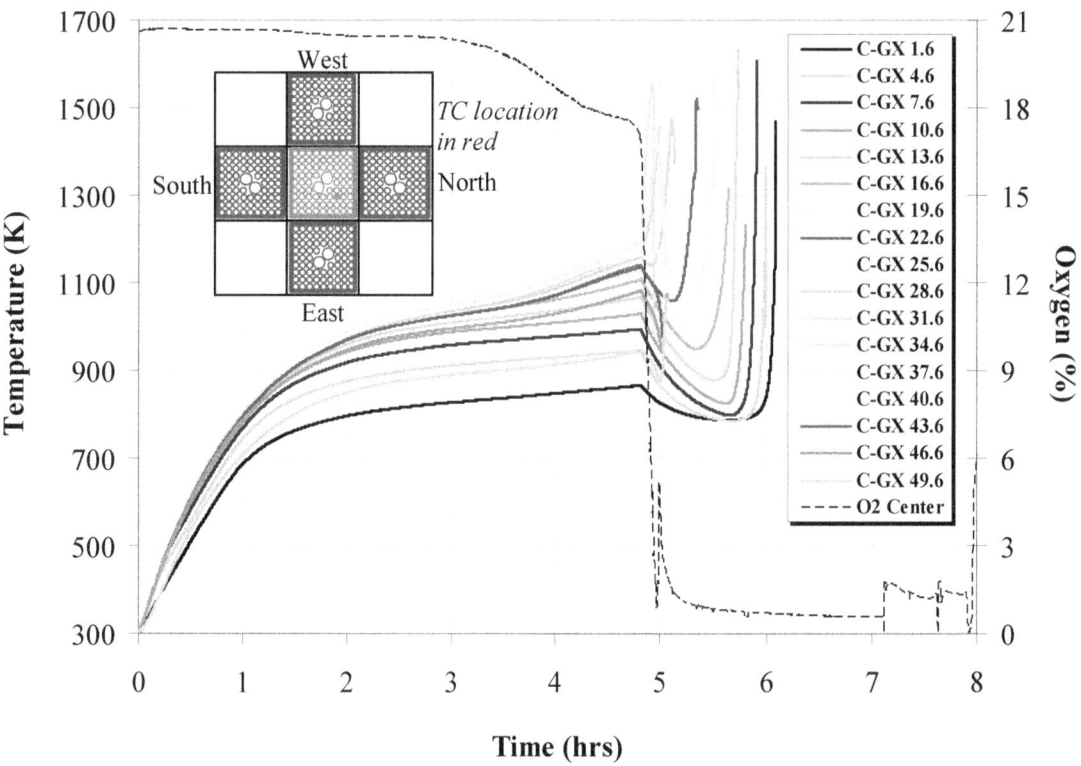

Figure 8.15 **Temperature of the GX rod in the center assembly as a function of time for different axial locations**

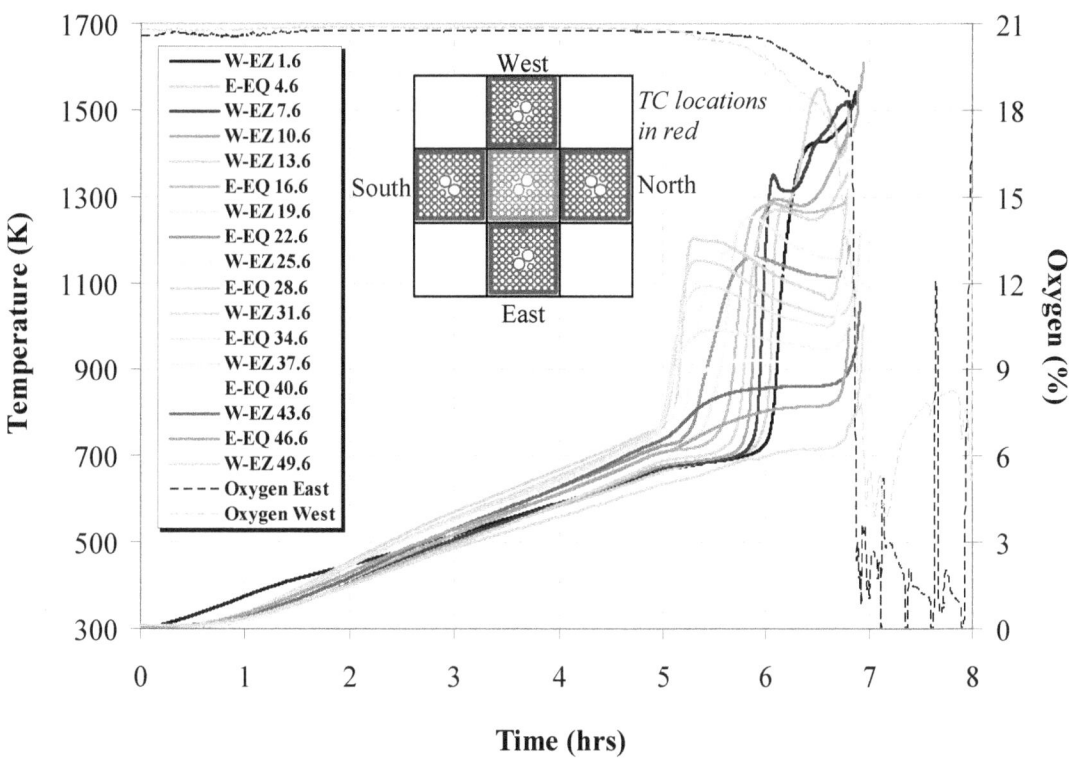

Figure 8.16 Temperature of the EZ and EQ rods in the west and east assemblies, respectively, as a function of time for different axial locations

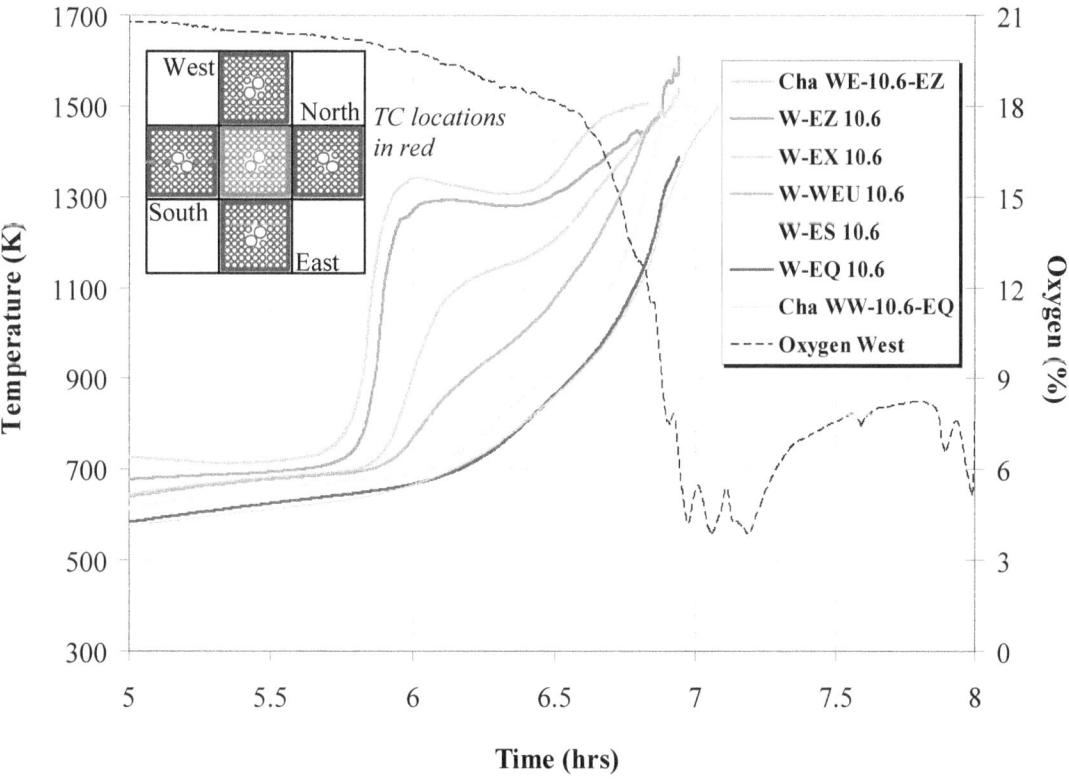

Figure 8.17 Detailed temperature of the peripheral assembly as a function of time at *z* = 10.6 in.

When ignition occurred in the center assembly at 4.83 hours (see Figure 8.15), the temperatures at levels 31.6, 34.6, and 37.6 in. increased sharply while the rest of the rod temperatures drop significantly. The drop in temperature was due to the expected heater rod failure once the rods ignited. At the time of ignition, the power input from Zircaloy oxidation is greater than twice that of the decay power supplied by the heaters therefore the impact of the heater rod failure is minimal, especially in the peripheral assemblies. The three locations that increased in temperature upon heater rod failure indicate the axial level where ignition occurred. The temperature at 34.6 in. increased at the fastest rate indicating this level was closest to the initiation point. The closest measurement points above and below this point, 37.6 and 31.6 in. respectively, increased at a slightly slower but equal rate. This suggests that the ignition initiation occurred across the 6-in. axial level from 31.6 to 37.6 in. The burn front then moved downward. Table 8.3 summarizes the time of ignition and apparent axial burn rate for all the axial levels. Comparison was made at 1220 K because this is the highest temperature available for all the axial levels. The burn front moved downward at an overall rate of 26 in/hr, which is half that measured in the full-length ignition test. This decrease in burn rate is most likely due to the transmission of energy to the colder peripheral assemblies via heat transfer.

As the burn front progressed downward, it is not likely that all of the Zirconium was consumed. The unconsumed Zirconium did not continue to oxidize due to lack of oxygen. When the burn front reached the bottom tie plate, it is likely that the burn would stall at this level until all of the Zirconium is consumed and then progress back upward burning the remaining unconsumed Zirconium. Since all the TCs were lost by this time, there is no data to measure the upward burn rate, however, video evidence shows that the center assembly burned long after the front reached the bottom.

Table 8.3   Measured burn rate at different axial heights based on an ignition criterion of 1220 K

Axial location (in)	Time to 1220K (hr)	Downward burn rate (in/hr)
34.6	4.820	--
31.6	4.880	50.0
28.6	5.020	21.4
25.6	5.145	24.0
22.6	5.275	23.1
19.6	5.453	16.9
16.6	5.623	17.6
13.6	5.705	36.6
10.6	5.806	29.7
7.6	5.896	33.3
4.6	6.055	18.9
1.6	6.090	85.7
	Overall	26.0

8.2.4 Radial Temperature Gradients

Figure 8.18 through Figure 8.23 show the temperature history of the central and peripheral assemblies at each of the three radial temperature arrays located at axial positions 10.6, 25.6, and 40.6 in. The series of temperatures shown parametrically span the apparatus radially at each of the axial levels as shown in the inset diagram. For each of the center assembly plots, the temperatures span from the center of the water rods, across three heater rods, the center channel box, pool rack, Boral neutron absorption plate, the peripheral channel box, and the inner most peripheral heater rod. In the peripheral plots, the temperatures span from the Boral plate across the peripheral channel box, two heater rods, the center of the water rods, two more heater rods, the back side of the channel box, and the back side of the peripheral pool cell. Note that the last three temperatures shown in the center assembly plots are the same as the first three temperatures shown in the peripheral plots.

Figure 8.19, Figure 8.21, and Figure 8.23 show the radial temperature histories in the peripheral assemblies at the 10.6-in., 25.6-in., and 40.6-in. levels, respectively. Ignition is only evident at the 10.6-in. level because once this level burned all the TCs leading to the two upper levels were lost. The temperatures in the peripheral assemblies responded to the burn front in the center assembly with heating progressing from the face adjacent to the center assembly and moving inward toward the back wall. Figure 8.21 shows the radial temperature histories at the 25.6-in. level. The burn front in the center assembly reached this level at 5.15 hours. The temperature of the canister and the first row of heater rods reached temperatures of 1200 to 1300 K and the oxygen concentration was still near 20%; however, the oxidation reaction was quenched once the burn front passed. The oxidation reaction could not be sustained because of radiative heat loss to the back half of the peripheral assembly where the heater rods and canister were only about 650 K.

Figure 8.19 shows the radial temperature histories at the 10.6-in. level in the west peripheral assembly. The burn front in the center assembly reached this level at 5.8 hours. The temperatures of the channel box and first row of heater rods reached 1300 K, but oxidation was not sustained because the back half of the assembly was still less than 700 K. The center burn front moved down past the 10.6-in. level and reached the bottom tie plate shortly after 6.1 hours. The burn front then likely stalled until all of the Zirconium was consumed and then started back upwards. This stationary burn front encouraged radial propagation to the peripheral assemblies. Based on data from the oxygen monitors, the west peripheral assembly was the first to ignite at 6.72 hours and ignition occurred uniformly across the 10.6-in. level of the west assembly at 6.9 hours. The timing of the radial propagation is not represented prototypically in this experiment due to the shortened assembly length. Based on the observed behavior of this experiment and the previous integral test, radial propagation in prototypic full length assemblies would likely have occurred only after the center assembly burn front reached the bottom tie plate and returned upwards to consume the remaining Zircaloy. This extended burn length could add at least 2 hours to the time of radial propagation in a prototypic situation.

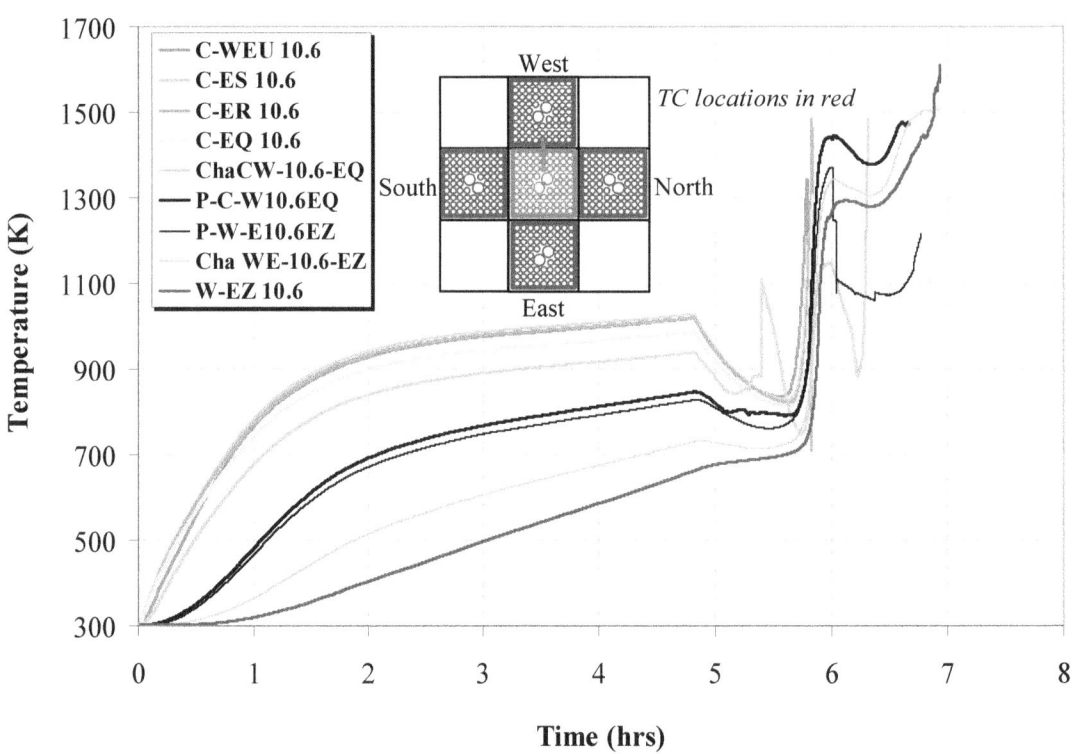

Figure 8.18   Temperature across the central portion of the apparatus as a function of time at z = 10.6 in.

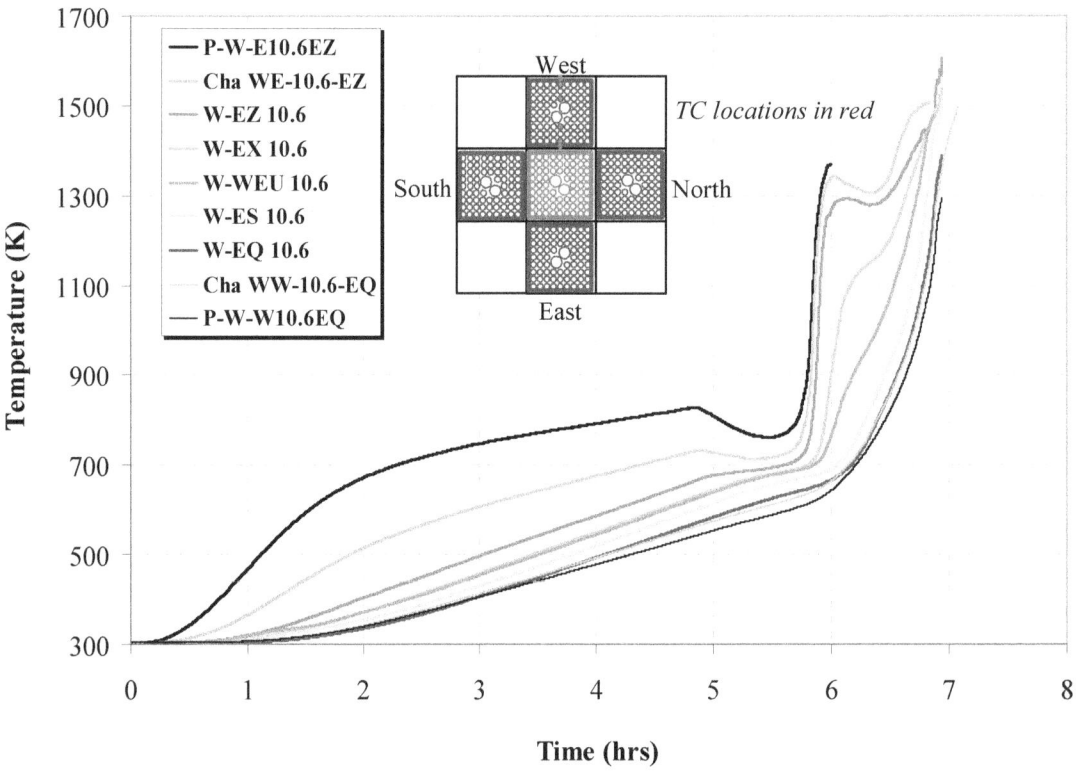

Figure 8.19   Temperature across the peripheral portion of the apparatus as a function of time at z = 10.6 in.

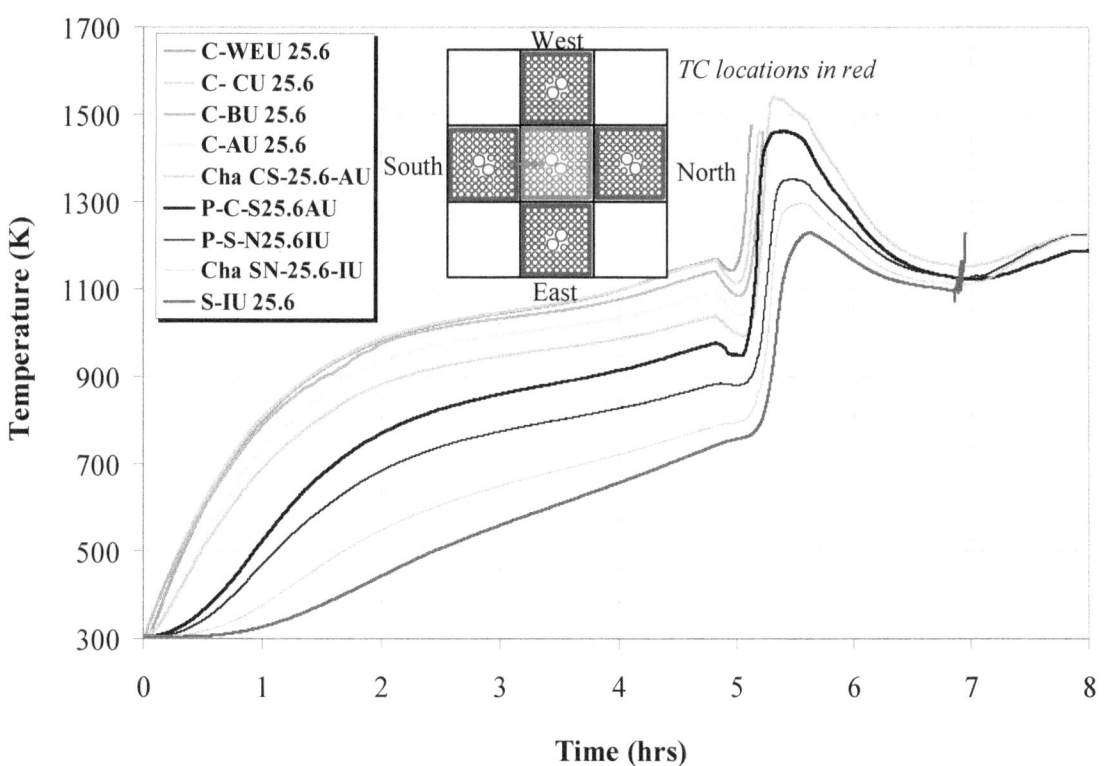

Figure 8.20   Temperature across the central portion of the apparatus as a function of time at $z = 25.6$ in.

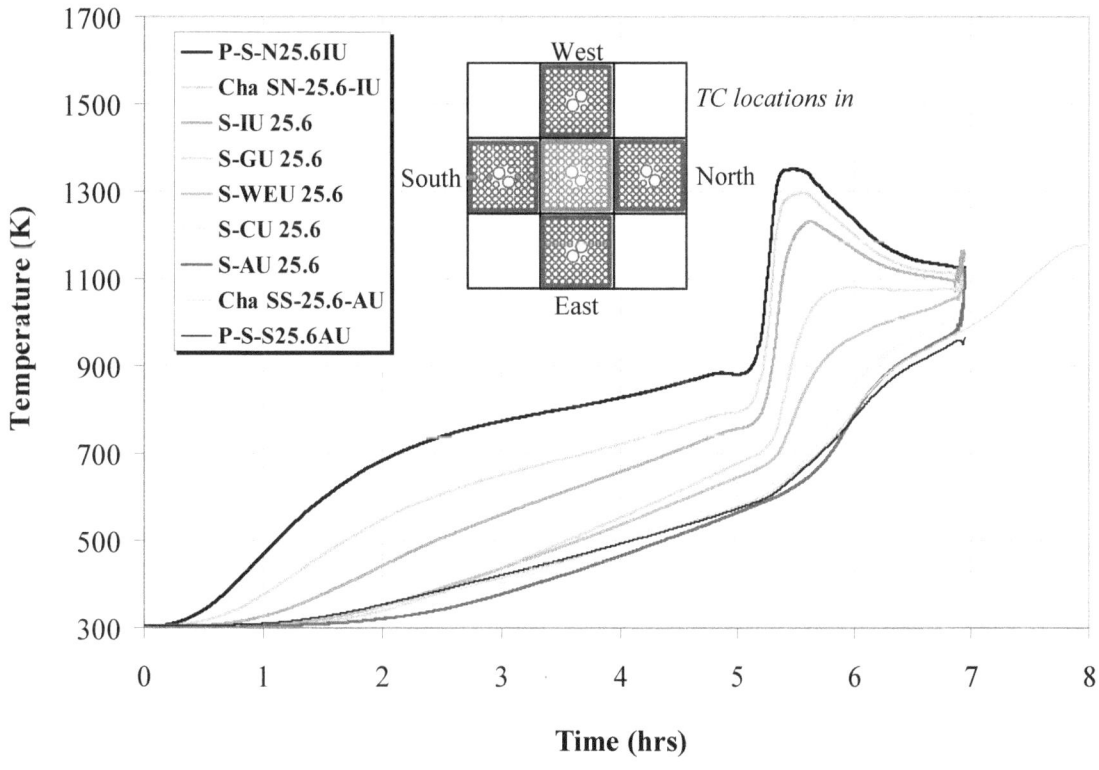

Figure 8.21   Temperature across the peripheral portion of the apparatus as a function of time at $z = 25.6$ in.

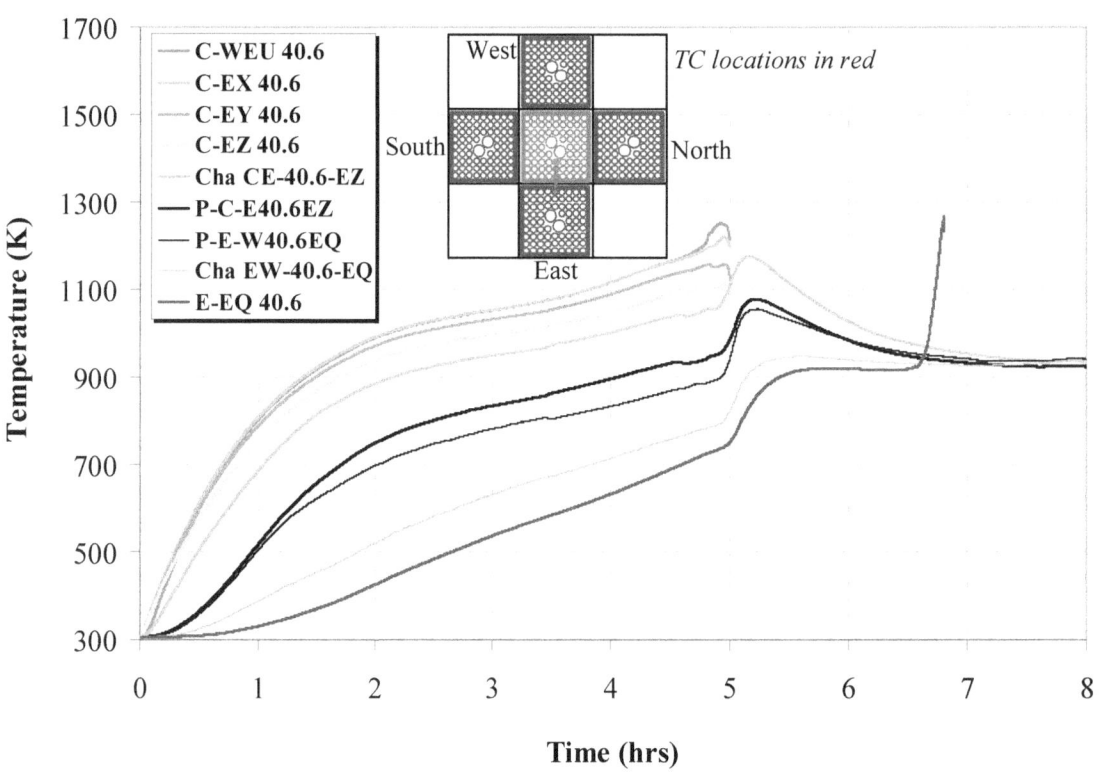

Figure 8.22   Temperature across the central portion of the apparatus as a function of time at *z* = 40.6 in.

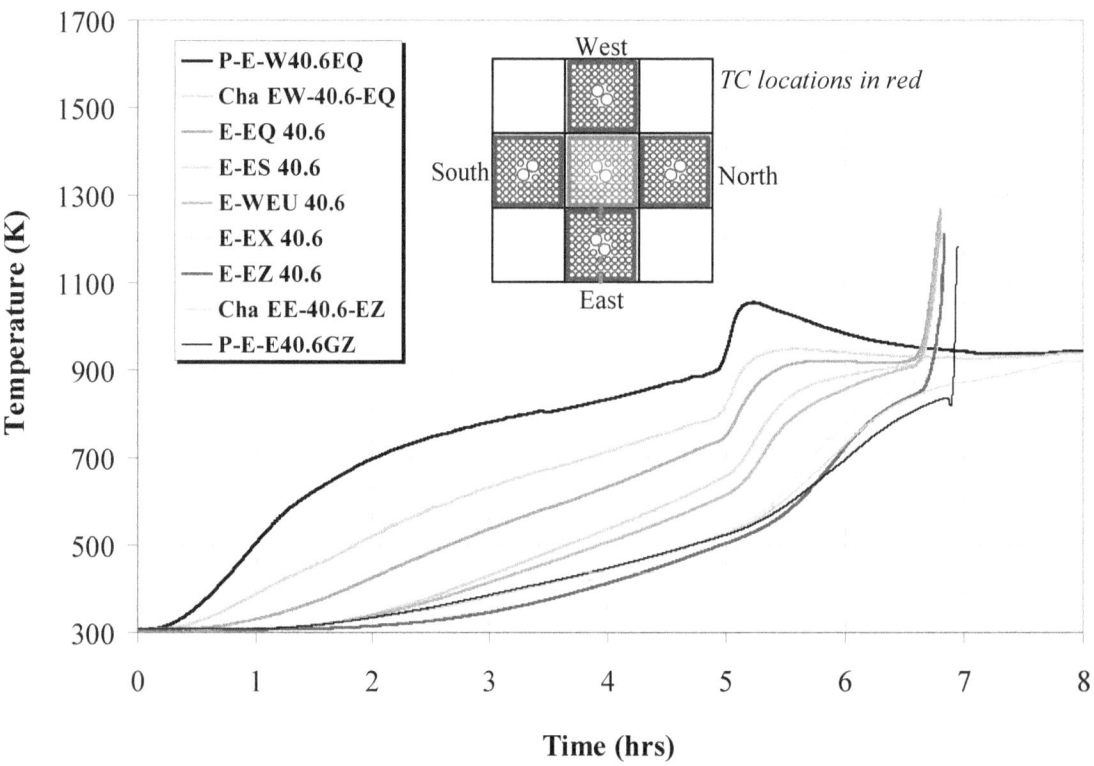

Figure 8.23   Temperature across the peripheral portion of the apparatus as a function of time at *z* = 40.6 in.

Table 8.4 summarizes the temperature and temperature drop between these radial locations at the time of initial ignition at 4.83 hours. The largest temperature drops (on the order of 100 K) were across the annular gaps between the pool cells and the adjacent channel box walls. The temperature drop across the pool cell Boral plate varied considerably from 19 K at the 10.6-in. level to 94 K at the 25.6-in. level. The radial temperature gradients are the smallest, and in some cases reversed on the back side of the peripheral assemblies. Thermal radiation and conduction results in the back pool cell and channel box walls reaching higher temperatures sooner than the adjacent heater rods.

Table 8.4   Measured temperatures across the apparatus at 4.82 hours

Radial Location	Temperature and radial temperature drop at 4.82 hours					
Axial Level =>	10.6 in.		25.6 in.		40.6 in.	
	Temp (K)	ΔT (K)	Temp (K)	ΔT (K)	Temp (K)	ΔT (K)
Center water rod	1018	-7	1170	6	1228	29
Center rod 1	1025	7	1164	24	1199	41
Center rod 2	1018	35	1140	47	1158	47
Center rod 3	983	44	1093	55	1111	54
Center channel	939	92	1038	61	1057	111
Pool cell - center side	847	19	977	94	946	57
Pool cell - peripheral side	828	98	883	95	889	104
Peripheral channel - center side	730	67	788	45	785	59
Peripheral rod 4	663	35	743	88	726	91
Peripheral rod 5	628	6	655	30	635	41
Peripheral water rod	622	24	625	44	594	82
Peripheral rod 6	598	30	581	35	512	25
Peripheral rod 7	568	8	546	-17	487	-25
Peripheral channel - back side	560	20	563	5	512	3
Pool cell - back side	540		558		509	

The axial temperature profiles placed initial ignition in the center assembly at the 34.6-in. level. This level is located between the 25.6-in. and 40.6-in. radial temperature array levels shown in Figure 8.20 and Figure 8.22, respectively, which help with determining the radial location of the initial ignition. At all three of the radial levels, the central water rod temperature and the adjacent heater rod temperatures tracked together as the hottest. Figure 8.22 shows the temperatures along the radial segment across the center assembly at the 40.6-in. level. Just after the time of ignition, the water rod temperature increased above that of the adjacent heater rod. Figure 8.20 shows the temperatures along the radial segment across the center assembly at the 25.6-in. level. After ignition the heater rod power was lost, and the water rod and adjacent heater rod temperatures dropped. However, the heater rod temperature dropped more, leaving the water rod as the hottest location. When the burn front reached this level, the water rod was first to ignite at 5.06 hours, almost 4 minutes before the adjacent heater rod. This strongly suggests that the point of ignition in the center assembly was in the very center at the water rods at the 34.6-in. level. By the time the burn front reached the 10.6-in. level the radial temperature gradients had flattened. Here, the heater rods and water rods ignited simultaneously as shown in Figure 8.18.

8.2.5 Post-Mortem of the Test Apparatus

Figure 8.24 shows two photographs of the exterior of the apparatus after the ignition test with the insulation removed. On the left, the extensive damage to the south and east peripheral assemblies is clearly evident. The south and southeast cells have been breached at the middle axial level. Indeed, the entire pool rack shows signs of warpage and buckling. Note the exposed channel cans at the top of the apparatus. The original construction placed the top of the assemblies and pool rack at the same level (see Figure 8.3a). The photo on the right depicts the damage incurred in the north and west assemblies. Similar to the south and east, the pool rack shows signs of melting and geometric failure.

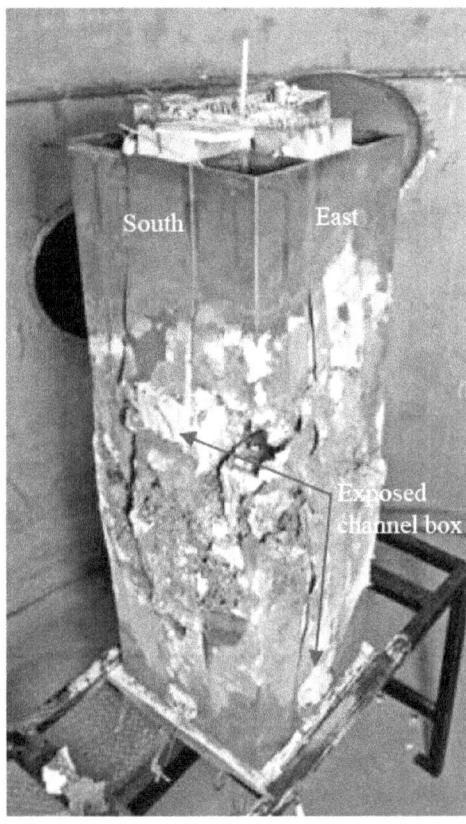

Figure 8.24   **Post-ignition test photographs of the apparatus with the insulation removed**

Figure 8.25a shows an overhead of the apparatus and further demonstrates the distortion caused during the ignition. Water rods in the center and east assemblies unmistakably protrude through the top tie plate indicating the slumping of the other structural components (bundle, channel box and pool rack). Figure 8.25b shows the air inlet flow boxes after their removal from the apparatus. The south assembly burnt through the bottom flow plate and deposited rod debris into the flow box. The east assembly deposited approximately 1 to 1.5 in. of aluminum from the Boral panels into its flow box. Remarkably, post-ignition inspection of all flow paths in the bundles and annuluses did not reveal any catastrophic blockages.

 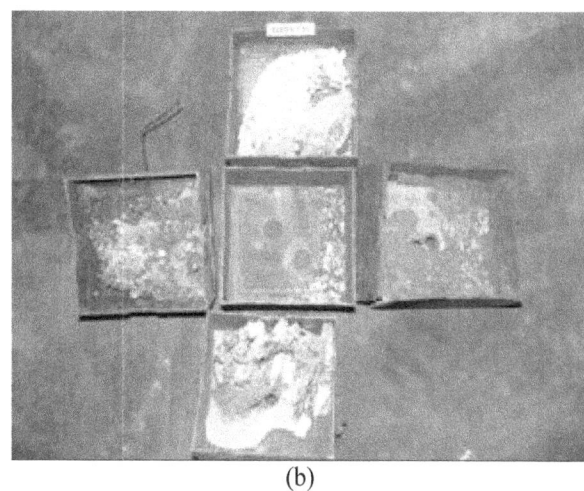

(a)   (b)

Figure 8.25   **Post-ignition overhead photographs of (a) the apparatus and (b) the inlet air flow boxes**

8.3   MELCOR Baseline Case Comparison

The MELCOR baseline model was built from the previous 1×4 efforts (see Section 6.2.1). For this case the flow paths were open, allowing air flow rates and temperatures to be defined as inlet conditions. The nodalizations described in the earlier section are accurate.

Recall that MELCOR does not account for radial temperature gradients in the fuel assembly tube bundle (as discussed in Section 6.2.2). In spite of this limitation, the experiments described in Section 6 demonstrate that MELCOR can adequately represent the radative coupling between the center and peripheral assemblies especially with respect to the impact on the center assembly. The primary objective of MELCOR whole pool calculations is to accurately predict when the first assembly ignites; therefore, accurately representing the center assembly is of paramount importance.

8.3.1   Model Input Parameters

The model input parameters were chosen based on earlier modeling and testing efforts. Figure 8.26 shows the inlet air flow rates during the ignition test and the corresponding MELCOR model values. Again, the flow rate shown for the peripheral annuluses is twice the flow rate entering a single annulus. Measured and modeled values are all within experimental error. At the point of ignition in the center assembly the test flow program was manually interrupted to begin a ramp down in flow rate.

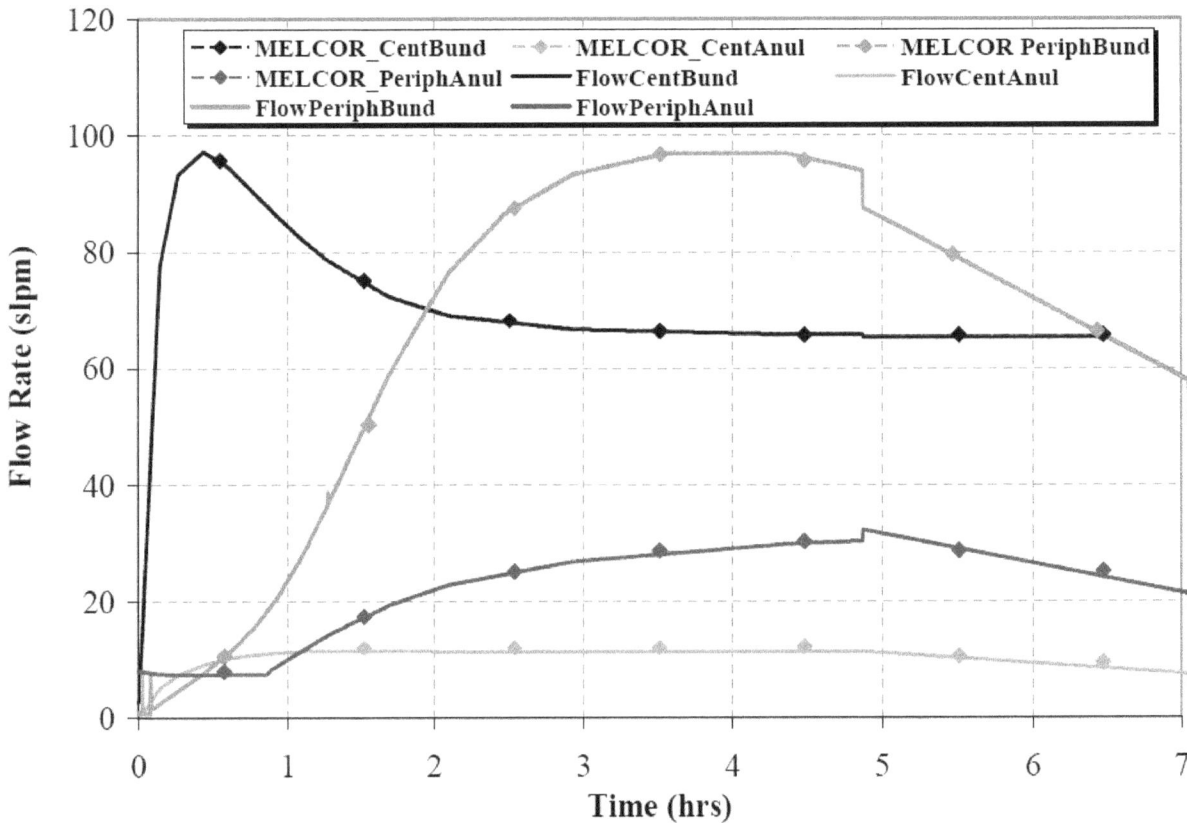

Figure 8.26 Comparison of air flow rates for the different channels inside the test apparatus and the MELCOR model

Inlet air temperatures for the ignition test and the MELCOR model are plotted in Figure 8.27. The model values are consistent with the measured temperatures up until the ignition front neared the bottom of the apparatus. At this time, the TCs registered the heat-up of the stainless steel base plate. The air temperature measurement of the heated air into the north and west annuluses is not shown. This measurement was anomalously much higher than its counterpart in the south and east annuluses or bundles and was therefore disregarded. Peripheral annulus temperatures in the MELCOR model were assumed to track with the south and east annulus air temperature measurement until they reached the average temperature of the peripheral bundle air temperature. From this point on, the MELCOR peripheral annulus air temperature tracked with the MELCOR peripheral bundle air temperature.

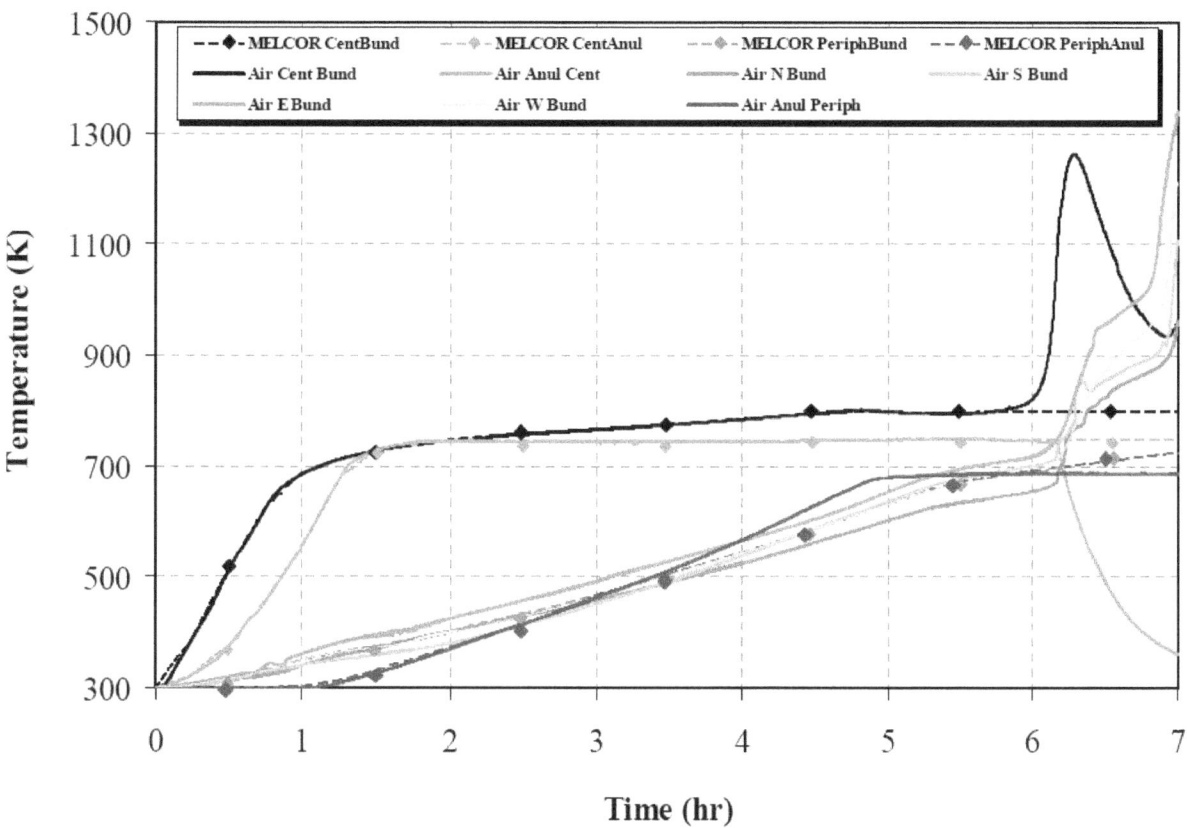

Figure 8.27 **Comparison of air inlet temperatures for the different channels inside the test apparatus and the MELCOR model**

The default emissivity model for Zircaloy in MELCOR is shown graphically in Figure 8.28 [3]. The emissivity is assumed to be a function of oxide layer thickness, increasing sharply from 0.325 for unoxidized Zircaloy to a nearly constant value of ~0.8 for oxide thicknesses greater than 3.88 μm. For Zircaloy surfaces above a temperature of 1500 K, the emissivity is modified by a multiplicative factor decreasing the emissivity back to the initial value of 0.325 at ~1800 K.

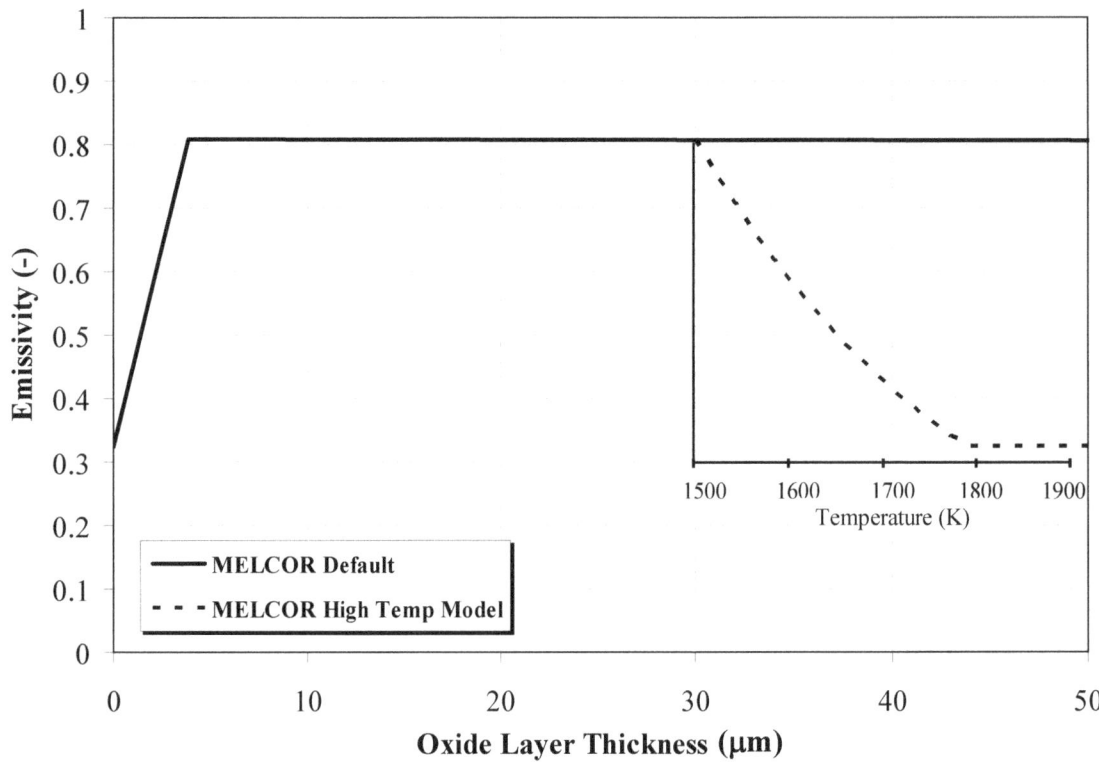

Figure 8.28   Default emissivity model in MELCOR with high temperature correction

8.3.2   Comparison of Results

Figure 8.29 gives the measured and predicted oxygen concentrations in the center and peripheral assemblies. Based on a 15% concentration level, the MELCOR model overestimated the time to reach ignition in the center by 31 minutes. The ignition of the peripheral assemblies was predicted to within 3 and 11 minutes for the west and south, respectively.

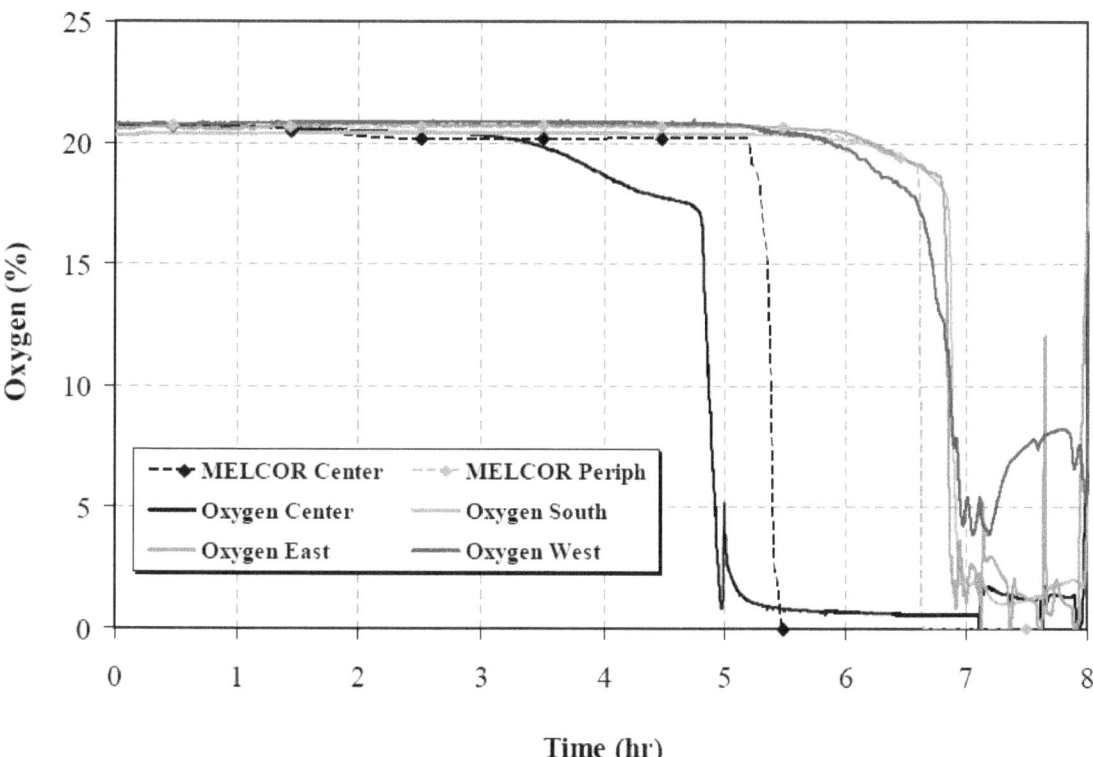

Figure 8.29 **Comparison of oxygen concentrations for the different assemblies inside the test apparatus and the MELCOR model**

Figure 8.30 shows the measured and modeled temperatures in the center assembly between $z = 4$ to 14 in. as a function of time. Note that the center GX rod was about 30 K hotter than the average of radial temperatures and 20 K cooler than PCT (C-DT and C-FV) at the same axial level. Therefore, the MELCOR model conservatively represents a temperature hotter than the average rod temperature. This trend was also evident in the 1x4 Incoloy experiments discussed in Section 6. The model heat-up matched the experiment within 25 K for the first 4 hours. The apparatus then heated considerably faster than predicted. A possible explanation involves an increase in the reaction kinetics at temperatures around 1090 K and is discussed in a later sensitivity section. Heater failure occurred at 4.83 and 5.43 hours for the experiment and prediction, respectively. The MELCOR criteria for heater rod failure was a PCT of 1444 K.

Figure 8.31 compares the model and experiment temperatures in the center assembly for axial locations bounding the initial location of ignition in the test apparatus. Again, the initial heat-up compares well with the experiment heating at a faster rate above temperatures of about 1090 K. The data indicates that ignition occurred at 4.82 hours at the 34.6" level. The baseline MELCOR model accurately predicted the ignition location to be between 31" and 37" and the ignition time was 36 minutes later than the experiment. In agreement with the data, MELCOR also predicted the burn front would move downward.

The temperatures from the upper section of both model and experiment are plotted in Figure 8.32. These results are similar to those discussed earlier. The model captures the beginning heat-up accurately but under predicts the temperatures from 4 hours elapsed time to ignition.

MELCOR utilizes a lumped thermal analysis that inherently characterizes each axial level in any given assembly with a single average temperature. This approach represents the center assembly relatively well because it is radially symmetric and the thermal gradients are relatively small. Due to asymmetry and steep thermal gradients, the lumped approach does not characterize the peripheral assemblies as well. Figure 8.33 gives the temperatures in the peripheral assembly for the MELCOR model and experiment. The west EZ rod is situated just inside the channel can on the center-facing side and represents the hottest location in the peripheral bundle. The MELCOR model predicted lower temperatures during the heat-up, which is to be expected since the single temperature MELCOR uses to characterize the tube bundle is more indicative of an average temperature. The base case MELCOR model predicted the timing of the peripheral assembly ignition closely, being only about 10 minutes early. However, the location of the peripheral assembly ignition and the direction of the burn front movement were not well represented. The base case MELCOR model predicted the peripheral assembly would ignite at the 19" to 25" level and burn downward while the data showed that the peripherals ignited simultaneously at the 1.6-in., 4.6-in., and 7.6-in. levels at 6.8 hours and then burned upwards (see Section 8.2.3). Overall, the base case MELCOR representation of the center assembly ignition was very good. The representation of the peripheral assemblies was not as good due to the limitation of characterizing the tube bundle with a single temperature thus ignoring the complications of steep radial temperature gradients.

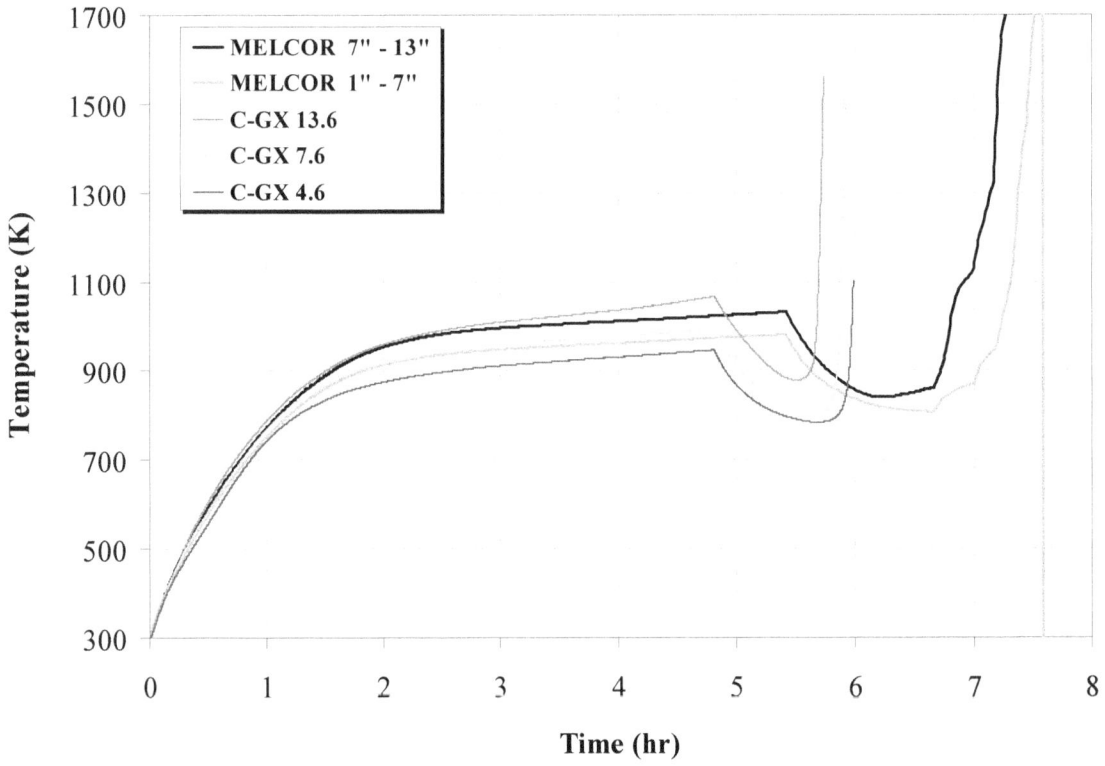

Figure 8.30   Comparison of temperatures inside the center assembly and MELCOR model for $z = 4$ to 14 in.

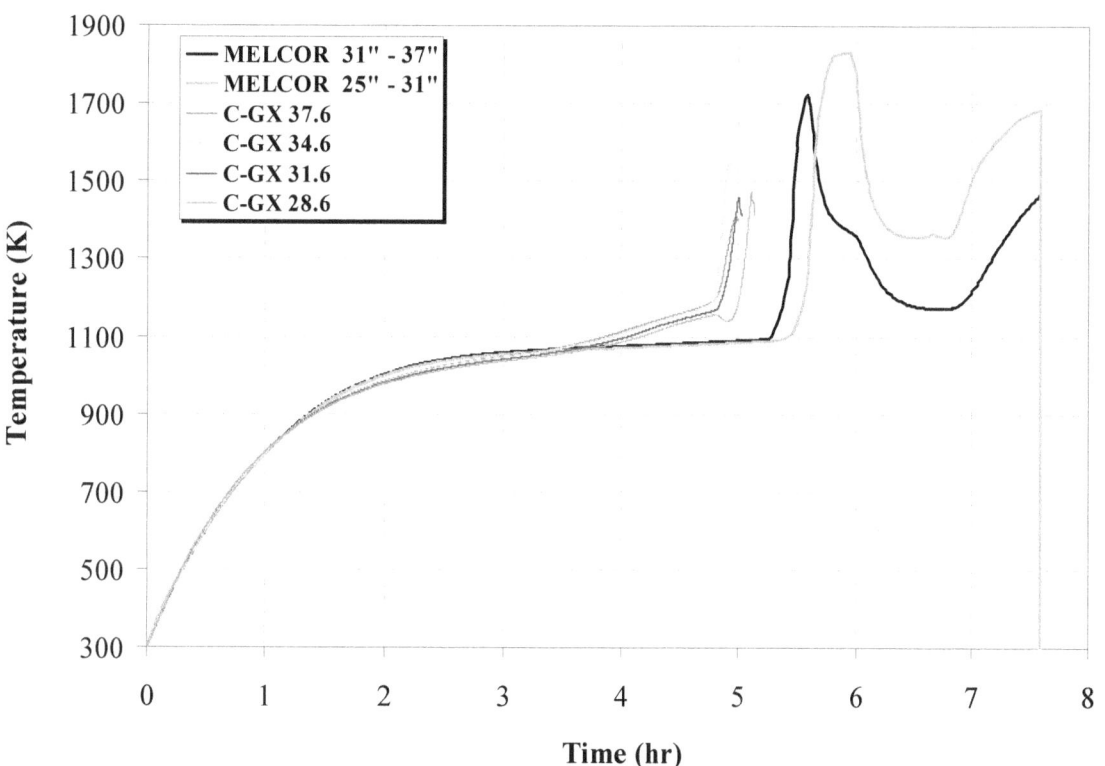

Figure 8.31   Comparison of temperatures inside the center assembly and MELCOR model for z = 28 to 38 in.

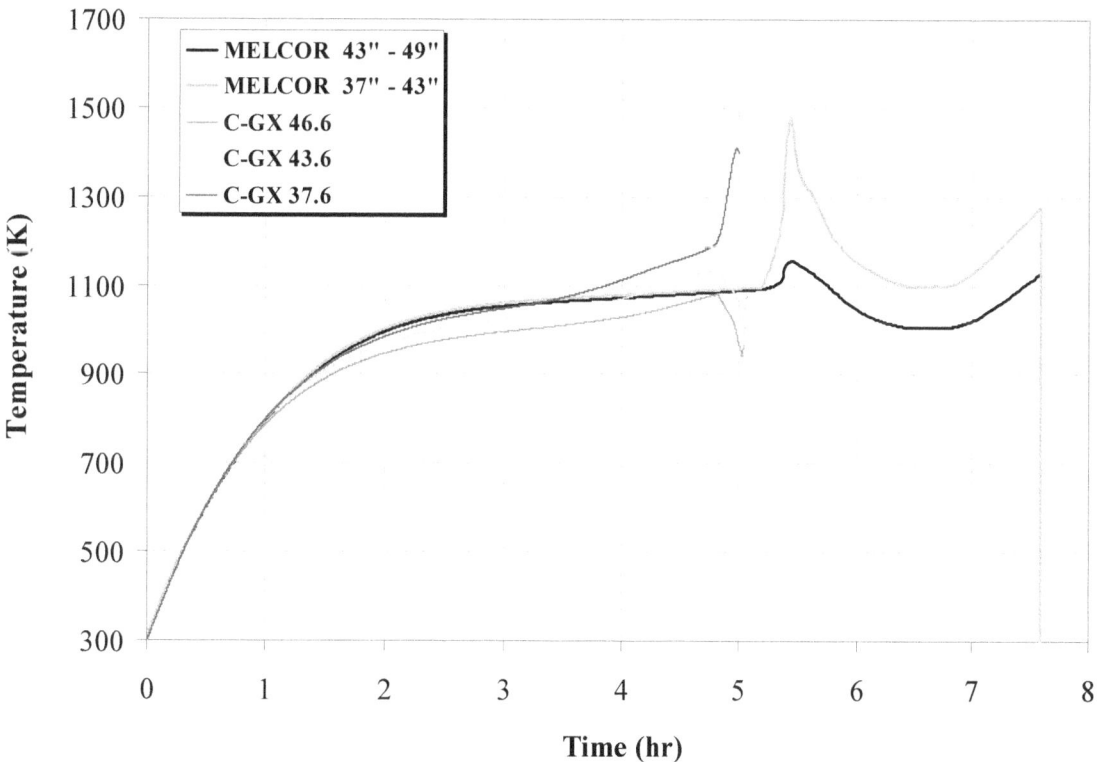

Figure 8.32   Comparison of temperatures inside the center assembly and MELCOR model for z = 37 to 47 in.

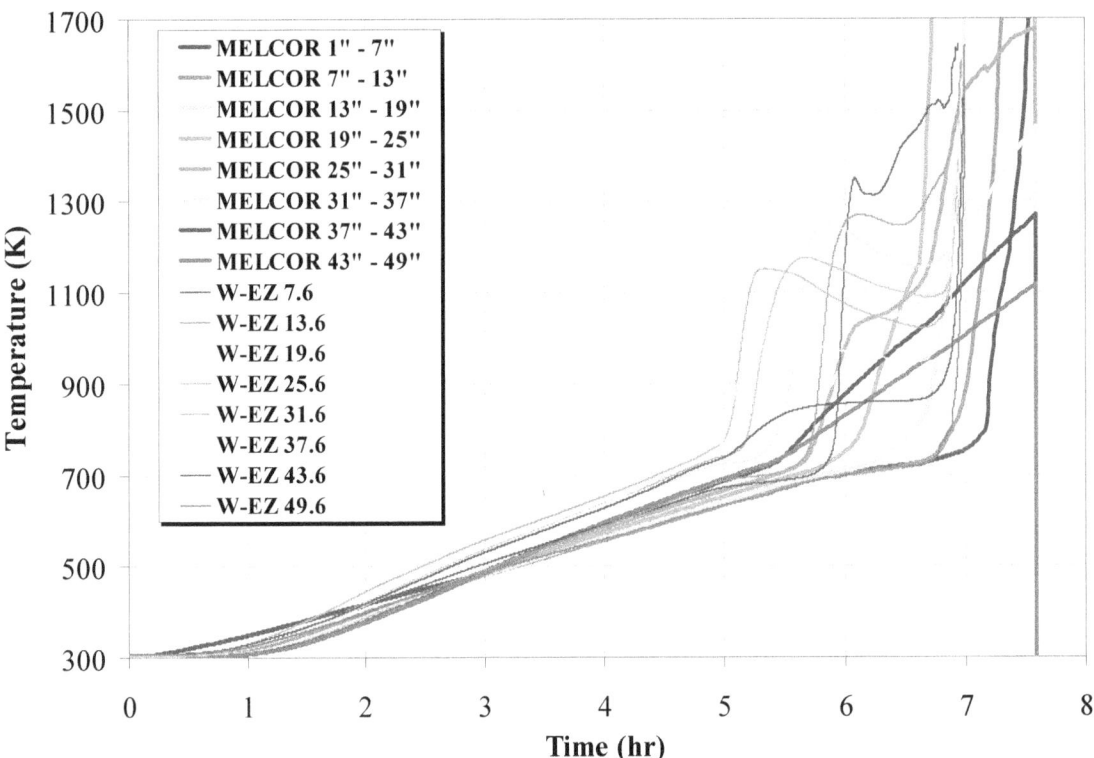

Figure 8.33   Comparison of temperatures inside the peripheral assembly and MELCOR model

8.4   MELCOR Sensitivity Study

While the representation of the center assembly ignition by the base case MELCOR model was relatively good, the time to ignition was slightly over predicted. Since this is not conservative, a sensitivity study was conducted to determine which parameters could explain the discrepancy. In the hour before ignition, radiation was the most important heat transfer mechanism and zirconium oxidation was increasing in importance as an energy input. The two parameters considered were the emissivity of the zirconium oxide surfaces, which effects the radiation heat transfer, and the Zircaloy oxidation kinetics transition to breakaway, which effects the energy input from zirconium oxidation. Although these two parameters are not the only possible explanation for the differences between the MELCOR simulations and the test results, they do appear to be the variables with the greatest influence to reconcile the MELCOR results.

As mentioned in the previous section, an increased heat-up rate was observed during the ignition test at temperatures around 1090 K. Adjustment of the zirconium oxide emissivity could be used to bring the MELCOR calculation in agreement with the increased heat-up rate observed; however, an aphysically low emissivitiy (~0.10) was required. Furthermore, the agreement between the MELCOR oxygen and the oxygen data was not significantly improved. Therefore, error in the zirconium oxide emissivity is not considered a viable explanation for MELCOR missing the increased heat-up rate after 1090 K and thereby slightly over predicting the time to ignition.

Modification of the transition from pre- to post-breakaway Zircaloy oxidation kinetics brought the MELCOR calculation into agreement with both the increased heat-up rate and the oxygen concentration history observed. There is also great significance to temperatures around 1090K because this happened to be coincident with an alpha to beta phase change in Zircaloy that mechanistically could result in an oxidation kinetic rate change. The remainder of this section details the changes made to the baseline model to explore this sensitivity.

8.4.1 Modification to Kinetics Model

As previously asserted, the increased oxidation is assumed to be linked to the alpha to beta phase change occurring at ~1090 K. Figure 8.34 shows the circumferential coefficient of thermal expansion (CTE) for Zircaloy as a function of temperature [4]. The MATPRO manual further lists the transition for the phase change occurring from 1083 to 1244 K. The sharp change in the circumferential, as well as the axial, CTE leads surprisingly to decrease in volume as the material changes from the alpha to beta phases. This volume change is hypothesized to provide a mechanism for increased oxygen diffusion through the outer crust of Zirconium oxide via fracture and rupture of this external surface.

Figure 8.35 shows the pre- and post-breakaway oxidation correlations from the ANL Zircaloy oxidation study [2]. For the default MELCOR model, the oxidation reaction follows the pre-breakaway correlation until an auxiliary time-at-temperature function reaches a preset value. The reaction rate then jumps to the post-breakaway correlation. Note that in the ANL study, the kinetic rates determined were based on pre- and post reaction weight measurements and the reaction was conducted under isothermal conditions. Furthermore, only one temperature considered (1173 K) was above 1090K. This means that the ANL study provides no information on how the kinetics transition from pre- to post-breakaway and would not have detected any effects resulting from the alpha to beta phase change.

As shown in Figure 8.35, the MELCOR oxidation correlations were modified for this study to ramp up to 5.3 times the pre-breakaway rate, from 1065 to 1115 K. The same time-at-temperature function from the default model is still tracked and determines the shift from the modified kinetics curve to the post-breakaway correlation. The bounding temperature values of 1065 to 1115 K were chosen based on the phase change temperature (1090 K) and the difference between the peak and average temperatures (50 K or ± 25 K) in the center assembly near the location of the initial ignition (see Figure 8.36). The reaction rate increase of 5.3 times was chosen to best match the heat-up rate after 1090K as well as the time of ignition. Although empirical, this modification is not considered to be unreasonable and is fully consistent with the ANL kinetics.

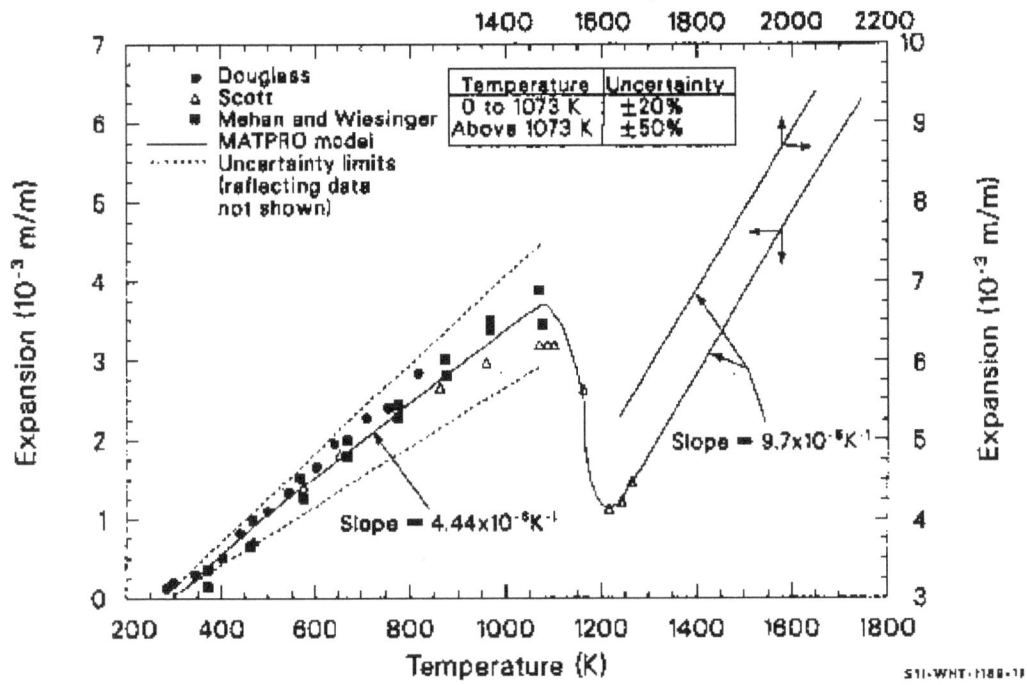

Figure 8.34 **Comparison of CTE prediction with Kearns' model for Zircaloy in the circumferential direction**

Note: Reproduced from MATPRO manual [4].

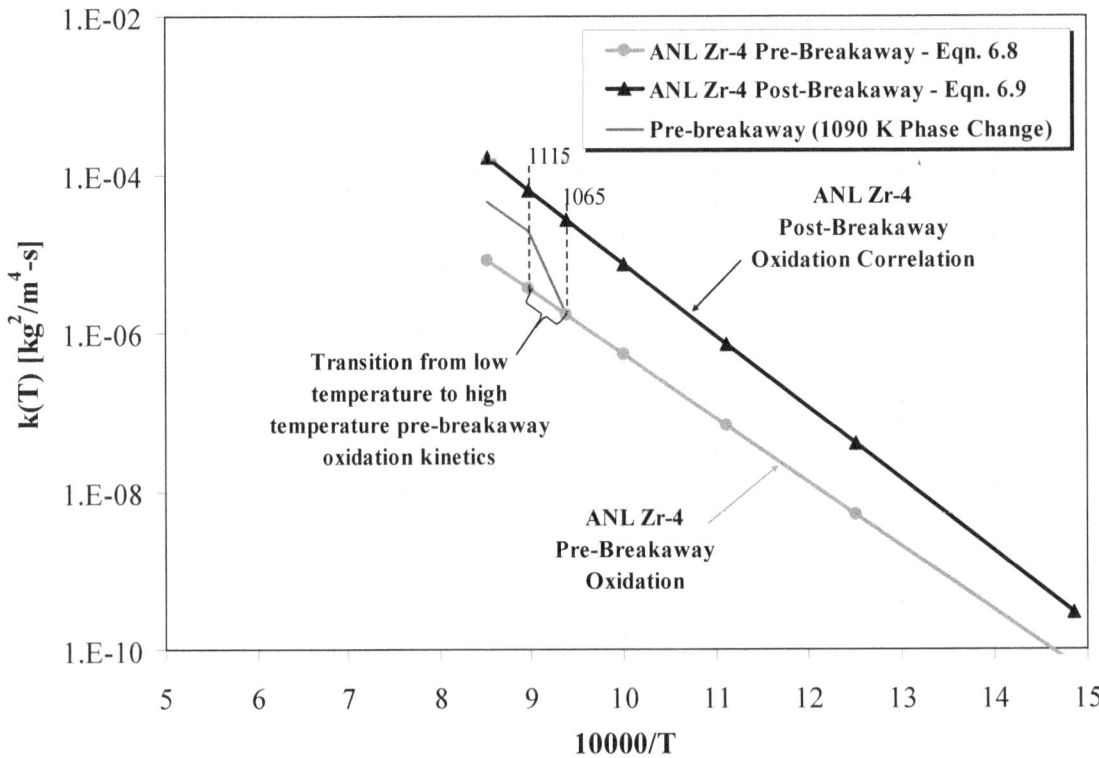

Figure 8.35 **Modification of the default MELCOR oxidation kinetics model**

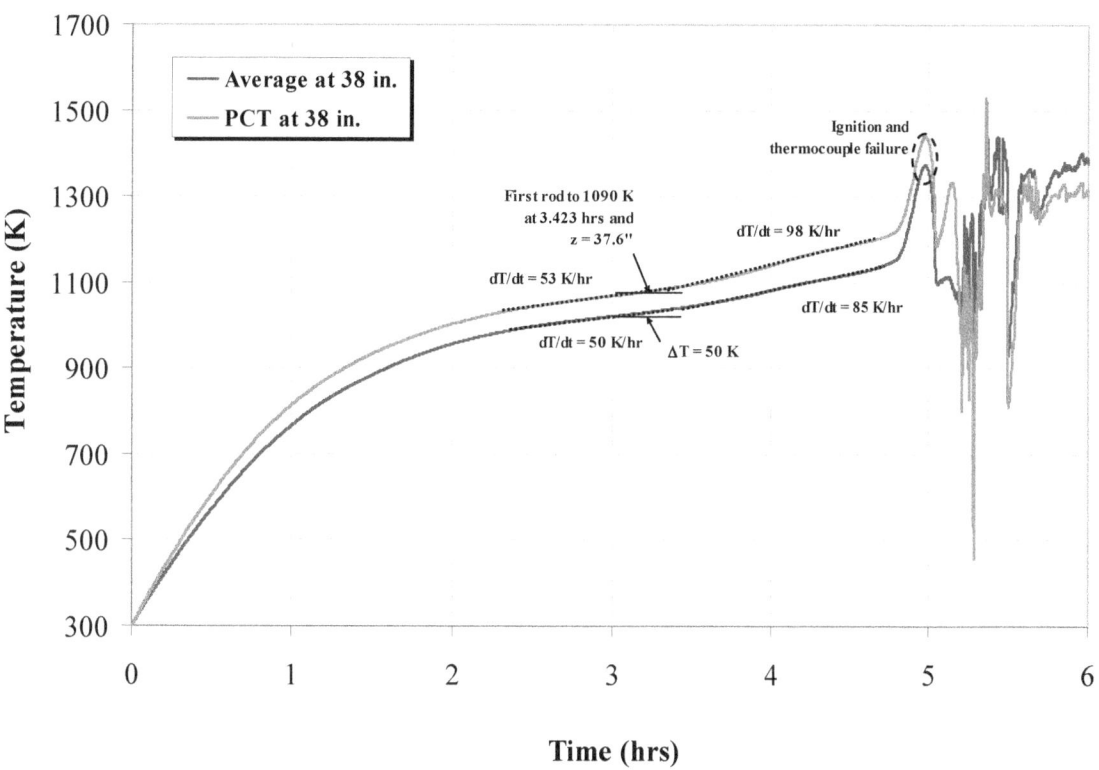

Figure 8.36   Average and PCT as a function of time in the center assembly at $z = 37.6$ **in.**

8.4.2   Comparison of Results

Figure 8.37 compares the oxygen monitor results with the predicted values from the oxidation sensitivity case. The depletion in the center assembly was modeled to within 1 minute based on a 15% criterion. The sensitivity case also captured the moderate decrease in oxygen concentration from 3 to 4.83 hours. The peripheral assemblies ignited just over 1 hour faster in the simulations. This discrepancy is likely due to the inability of the model to capture the radial temperature gradient in the peripheral assembly.

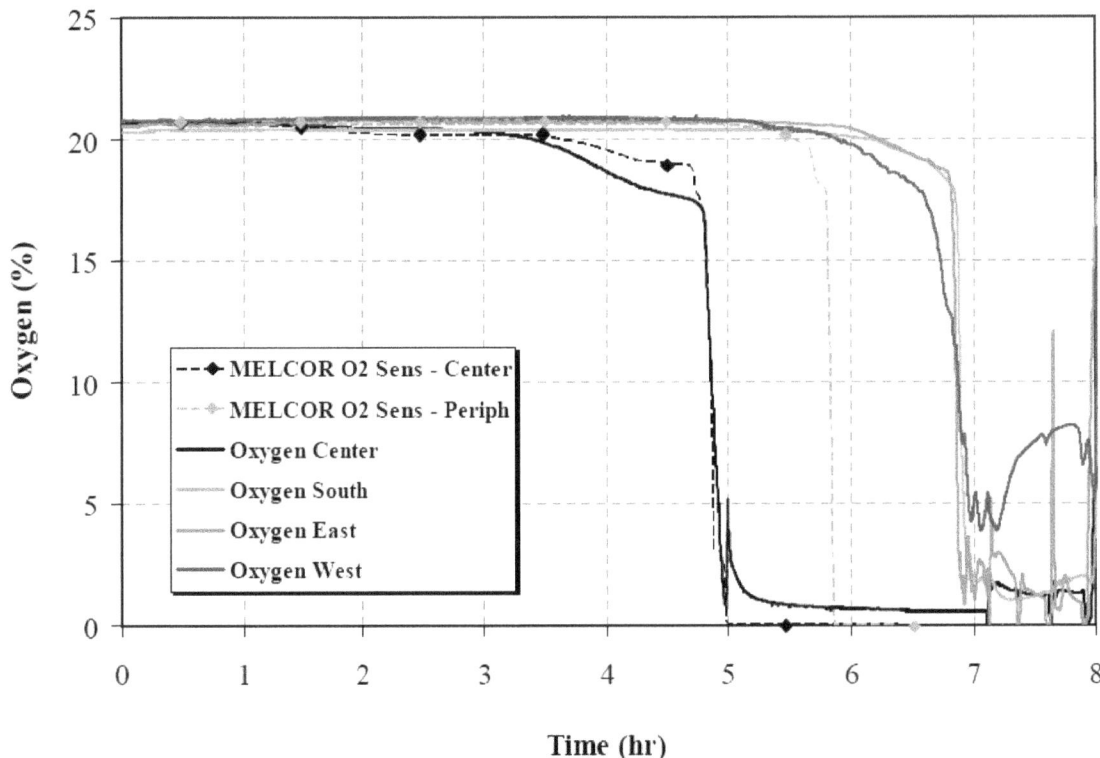

Figure 8.37 Comparison of oxygen concentrations for the different assemblies inside the test apparatus and the MELCOR sensitivity model

Figure 8.38 shows the thermal response of both MELCOR models and two data series, the axial maximum of all radial PCTs and the axial maximum of all radially-averaged temperatures. The MELCOR baseline and sensitivity models track with the maximum of the PCTs up until 3 hours, which coincides with model temperatures of 1065 K. At times past 3 hours, the sensitivity model begins to deviate from the baseline and falls between the two data curves. The time of ignition was again predicted to within 1 minute, here based on a criterion of 1220 K.

Similar to Figure 8.38, the thermal response of the peripheral assemblies channel box and bundle are shown in Figure 8.39 and Figure 8.40, respectively. The models trend with the axial maximum of the radially-averaged temperatures until ignition in the center assembly, or 4.83 hours. The sensitivity model then follows the general trend of the lower data series but ignites 1 hour earlier than measured. This difference in ignition times is attributed to the lack of the model to fully appreciate the radial temperature gradient. Therefore, the model could not capture the quenching effect of radiative heat loss within the peripheral assemblies on the oxidation reaction.

Finally, the modified kinetics model was applied to the full-scale ignition test to determine its applicability to other situations. Figure 8.41 shows the thermal response of the full-scale apparatus and both models. Using the modified kinetics leads to an ignition time within 10 minutes of the measured response, or a 2.3% error. The impact of the sensitivity case appears to have minimal impact on the overall response of the full-scale ignition test.

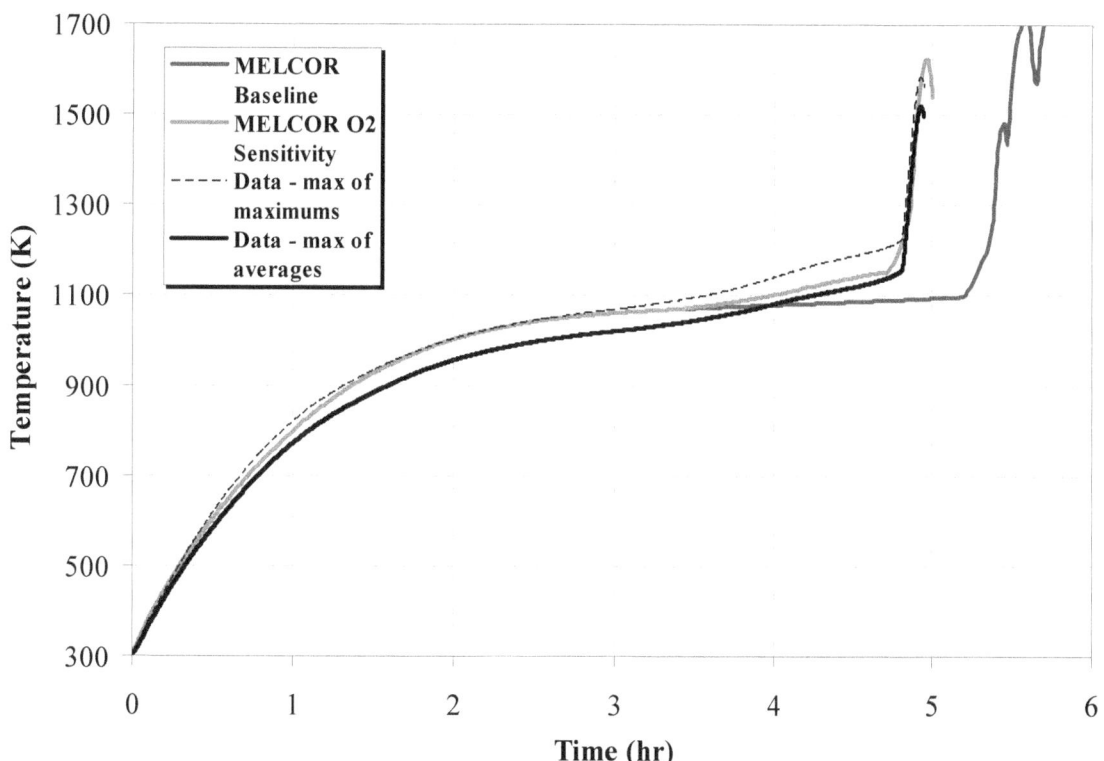

Figure 8.38   Comparison of the maximum and average temperatures in the center assembly to the MELCOR baseline and sensitivity models

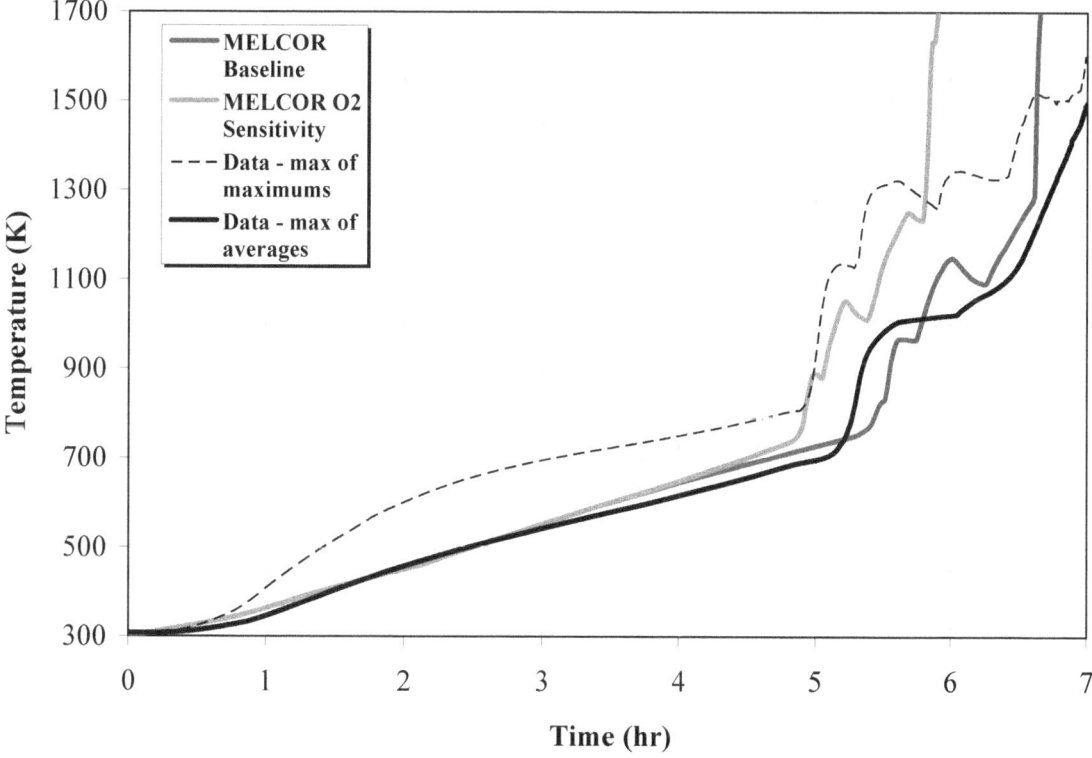

Figure 8.39   Comparison of the maximum and average temperatures in the peripheral canister to the MELCOR baseline and sensitivity models

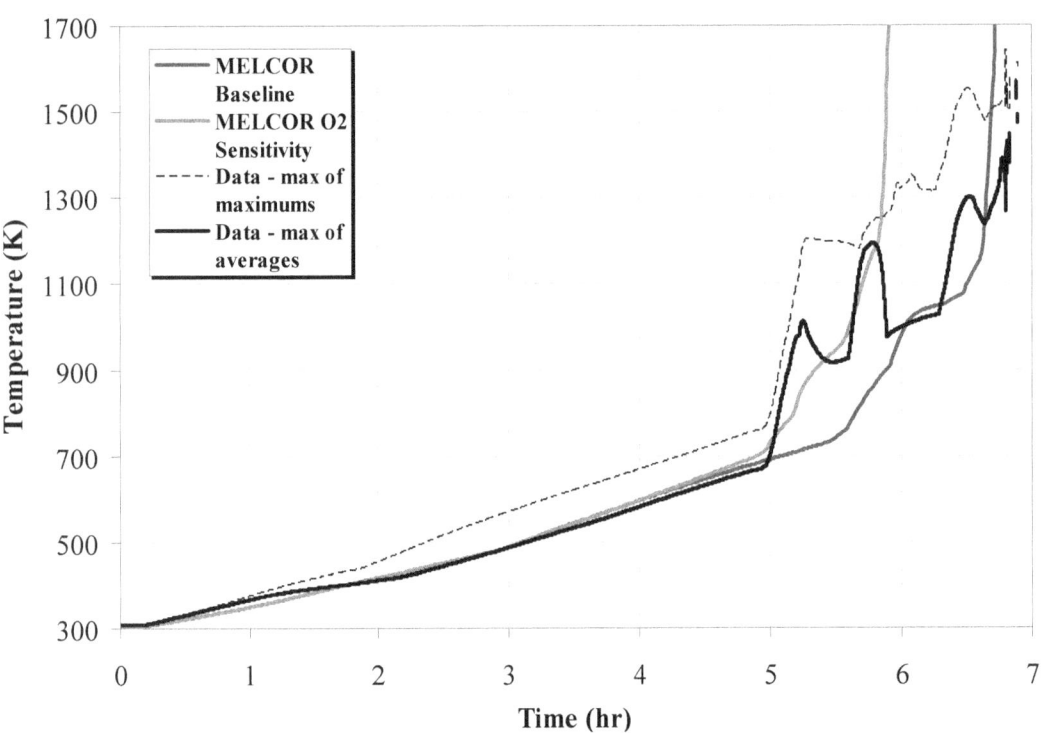

Figure 8.40   Comparison of the maximum and average temperatures in the peripheral assemblies to the MELCOR baseline and sensitivity models

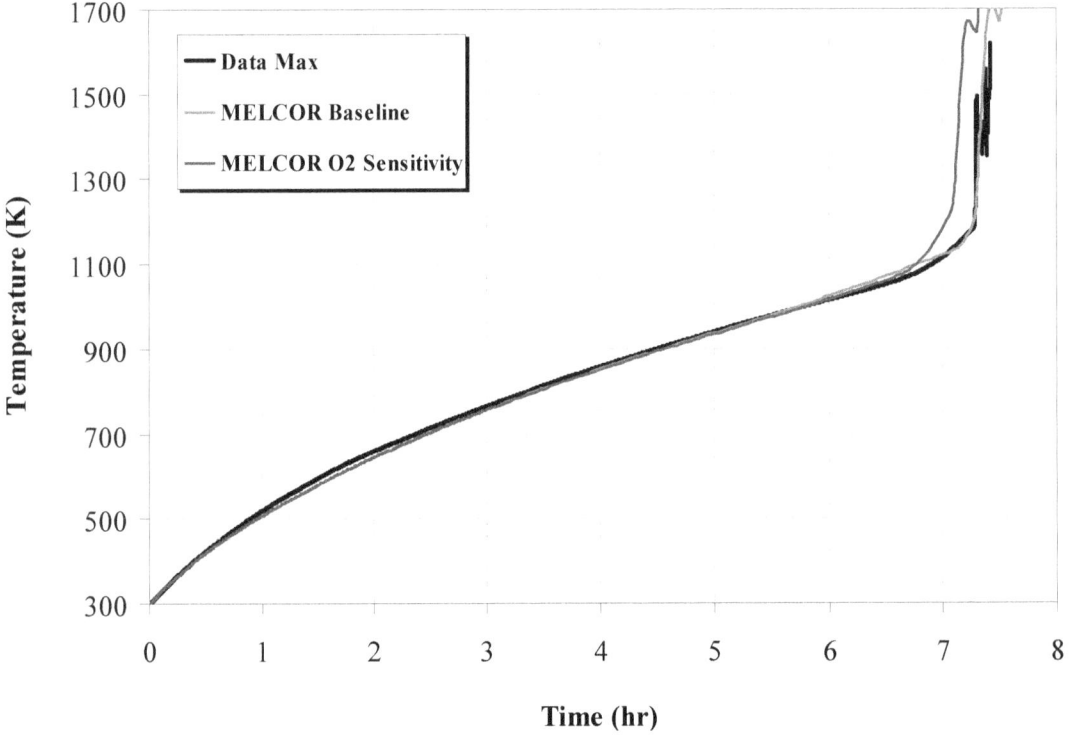

Figure 8.41   Effect on temperature response in the full-scale Zircaloy MELCOR model to the modified oxidation kinetics model

8.5   Summary

The 1×4 ignition experiment represents the second integral effects test of the Zircaloy ignition phenomena and propagation program. This test utilized an array of five electrically-heated prototypic partial-length assembly segments. In addition, the assemblies were constructed of Zircaloy-clad heater rods and allowed for flow through the bundle and annular regions of each assembly. The objective of the test was to study the thermal history, ignition, and radial burn propagation behavior of a central heated assembly when adjacent to cold neighbors. The experimental approach was designed to represent a one-third axial piece of prototypic assemblies located in the upper portion of the fully populated bundle region. Each assembly was supplied with metered, heated air. The results of the full-length ignition test and MELCOR modeling were used to determine the flow and temperature histories to impose as boundary and inlet conditions.

The center assembly was heated at 4.87 kW, which is the full-length equivalent of 17.5 day old fuel. This hot assembly was surrounded by four peripheral assemblies representing fuel at the average pool background power, or 0.2 kW. The center assembly ignited at the 34.6-in. level after 4.83 hours and burned down to the bottom tie plate by 6.1 hours. The peripheral assembly ignited near bottom at 6.8 hours and burned upwards. Post-mortem inspection revealed severe damage to both the pool rack and fuel assemblies.

The initial MELCOR model estimated the time to ignition in the center assembly within 31 minutes based on a 15% oxygen concentration criterion. Propagation into the peripheral assemblies was predicted to within 3 and 11 minutes for the west and south, respectively. The model heat-up matched the experiment within 25 K for the first 4 hours. The apparatus then heated considerably faster than predicted, perhaps due to a change in the oxidation kinetics.

A sensitivity case was created in MELCOR to study the effect of an increased reaction rate at temperatures of a known alpha to beta phase change in Zircaloy. Many other parameters were considered and while consideration of the reaction kinetic parameter is reasonable it is not the only parameter that could be used to explain the small discrepancy in the MELCOR's prediction of the center ignition time. The default MELCOR reaction kinetics were increased from the pre-breakaway correlation by a factor of 5.3 times over the temperature range of 1065 to 1115 K. The kinetic rate increase was chosen such that the ignition in the center assembly was predicted to within 1 minute based on criteria of both 15% oxygen concentration and temperature of 1220 K. The sensitivity model predicted propagation of ignition in the peripheral assemblies 1 hour earlier than measured. This difference in ignition propagation time is attributed to the lumped analysis approach inherent to MELCOR that results in a lack of the model to fully capture the radial temperature gradients in the peripheral bundles. As a result, the model could not capture the quenching effect of radiative heat loss within the peripheral assemblies on the oxidation reaction.

9   REPORT SUMMARY

The Spent Fuel Pool Heatup and Propagation Phenomena Experimental program was conducted from April 2004 until November 2006; over which time seven experimental apparatuses were tested. The objective of this project was to provide basic thermal-hydraulic data associated with a SFP complete loss-of-coolant accident. The accident conditions of interest for the SFP were simulated in a full-scale prototypic fashion (electrically-heated, prototypic assemblies in a prototypic SFP rack) so that the experimental results closely represent actual fuel assembly responses. A major impetus for this work was to facilitate code validation (primarily MELCOR) and reduce questions associated with interpretation of the experimental results.

Table 9.1 summarizes the phased experimental approach employed to study Zircaloy fires in prototypic spent fuel assemblies. Three basic types of experiments were used to study these accident situations. As a proof of concept, two heater design tests were first performed to determine the suitability of the electrically-heated, Zircaloy-clad spent fuel rod simulators. Next, three separate effects tests were conducted to study and understand specific phenomena independently. Finally, two prototypic assemblies were heated to ignition in the integral effects test series.

Table 9.1   Summary of the testing elements in the experimental program

Description	Purpose	Assembly	Rod material
Heater Design	Test electrical heater performance, preliminary data on Zircaloy fire	12 rod bundles	Zircaloy
Separate Effects	Hydraulics – Determine form loss and laminar friction coefficients	Prototypic	Stainless Steel
Separate Effects	Thermal hydraulics – Measure naturally induced buoyancy driven flow rates.	Prototypic	Incoloy
Separate Effects	Thermal radiation – Radiation coupling in a 1×4 arrangement.	Prototypic – Partial length	Incoloy
Integral Effects	Axial ignition – Temperature profiles, induced flow, axial O_2 profile, nature of fire.	Prototypic	Zircaloy
Integral Effects	Radial propagation – Determine nature of radial fire propagation.	Prototypic – Partial length	Zircaloy

9.1   Heater Design Tests

The main objectives of the two heater design tests were to prove that electrical heater rods could simulate spent fuel under accident conditions. The internal Nichrome heating element was able to successfully bring the Zircaloy cladding up to air ignition temperature before failure. In addition, the thermal mass of the magnesium oxide insulation packing was in excellent agreement with that spent fuel over a wide temperature range. As a result, the initial heat-up rates closely represented the heating of spent fuel rods.

9.2   Separate Effects Tests

For the first separate effects test series, a highly prototypic BWR assembly was constructed using prototypic components and stainless steel rods. The commercial components included the top and bottom tie plates, seven spacers, two water rods, the channel box, and all the nuts, springs, and washers used in the assembly process. The fuel rods were constructed of 7/16-in. stainless steel tubing that was within 0.002 in. of the diameter of a prototypic fuel rod. The pins were machined from stainless steel based on vender drawings and welded to the tubing ends. Extensive hydraulic characterization was conducted to determine the appropriate viscous and form loss coefficients for use with the MELCOR severe accident analysis code. The parameters determined were different from previously used and accepted textbook parameters. The experimental range spanned the laminar region with volumetric flow rates of 1.1×10^{-3} to 1.3×10^{-2} m^3/s. The resulting Reynolds numbers based on the bundle velocity and hydraulic diameter were 70 to 900, respectively. These experiments included testing with the water rods in the assembly, both blocked and unblocked.

For the second separate effects testing series, benchmark experiments were conducted with a highly prototypic electrically-heated BWR spent fuel assembly to measure thermal and hydraulic response. Peak cladding temperatures for both experiments and MELCOR simulations were within 5% error for all flow configurations and for all assembly power inputs, which ranged from 200 W to 2500 W. Naturally-inducted flow rates were within 10% error over the entire power input range between tests and MELCOR. The overall excellent agreement of the MELCOR calculations with the experimental data is attributed to the direct application of prototypic viscous and form loss coefficients determined in previous, unheated hydraulic characterization.

For the third separate effects testing series, an array of five electrically-heated prototypic partial length assembly segments were used to study thermal radiation effects. The experimental approach was designed to represent a one-third axial length from a grouping of five prototypic assemblies. The grouping arrangement studied was a central heated assembly with an adjacent assembly on each side. The assembly segments were located in an analogously shortened but otherwise highly prototypic 3×3 pool rack fabricated by Holtec Inc. Flow into the bottom and out of the top of each assembly was blocked in order to minimize convection and maximize the significance of thermal radiation heat transfer. The center assembly was rapidly heated to temperatures of 900 K and abruptly stopped. The temperature histories of the center and peripheral assemblies were analyzed with two codes, MELCOR and COBRA. Both codes were in very good agreement with the experimental data. MELCOR predicted the center assembly response better than the peripheral assembly response. This is due to assumptions built into MELCOR that allow axial temperature gradients in the fuel assemblies but no radial temperature gradients. Since prediction of initial ignition in the center assembly is of primary importance, the small deficiency in the peripheral assembly is acceptable.

9.3   Integral Effects Tests

In the first integral effects test, a single, highly-prototypic full length assembly was allowed to heat up until the Zircaloy-clad heater rods ignited. The objective was to measure the thermal history leading to ignition, the axial burn front advance rate, the buoyancy driven flow leading up to and during the burn, and the axial oxygen concentration profile. The results of these measurements were used to set boundary conditions in the second integral effects test. Ignition occurred after 7.3 hours of heating near the top of the full populated bundle region 96 in. above the bottom tie plate and 60 in. below the top tie plate. The MELCOR model predicted the peak cladding temperature of the assembly to within 40 K at all times and the ignition time to within 5 minutes.

The 1×4 ignition experiment represents the second integral effects test of the Zircaloy ignition phenomena and propagation program. This test utilized an array of five prototypic partial length assembly segments. The assemblies were constructed of Zircaloy-clad heater rods and allowed for flow through the bundle and annular regions of each assembly. The objective of the test was to study the thermal history, ignition, and radial burn propagation behavior of a central heated assembly when adjacent to cold (unheated) neighbors. The experimental approach was designed to represent a one-third length axial piece of prototypic assemblies located in the upper portion of the fully populated bundle region. Each assembly was supplied with metered, heated air. The results of the full-length ignition test and MELCOR modeling were used to determine the flow and temperature histories to impose as boundary and inlet conditions. The center assembly ignited at the 34.6-in. level after 4.83 hours and burned down to the bottom tie plate by 6.1 hours. The peripheral assemblies ignited near bottom at 6.8 hours and burned upwards. Initial MELCOR modeling simulated the time to ignition in the center assembly to within 0.5 hours. Sensitivity of the model to the default oxidation kinetics correlations was also investigated, resulting in code predictions within 1 minute of the measured ignition in the center assembly.

9.4 Key Findings

The Spent Fuel Pool Heatup and Propagation Phenomena project was conducted from April 2004 until November 2006; over which time seven unique experimental apparatuses were tested at Sandia National Laboratories in Albuquerque, New Mexico. In addition to these experiments, extensive simulation efforts were undertaken with the MELCOR severe accident modeling code to understand and predict the behavior observed in the tests. The key findings from this integrated experimental and simulation program are:

- Electrically heated spent fuel rod simulators can be fabricated with Zircaloy cladding to accurately represent the decay heat, thermal mass and Zircaloy reactivity of a prototypic spent fuel rod.

- The measured form and friction loss coefficients of a prototypic BWR assembly were significantly different from generally accepted values. Use of the measured coefficients was vital for accuracy when calculating (with MELCOR) the naturally induced flow in a heated, prototypic BWR assembly.

- Incorporation of "breakaway" Zircaloy oxidation kinetics into MELCOR was vital for accurately capturing the Zircaloy heat-up to ignition and oxygen consumption.

- For the full length ignition test, the MELCOR model predicted the peak cladding temperature (PCT) of the assembly to within 40 K at all times and the time of ignition to within 5 minutes.

- For the 1×4 ignition experiment, the standard MELCOR model predicted ignition in the center and peripheral assemblies to within 30 and 15 minutes, respectively. The error in ignition timing between the simulations and experiment is approximately 10%. The difference in timing is likely due to the inability of the lumped parameter approach used in MELCOR to account for steep radial temperature gradients.

- Post-mortem examination of the integral test assemblies revealed gross distortion of the pool rack and channel box, rubblization of the tubing bundle and accumulation of debris on the bottom tie plate that resulted in flow blockage. Flow blockage was also evident from molten aluminum (originating from Boral plates built into the pool rack) that collected on and below the bottom tie plates.

9.5   Uncertainties and Potential Mitigating Factors

In the course of this research, some uncertainties and approaches to mitigating the postulated accident were noted. These potential mitigating factors were not further evaluated as they were outside the study's scope.

- Characterization of a BWR assembly with the channel box removed was considered as a possible mitigation strategy. Removal of the BWR canister could increase the naturally induced flow rate and thus improve the coolability of the assembly. This hypothesis was noted on this research but was not quantified or confirmed.

- The independent flow paths evaluated as part of this study (i.e. bundle, water rods, and bypass) were quantified by indirect methods. Direct measurements using techniques such as laser Doppler anemometry were not used, so this increased uncertainty in the test results. The partitioning of flow is of much greater importance in a PWR assembly, where the flow can exit the fuel bundle into the annular region along the entire length of the assembly.

- The presence of both viscous and form loss in pure bundle runs indicates possible entry or exit effects from the spacers. Direct flow velocity measurements in the bundle region to confirm the presence of these disturbances or other unexpected flow structures were not made.

- Water sprays may represent a possible mitigation approach for a complete-loss-of-coolant accident. If the supply of makeup water is not adequate to keep the pool from draining, spraying the available water onto the assemblies may be effective cooling strategy. Currently, the amount of water needed and the best delivery approach is uncertain.

- Flooding the pool with an inert gas is another possible mitigation strategy to prevent or halt the combustion of Zircaloy components. The inert gas could displace the oxygen and thereby shut down the oxidation reaction. Cooling would still need to be considered to avoid releases due to ballooning of the fuel cladding and other high temperature effects.

10 REFERENCES

1. NUREG-1738, "Technical Study of Spent Fuel Pool Accident Risk at Decommissioning Nuclear Power Plants," U.S. Nuclear Regulatory Commission, Washington, DC, February 2001.

2. NUREG/CR-6846, "Air Oxidation Kinetics for Zr-Based Alloys," prepared by Argonne National Laboratory, U.S. Nuclear Regulatory Commission, Washington, DC, July 2004.

3. NUREG/CR-6119, "MELCOR Computer Code Manuals, Reference Manual, Version 1.8.5, Vol. 2, Rev. 2," U.S. Nuclear Regulatory Commission, Washington, DC, September 2001.

4. NUREG/CR-6150, "SCDAP/RELAP5/MOD 3.2 Code Manual," U.S. Nuclear Regulatory Commission, Washington, DC, July 1998.

APPENDIX A. SFP HEATER DESIGN TEST 1 DATA SUMMARY

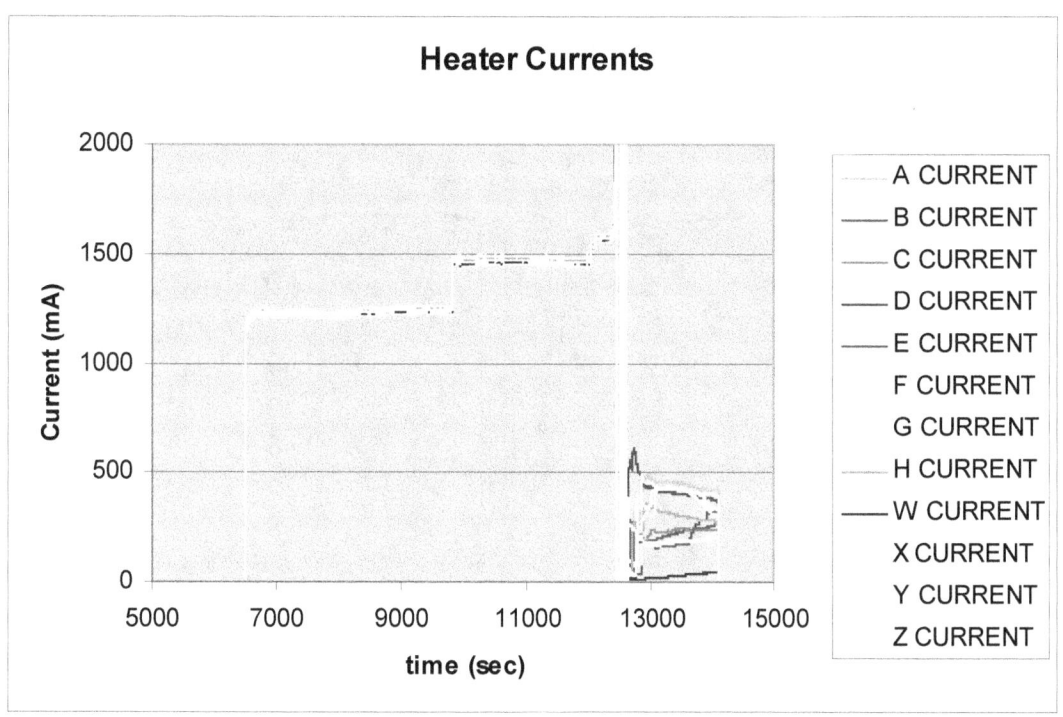

Figure A1

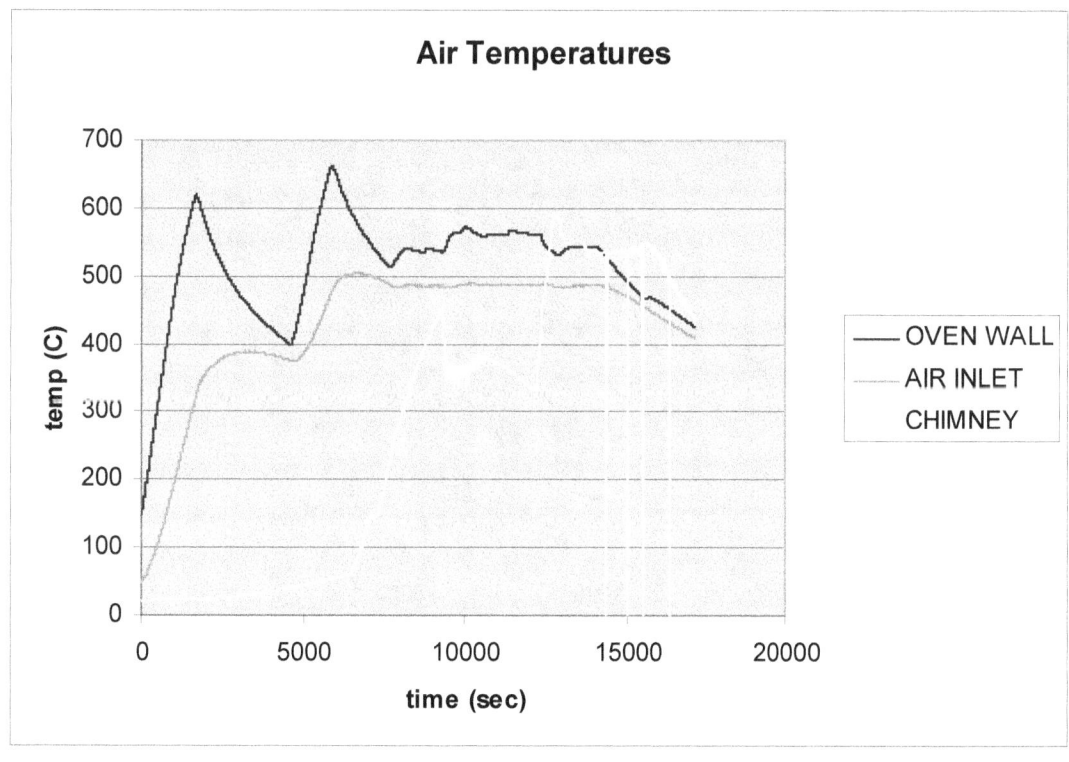

Figure A2

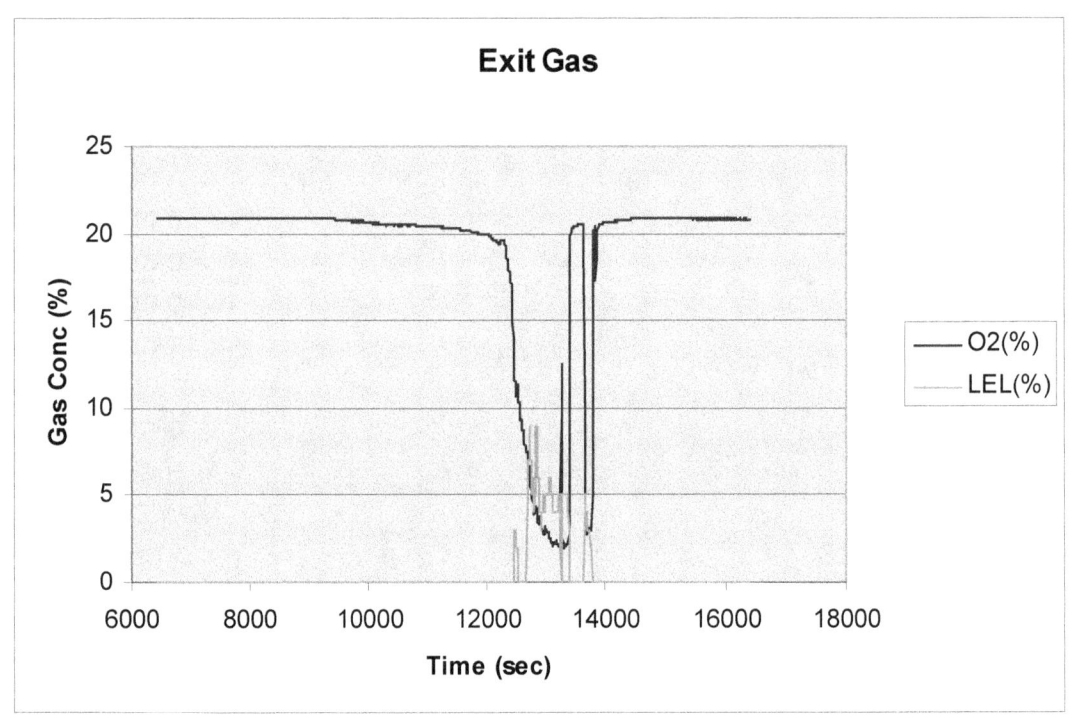

Figure A3

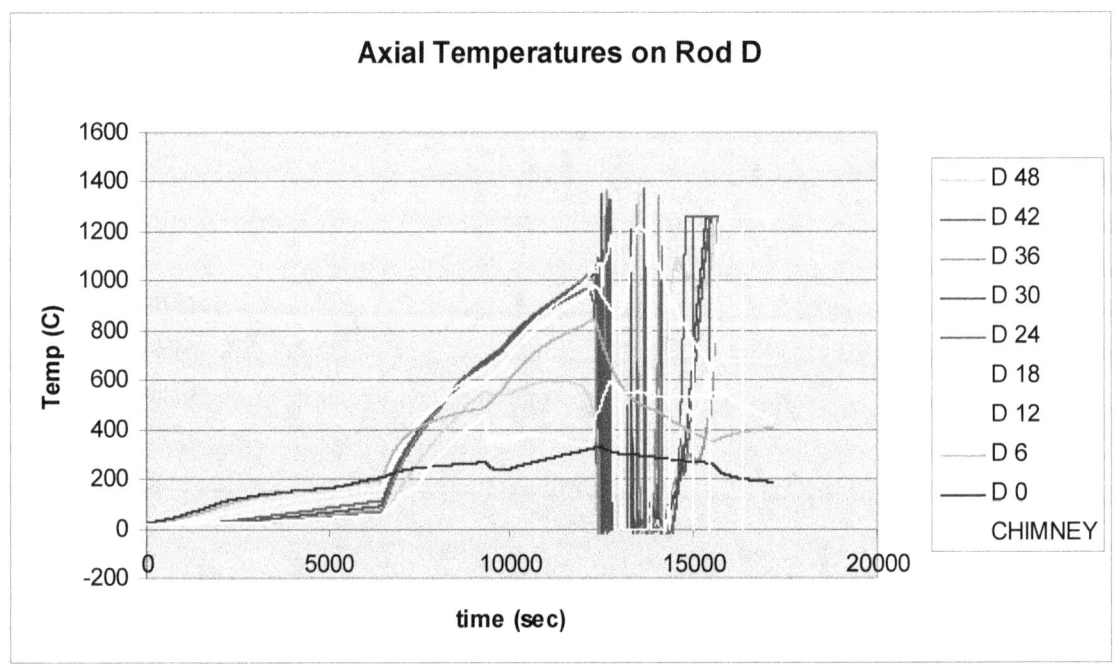

Figure A4

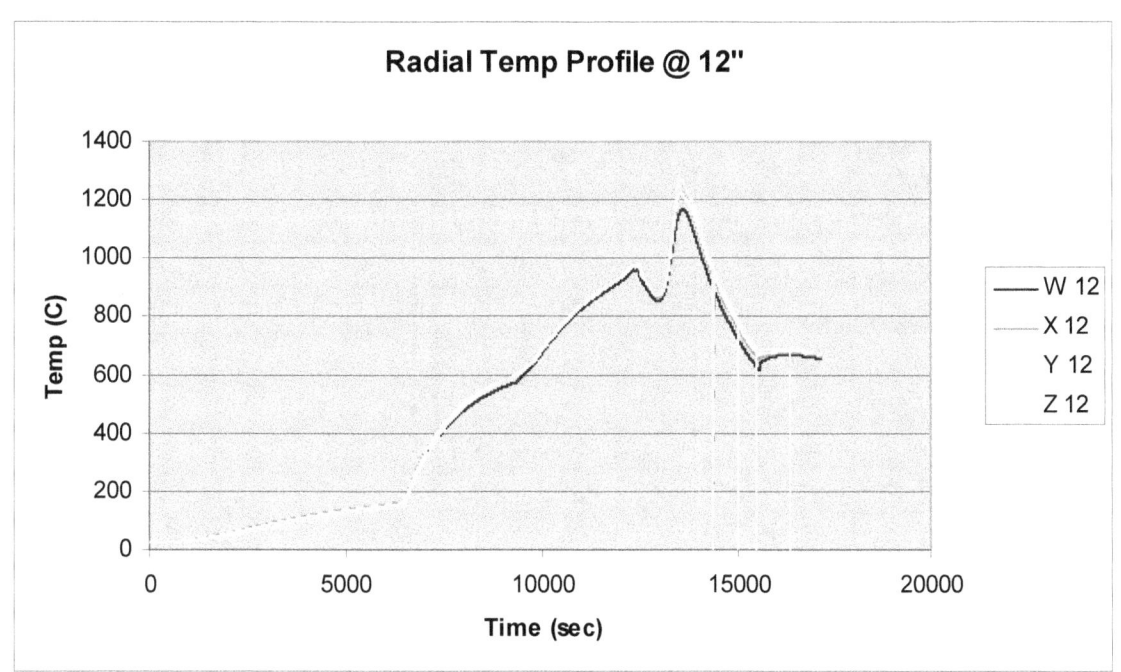

Figure A5

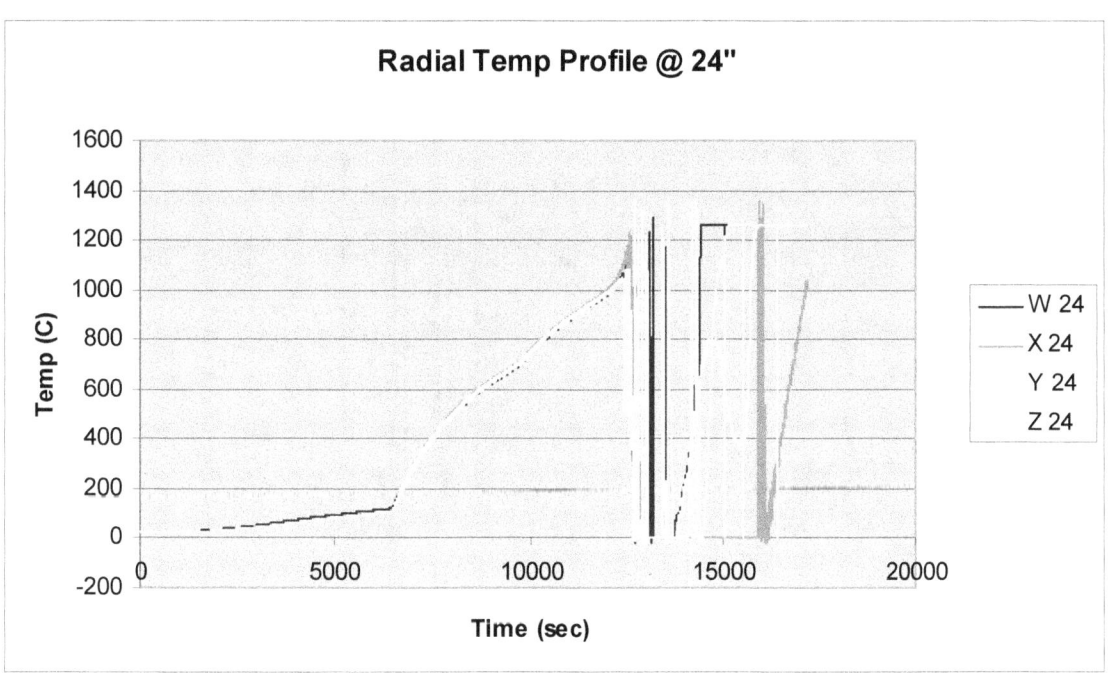

Figure A6

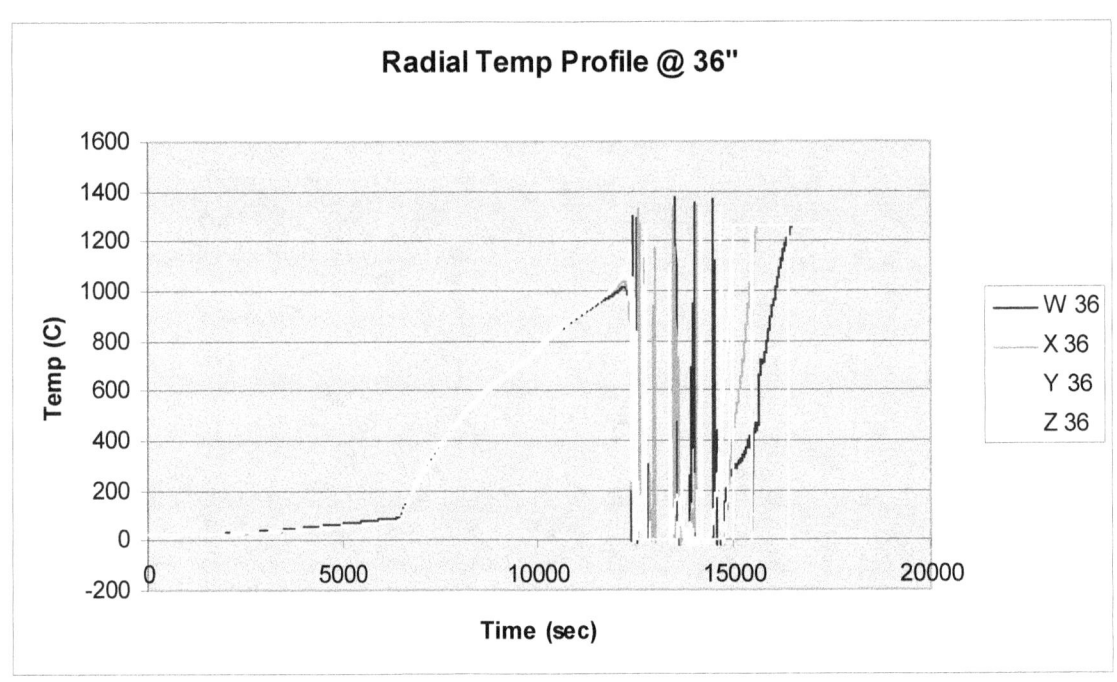

Figure A7

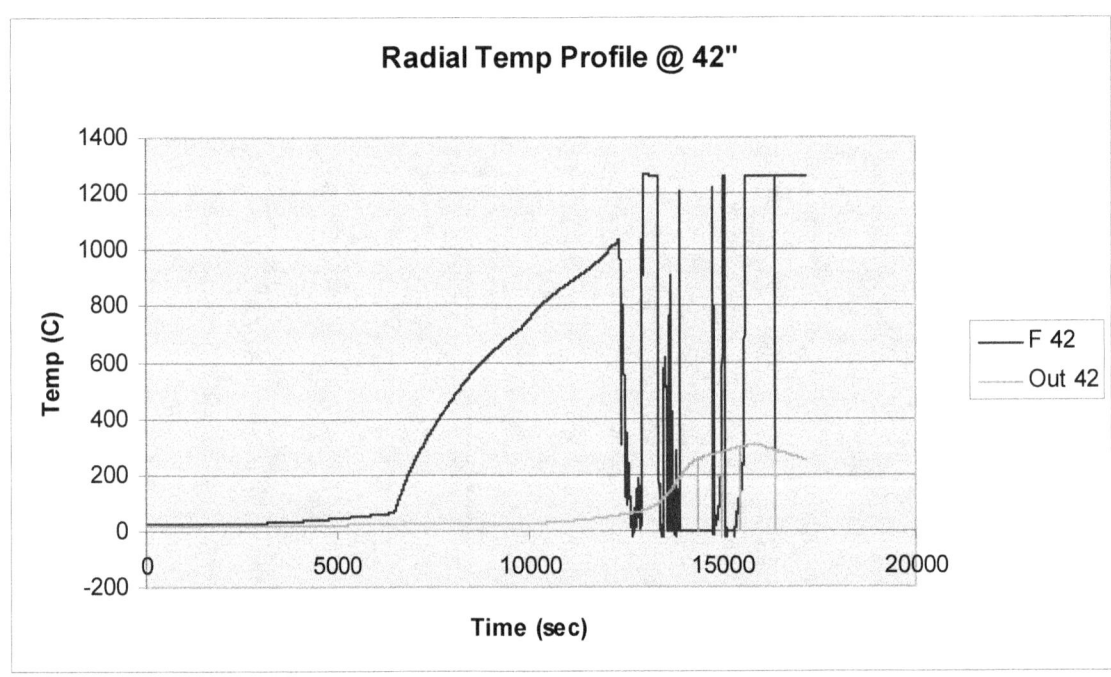

Figure A8

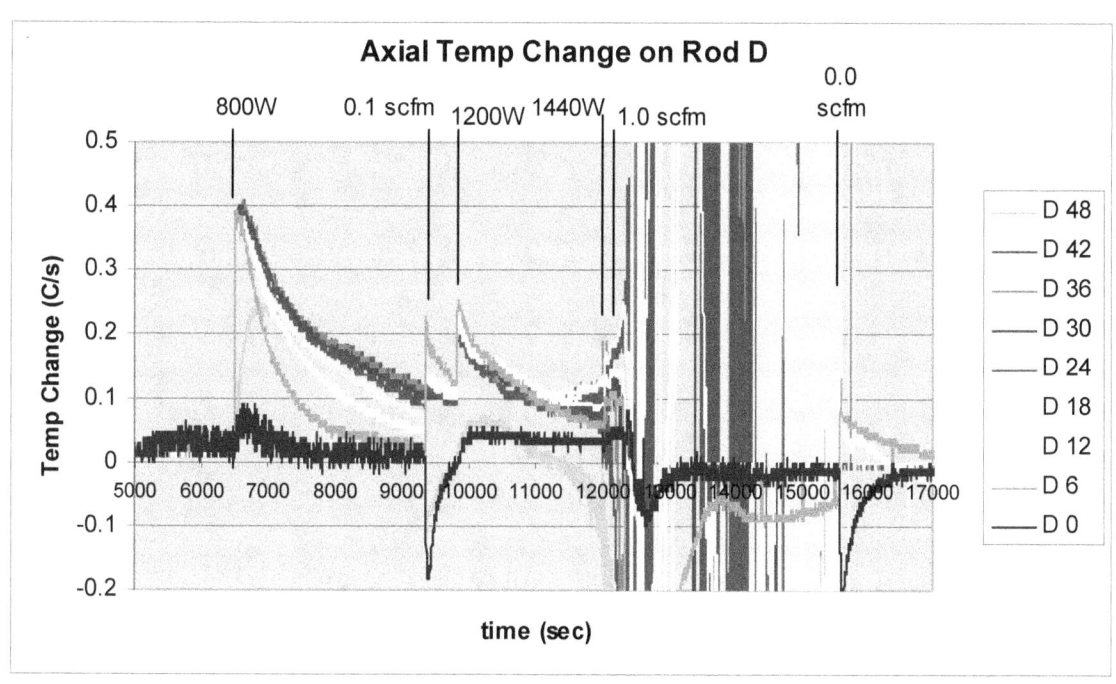

Figure A9

APPENDIX B. ERROR ANALYSIS

The error and uncertainty inherent to an experimental result are critical to the accurate interpretation of the data. Therefore, the uncertainties in the experimental measurements are estimated in this section. Results of this analysis are given, followed by a general description of the method used and a brief explanation of the source of each reported measurement uncertainty.

The overall standard uncertainty of an indirect measurement y, dependent on N indirect measurements x_i, is defined in Equation 1. The standard uncertainty associated with an indirect measurement is analogous to the standard deviation of a statistical population.

$$u^2 = \sum_{i=1}^{N} \left(\frac{\partial y}{\partial x_i} u_i \right)^2 \qquad\qquad 1$$

Here, u is used to define the standard uncertainty of a measurement.

The expanded uncertainty U is reported in this appendix and defines the bounds that include 95% of the possible data. The expanded uncertainty is usually defined as some multiple of the standard uncertainty. Equation 2 shows the definition of the expanded uncertainty as used in the following sections.

$$U = 2 \cdot u \qquad\qquad 2$$

B.1 Uncertainty in Bundle Velocity for the Apparatus

The uncertainty in the bundle velocity was determined using error propagation analysis (EPA) for the blocked water rod measurements. Errors associated with auxiliary flows through the bypass and water rods are not considered. However, the magnitude of these flows is small compared with the highest, measured bundle flow rate. The bundle velocity was determined from Equation 3 in which Q_i is the volumetric flow rate in slpm for each flow controller, A_{bundle} is the bundle area, R is gas constant for air, T is the ambient air temperature, and P is the ambient air pressure. The first term in the equation represents the conversion from slpm to kg/s.

$$V_{bundle} = \left(\frac{0.001 \, m^3/1 \cdot \rho_{STP}}{60 \, s/min} \right) \frac{(Q_1 + Q_2)}{A_{bundle}} \left(\frac{R \cdot T}{P} \right) \qquad\qquad 3$$

Equation 4 gives the relation between the overall uncertainty of V_{bundle} and the contributions from the measurement uncertainties of Q, T, and P.

$$u_{V_{bundle}}^2 = \left(\frac{0.001 \, m^3/1 \cdot \rho_{STP}}{60 \, s/min \cdot A_{bundle}} \right)^2 \left[\begin{array}{c} \left(\frac{RT}{P} u_{Q_1} \right)^2 + \left(\frac{RT}{P} u_{Q_2} \right)^2 + \left(\frac{(Q_1 + Q_2)R}{P} u_T \right)^2 \\ + \left(\frac{(Q_1 + Q_2)RT}{P^2} u_P \right)^2 + \left(\frac{(Q_1 + Q_2)RT}{PA_{bundle}^2} u_{A_{bundle}} \right)^2 \end{array} \right] \qquad 4$$

Table B.1 summarizes the values used to determine the overall uncertainty of the bundle velocity. The overall uncertainty in V_{bundle} was found for the two highest volumetric flow rates, $Q_1 = 300$ slpm and $Q_2 = 300$ slpm, at a standard ambient condition of $T = 298$ K, $P = 83,400$ Pa, and $A_{bundle} = 9.8 \times 10^{-3}$ m^2. The standard uncertainty was determined to be $u_{V_{bundle}} = 0.01$ m/s. The uncertainty was most affected by volumetric flow rate, Q, which contributed 80% of the overall uncertainty.

Table B. 1   Measurement uncertainties and intermediate calculations for V_{bundle}

Measurement, x_i	Standard Uncertainty, u_i	Influence Coefficient $\dfrac{\partial(V_{bundle})}{\partial x_i}$	Section Containing Explanation
Volumetric Flow Rate, Q_1	3.0 slpm	$\dfrac{RT}{P}$	B.1.1
Volumetric Flow Rate, Q_2	3.0 slpm	$\dfrac{RT}{P}$	B.1.1
Ambient Air Temperature, T	1.1 K	$\dfrac{(Q_1 + Q_2)R}{P}$	B.1.2
Ambient Air Pressure, P	110 Pa	$\dfrac{(Q_1 + Q_2)RT}{P^2}$	B.1.3
Bundle Hydraulic Area, A_{bundle}	3.4×10^{-5} m^2	$\dfrac{(Q_1 + Q_2)RT}{PA_{bundle}^2}$	B.1.4

B.1.1 Uncertainty in Volumetric Flow Rate Q

The volumetric flow rate was controlled with two MKS volumetric flow controllers operated in parallel (Model # 1559A-24174). The uncertainty of the volumetric flow rate was determined from the stated manufacturer's upper uncertainty of 1% of full scale. The uncertainties in flow rate were 3 slpm for flow controllers 1 and 2. The value shown in Table B.1 represents these standard uncertainties associated with the volumetric flow rate.

B.1.2 Uncertainty in Ambient Air Temperature

The air temperature was measured with a standard k-type TC. The standard uncertainty for this type of TC is $u_T = 1.1$ K.

B.1.3 Uncertainty in Ambient Air Pressure

The air pressure was measured with a Setra Systems barometer (Model 276). The uncertainty of the ambient air pressure was taken from the manufacturer's calibration sheet, which indicated an uncertainty in the instrument of ±0.1% of full scale (110,000 Pa). Therefore, the standard uncertainty in the pressure reading is $u_P = 110$ Pa.

B.1.4 Uncertainty in Bundle Hydraulic Error

The inner dimension of the channel box was measured to within ± 0.127 mm (0.005 in). This tolerance leads to a standard uncertainty of 3.4×10^{-5} m^2 in the hydraulic area. The uncertainty in the inner channel box dimension dominates the uncertainty in the hydraulic cross-sectional area, including variations of ± 0.0254 mm (0.001 in) in the outer diameter of the rods.

B.2 Uncertainty in Pressure Drop Measurements

The manufacturer of the Digiquartz pressure transducers used in these experiments lists a *static error band* of ±0.02% of full scale. This error band includes repeatability, hysteresis, and conformance. Furthermore, these error bands consider the *zero-drift* of the instrument over periods of up to 14 years. Conversations with the manufacturer indicate the experimental procedure followed for these investigations, namely the zero flow measurements to correct any zero drift and the relatively short experimental data collection times (~ 2 minutes), should place the uncertainty in any pressure data closer to the resolution of the instrument, or 1 part per million of full scale. The observed noise level in the zero flow measurements was approximately twice the instrument resolution. This noise-based uncertainty is approximately 0.04 Pa, which is smaller than the plotted symbols in this report. Any spread in the data outside this range is believed to be caused by positioning errors associated with the installation and removal of the ported canister between experimental runs.

B.3 Uncertainty in S_{LAM} and k Coefficients

The following procedure was adopted to determine the uncertainty in the S_{LAM} and k coefficients. Because the greatest experimental uncertainty comes from the bundle velocity, the influence of the velocity on the quadratic curve fits was examined. The pressure drops across 2–8 and 8–17 for the blocked water rod assembly were curve fit as a function of $V_{bundle} \pm u_{V_{bundle}}$.

Figure B. 1 shows the resulting curve fits to the pressure drop data across 8–17. Using these curve fit coefficients, the error associated with the S_{LAM} and and k coefficients may now be determined. This procedure was also followed for the derived pressure drop across 2–17.

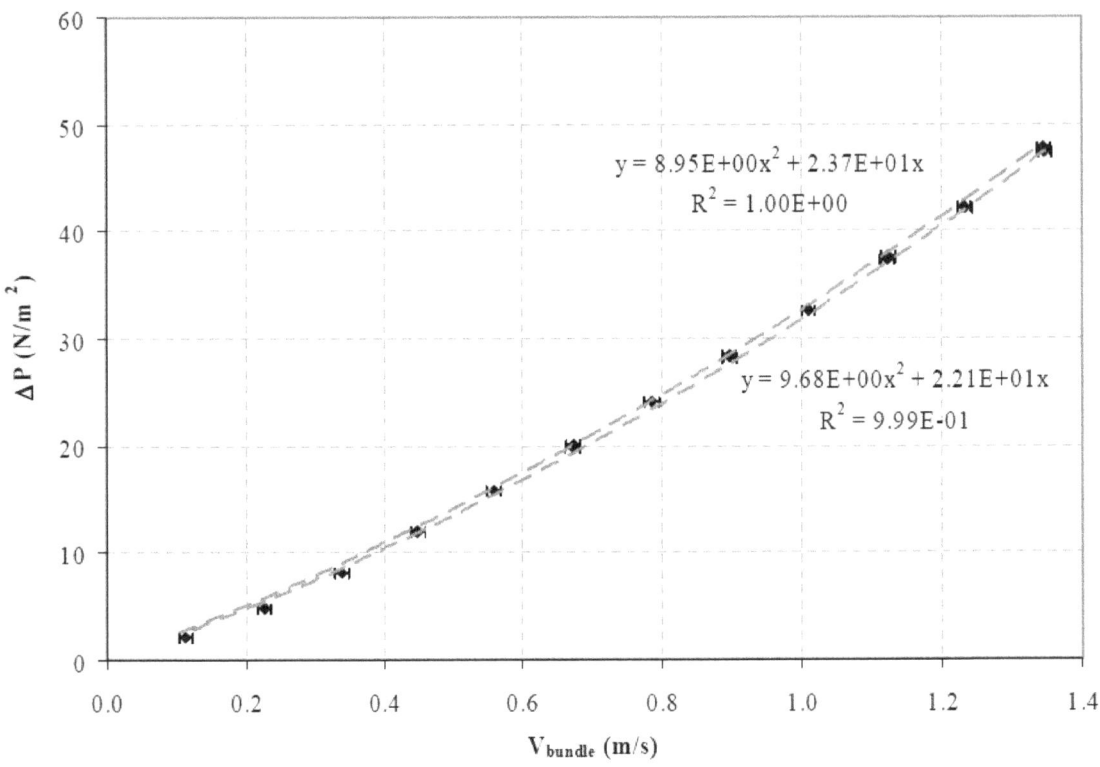

$$y = 8.95E+00x^2 + 2.37E+01x$$
$$R^2 = 1.00E+00$$

$$y = 9.68E+00x^2 + 2.21E+01x$$
$$R^2 = 9.99E-01$$

Figure B. 1   Pressure drop across 8–17 as a function of bundle velocity

The two dashed curves represent the quadratic fits to the data shifted by $\pm u_{V_{bundle}}$.

Table B. 2 summarizes the S_{LAM} and k coefficients determined from this error analysis. The hydraulic diameter and bundle velocity used for 2–17 was the same as 8–17.

Table B. 2 S_{LAM} and k coefficients showing the effect of the uncertainty in bundle velocity

Fitting velocity (m/s)	2–8		8–17		2–17	
	S_{LAM}	Σk	S_{LAM}	Σk	S_{LAM}	Σk
V_{bundle}	88	7.3	138	19.0	109	25.3
$V_{bundle} + u_{V_{bundle}}$	92	7.1	143	18.2	113	24.3
$V_{bundle} - u_{V_{bundle}}$	84	7.6	133	19.7	105	26.3

The uncertainty for S_{LAM} and k coefficients appears to be dependent on bundle location of the pressure data. The maximum differences in the S_{LAM} and k coefficients in Table B. 2 are taken to be the uncertainty, $u_{S_{LAM}} = 5$ and $u_k = 1$. These values represent the conservative limit.

B.4 Uncertainty in Thermal-Hydraulic Measurements

The uncertainty in volumetric flow rate was determined using EPA. The flow rates were determined from Equation 5 in which Q_i is the volumetric flow rate in slpm for each flow controller.

$$Q_{tot} = Q_1 + Q_2 \qquad\qquad 5$$

Equation 6 gives the relation between the overall uncertainty of Q_{tot} and the contributions from the measurement uncertainties of Q_i.

$$u_{Q_{tot}}^2 = \left[\left(u_{Q_1} \right)^2 + \left(u_{Q_2} \right)^2 \right] \qquad\qquad 6$$

Table B. 3 summarizes the values used to determine the overall uncertainty of the volumetric flow rates. The overall uncertainty in Q_{tot} was found to be 4.24 slpm. For situations where only one flow controller was used, the uncertainty was simply the stated manufacturer limit of 1% of full scale, or 3 slpm.

Table B. 3   Measurement uncertainties and intermediate calculations for Q_{tot}

Measurement, x_i	Standard Uncertainty, u_i	Influence Coefficient $\dfrac{\partial \left(Q_{tot} \right)}{\partial x_i}$	Section Containing Explanation
Volumetric Flow Rate, Q_1	3 slpm	1	A.1.1
Volumetric Flow Rate, Q_2	3 slpm	1	A.1.1

B.4.1 Uncertainty in Omega Hot Wires

The 95% uncertainty in the Omega FMA-900-V-R hot wires was stated by the manufacturer as 3% of full scale. This translates to an expanded uncertainty of 0.01524 m/s, or 3 sfpm.

B.4.2 Uncertainty in Measured Temperatures

The standard limits of error of the k-type TCs are quoted by the manufacturer as ± 1.1 K. This uncertainty in temperature is smaller than the data symbols and does not display on the graphs presented in the text.

APPENDIX C. TSI HOT WIRE ANEMOMETER MEASUREMENTS

In addition to the integrated hot wires 1 and 3 (HW1 and HW3), a standalone TSI constant temperature hot wire system (Model IFA 300) was instrumented as a backup measurement to determine the flow rate in the annulus. During heated tests, this instrument was located at $z = 0.305$ m (12 in.) inside the annulus formed by the channel box and pool cell. The lower threshold of this instrument was 0.2 m/s as determined from the lowest achievable external calibration velocity, described next.

The single sensor, end flow probes (Model 1210-20) were externally calibrated prior to insertion into the assembly annulus. Figure C.1 shows the TSI calibration equipment (Model 1129). The probes are placed above the nozzle, which is capable of flows ranging between 0.2 and 5 m/s. Typical calibrations for these experiments are taken between 0.2 and 1 m/s.

The system records ambient pressure and temperature during the calibration. Once the calibration file was saved to the dedicated TSI system, the probe was placed into the assembly annulus. The pressure and temperature at the probe were input into the system during data acquisition. These values were referenced back to the original calibration file with appropriate temperature and pressure corrections applied. The pressure was taken to be local ambient, and the temperature was measured with a TC placed next to the probe in the annulus (Figure C.2). The probe was mounted to a linear stage to enable traverses of the annulus. However, the probe was placed in the middle of the annulus for most of the testing.

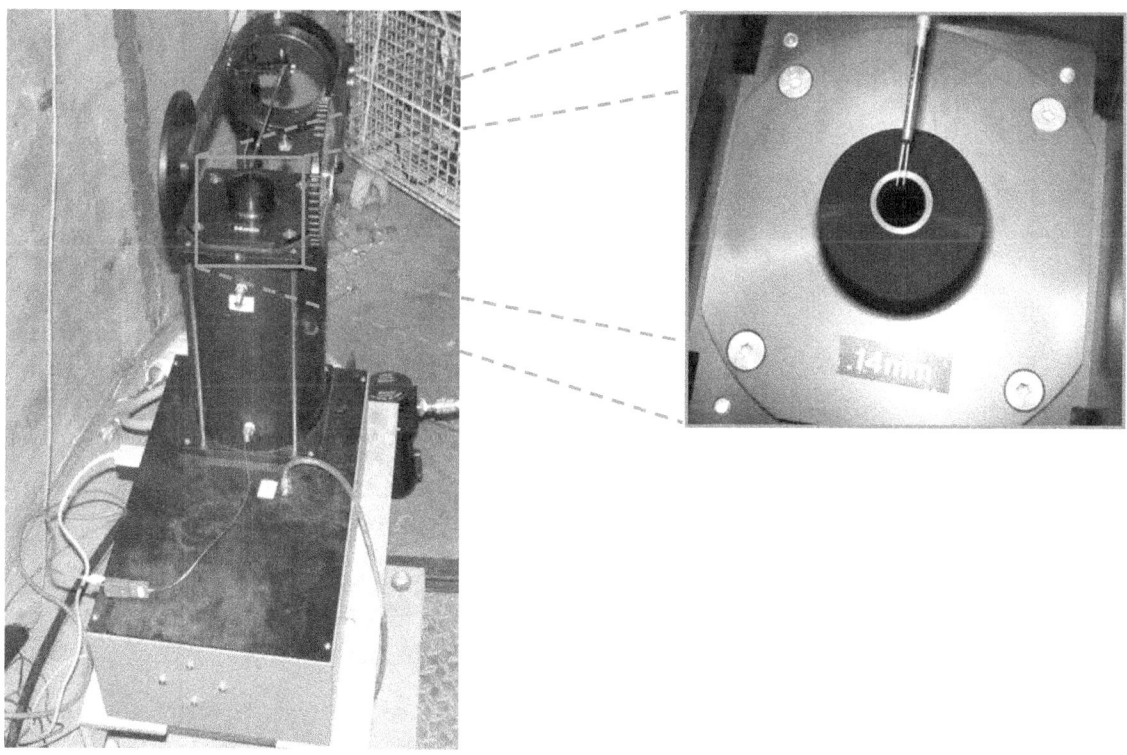

Figure C.1   **Photographs of the TSI calibration equipment**

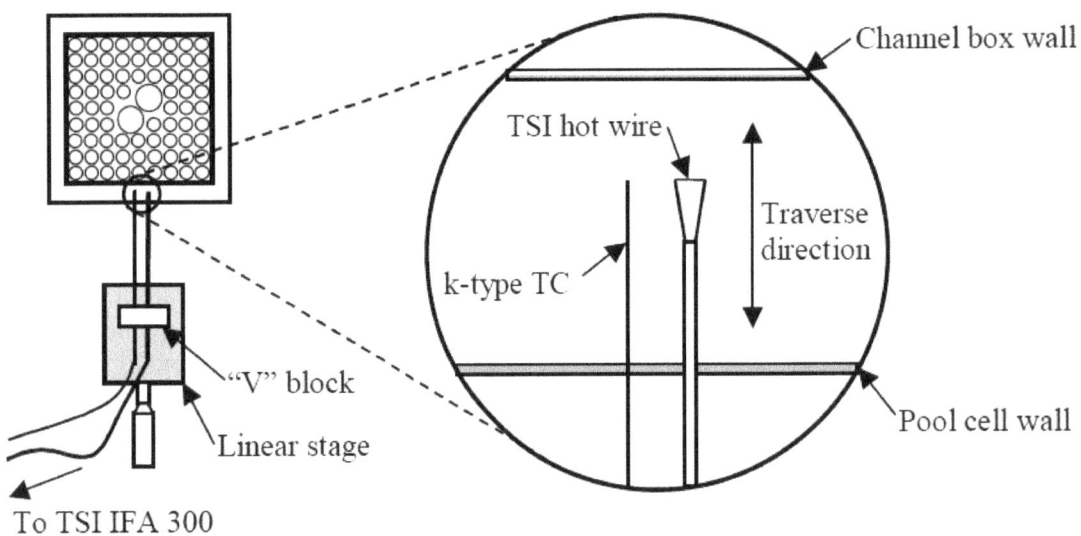

Figure C.2 **Schematic showing the arrangement of the TSI hot wire in the annulus**

Figure C.3 gives the response of the TSI anemometer to a given volumetric annular flow rate for calibrations performed on January 20 and January 31, 2006. For these calibrations the probe was located in the center of the annulus. The two data sets are within the experimental uncertainty, which is dominated by uncertainty in the flow rate.

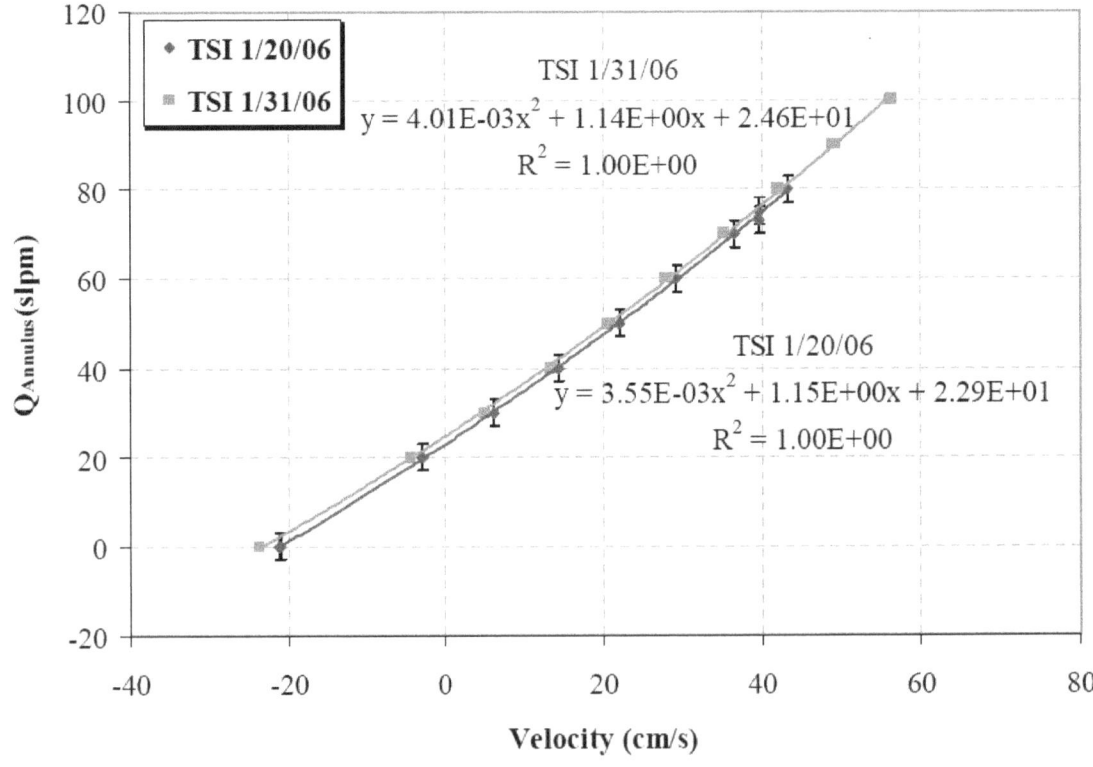

Figure C.3 **In situ calibration of the TSI hot wire located in the center of the annulus performed on 1/20/06 (blue diamonds) and 1/31/06 (red squares)**

Figure C.4 shows a velocity traverse across the annulus with the TSI hot wire probe. The solid line depicts the corresponding laminar velocity profile for a volumetric flow rate of 72 slpm passing through the annulus. This traverse was collected on January 17, 2006, during a heated test with 1370 W applied to the apparatus in the closed bypass/open drains configuration. The data were collected for this traverse at approximately 608 minutes elapsed test time. Typical experimental uncertainties are shown on the graph. Recall that calibrations were limited to a lower operating threshold of 20 cm/s. The flow through the annulus appears to be laminar in nature.

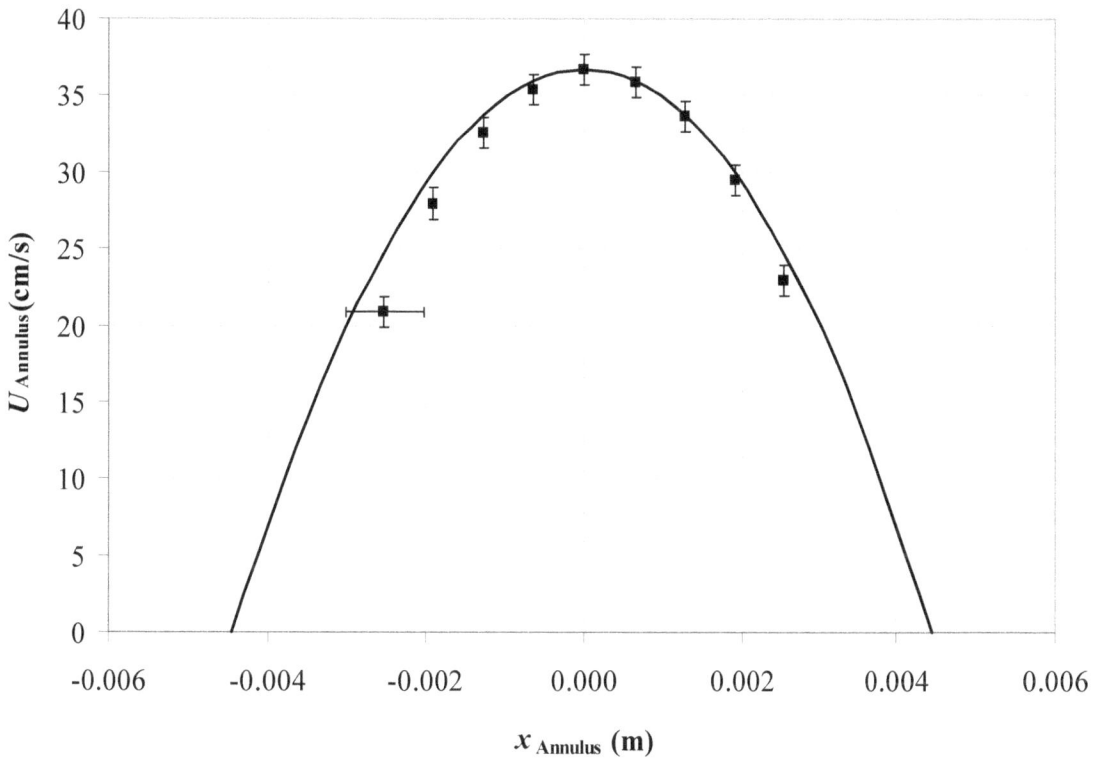

Figure C.4   Velocity traverse of the TSI hot wire across the annulus

Note: The solid line depicts the corresponding analytical laminar velocity profile.

APPENDIX D. ADDITIONAL DATA AND FINAL POST-TEST MELCOR RESULTS FOR THE FULL-SCALE ZIRCALOY IGNITION TEST

The following graphs represent additional data and final post-test MELCOR results from the full-scale ignition test not already presented in the report.

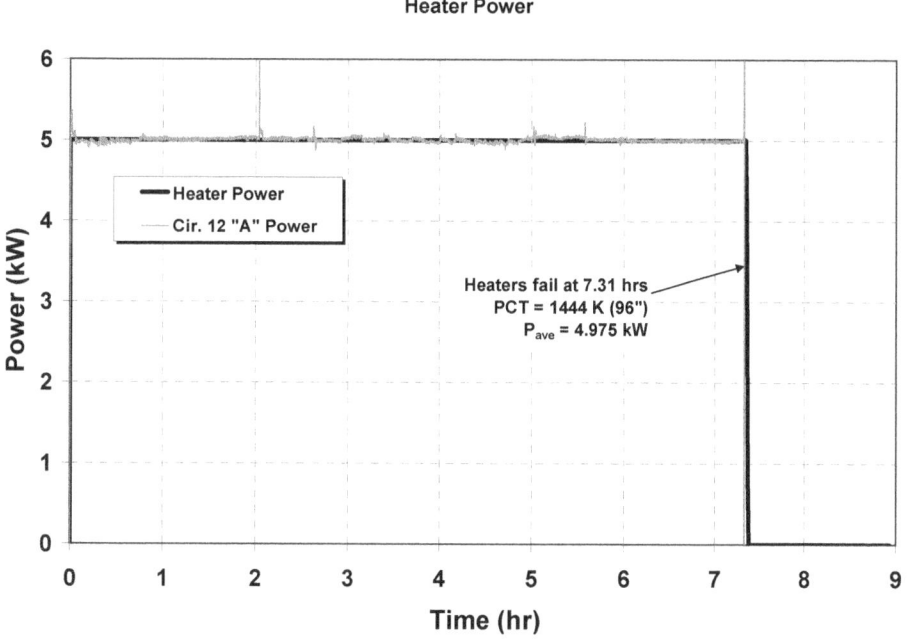

Figure D. 1   Input power for the experiment and the final post-test MELCOR model as a function of time

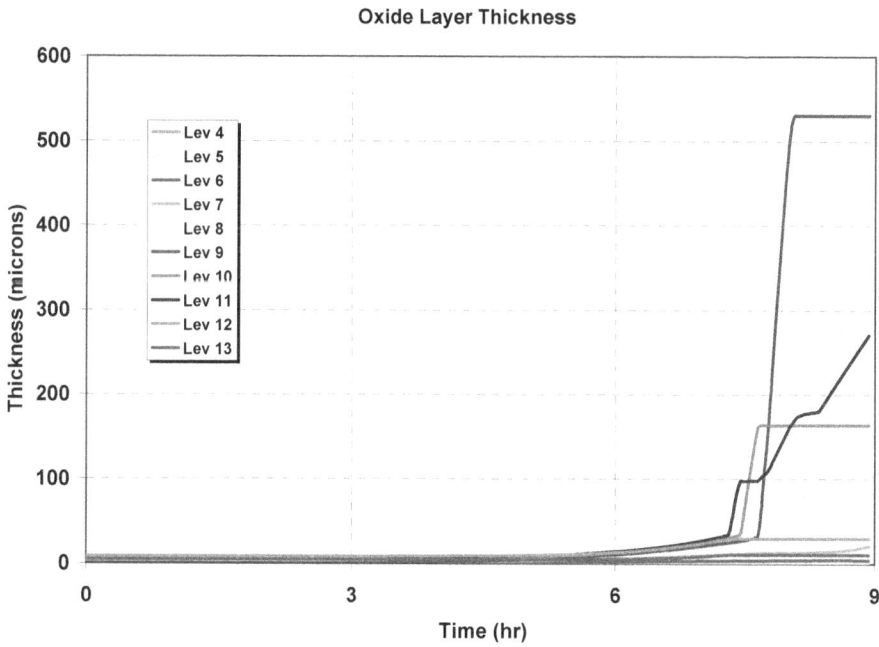

Figure D. 2   Cumulative oxide layer predictions from MELCOR at different axial levels

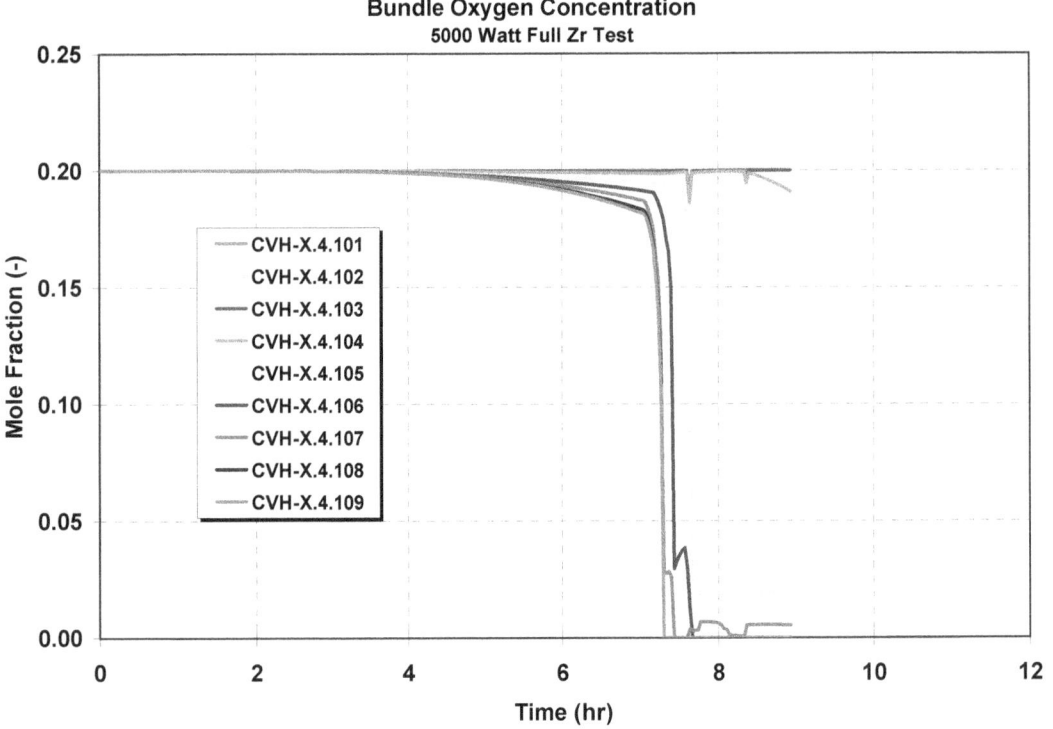

Figure D. 3   Oxygen concentration at different axial levels within the MELCOR model

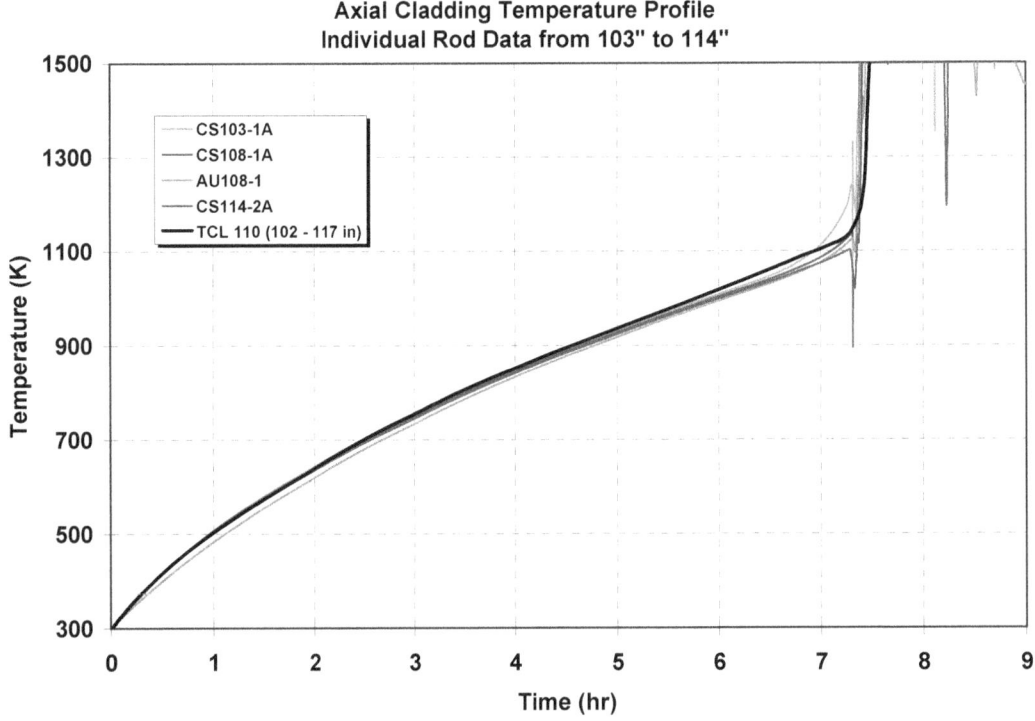

Figure D. 4   Cladding temperatures at $z = 103$ to 114 in. for both experiment and MELCOR

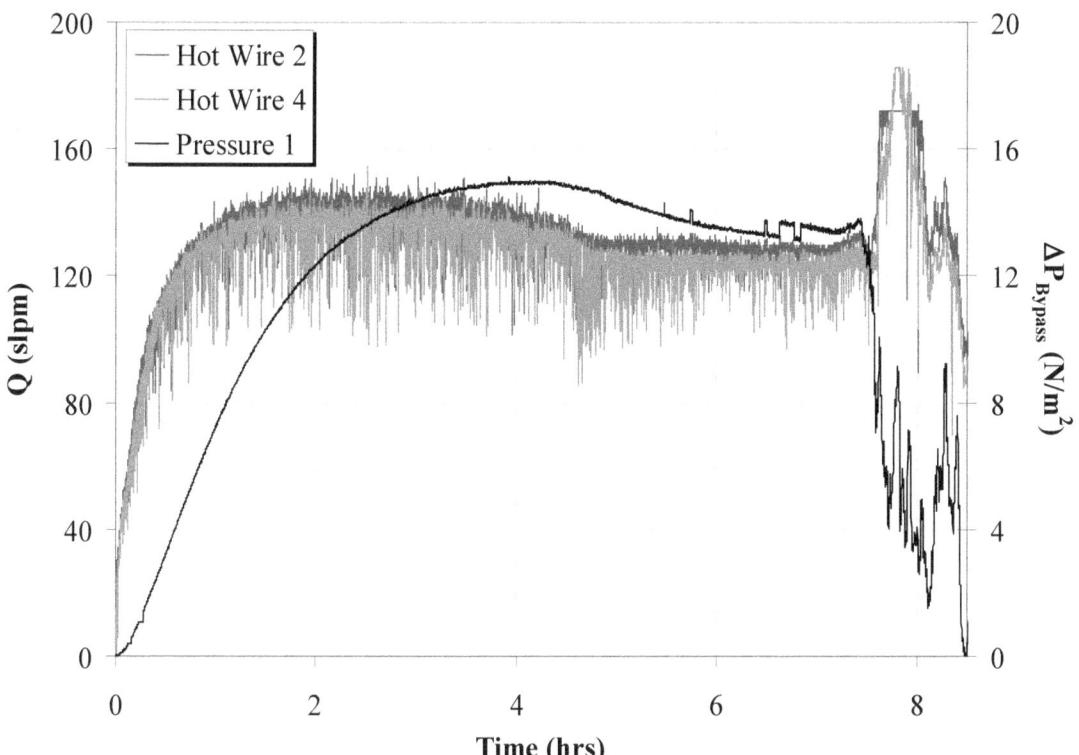

Figure D. 5   **Volumetric flow rate and bypass pressure drop as a function of time**

NRC FORM 335
(12-2010)
NRCMD 3.7

U.S. NUCLEAR REGULATORY COMMISSION

1. REPORT NUMBER
(Assigned by NRC, Add Vol., Supp., Rev., and Addendum Numbers, If any.)

NUREG/CR 7143

BIBLIOGRAPHIC DATA SHEET

(See instructions on the reverse)

2. TITLE AND SUBTITLE

Characterization of Thermal-Hydraulic and Ignition Phenomena in Prototypic, Full-Length Boiling Water Reactor Spent Fuel Pool Assemblies after a Postulated Complete Loss-of-Coolant Accident

3. DATE REPORT PUBLISHED

MONTH	YEAR
March	2013

4. FIN OR GRANT NUMBER

Y6758

5. AUTHOR(S)

E. R. Lindgren and S. G. Durbin

6. TYPE OF REPORT

Technical

7. PERIOD COVERED (Inclusive Dates)

March 2004 - September 2006

8. PERFORMING ORGANIZATION - NAME AND ADDRESS (If NRC, provide Division, Office or Region, U. S. Nuclear Regulatory Commission, and mailing address; if contractor, provide name and mailing address.)

Sandia National Laboratory
Albuquerque, NM 87185

9. SPONSORING ORGANIZATION - NAME AND ADDRESS (If NRC, type "Same as above", if contractor, provide NRC Division, Office or Region, U. S. Nuclear Regulatory Commission, and mailing address.)

Division of Systems Analysis
Office of Nuclear Regulatory Research
U.S. Nuclear Regulatory Commission
Washington, DC 20555-0001

10. SUPPLEMENTARY NOTES

G. A. Zigh, Technical Advisor and A. Velazquez-Lozada, Project Manager

11. ABSTRACT (200 words or less)

The NRC requires all nuclear power plants to have a spent fuel pool (SFP) where the used reactor fuels are allowed to cool for a number of years before being moved to interim or permanent storage. Spent fuel pools are robust structures with an extremely low likelihood of a complete loss of coolant under traditional accident scenarios. However, in the wake of the terrorist attacks of September 11, 2001, the SFP accident progression was reconsidered and reevaluated using best-estimate accident codes. NRC continued SFP accident research by applying best estimate computer codes to predict the severe accident progression following various postulated accident initiators. These code studies identified various modeling and phenomenological uncertainties that prompted a need for experimental confirmation.

Sandia National Laboratories (SNL) performed an experimental program to help NRC to address thermal-hydraulic issues associated with complete loss-of-coolant accidents in boiling water reactor SFPs. The objective of these experiments was to provide basic thermal-hydraulic data associated with a postulated SFP complete loss of coolant accident. The accident conditions of interest for the SFP were simulated in a full-scale prototypic fashion (electrically-heated rods in prototypic assemblies and SFP rack) so that the experimental results closely represent actual fuel assembly responses. A major impetus for this work was to facilitate code validation (primarily MELCOR) and reduce modeling uncertainties within the code.

12. KEY WORDS/DESCRIPTORS (List words or phrases that will assist researchers in locating the report.)

zirconium alloy fire, BWR spent fuel pool postulated accident, MELCOR analysis

13. AVAILABILITY STATEMENT

unlimited

14. SECURITY CLASSIFICATION

(This Page)

unclassified

(This Report)

unclassified

15. NUMBER OF PAGES

16. PRICE

Printed
on recycled
paper

Federal Recycling Program

UNITED STATES
NUCLEAR REGULATORY COMMISSION
WASHINGTON, DC 20555-0001

OFFICIAL BUSINESS

NUREG/CR-7143

Characterization of Thermal-Hydraulic and Ignition Phenomena in Prototypic, Full-Length Boiling Water Reactor Spent Fuel Pool Assemblies After a Postulated Complete Loss-of-Coolant Accident

March 2013

www.ingramcontent.com/pod-product-compliance
Lightning Source LLC
Chambersburg PA
CBHW080239180526

45167CB00006B/2342